Arrows

To The

Moon

Arrows To The Moon

Avro's Engineers and the Space Race

by

Chris Gainor

An Apogee Books Publication

Avro Arrow 201 in flight

We acknowledge the financial support of the Government of Canada through the Book Publishing Industry Development Program for our publishing activities.

Published by Apogee Books an imprint of Collector's Guide Publishing Inc., Box 62034, Burlington, Ontario, Canada, L7R 4K2
http://www.cgpublishing.com

Printed and bound in Canada

Arrows To The Moon - Avro's Engineers and the Space Race
by Chris Gainor

ISBN 1-896522-83-1
ISSN 1496-6921

Acknowledgements

This book springs from my lifelong fascination with space exploration, which began when I found my father, Don Gainor, watching TV unusually early one morning in 1962. Like millions of others, he was following the historic Mercury flight of astronaut John Glenn. I was soon hooked, and kept up the vigil for the space flights that followed, even after Dad and most others lost interest.

In June of 1966, I was 11 and the Gemini program was in full swing. My father was working at the annual meeting of the Canadian Medical Association, which that year took place in our home town of Edmonton, Alberta. One of the speakers was a physician from the Manned Spacecraft Center in Houston, Texas, named Dr. D. Owen Coons, and so my father arranged to get me into his talk. Dr. Coons answered my questions and kindly sent me some photos and leaflets from NASA that instantly became my most valued possessions.

Although other interests and aptitudes took my career in different directions, my interest in space travel never left me.

In May 1994, the Canadian Space Society and the National Space Society of the U.S. combined forces and held that year's International Space Development Conference in Toronto. The meeting included a reunion of what was called the "NASA Canadians" (although I later found that not everyone embraced this title). The members of this group were presented with special certificates, and they gathered in a room set aside for them to reminisce and talk with the other people at the meeting. The first member of the group I spotted was Dr. Coons, and soon I met other members of the group, including his former NASA medical colleague Dr. Bill Carpentier, and a man named Owen Maynard. In the months that followed the meeting, I thought about the many interesting stories I had heard from these space flight pioneers. Leafing through books on space travel, I found many references to the people who had come to NASA after the Canadian government scrapped the Avro Arrow program. They all said that these people had made major contributions to Mercury, Gemini and Apollo, but most didn't say how. That fall, I decided that the story of the people from Avro Canada at NASA needed to be told in full.

•

If a book transports the minds of its readers, it also takes the author on another journey altogether. The research for this book took me around Canada and the United States. During that time, I learned a great deal about the U.S. space program, the Canadian aircraft industry, the state of archives in Canada and the U.S., and many other things. Of course the best part was the friends I made along the way, many of whom provided invaluable help with this book.

First and foremost are the people who are the subject of this book and who generously agreed to talk to me and assist me. They include Bruce Aikenhead, Peter Armitage, Richard Carley, Dr. William Carpentier, Thomas Chambers, Robert Cockfield, Stanley Cohn, Dr. D. Owen Coons, Burton Cour-Palais, Bryan Erb, Dona Erb, Dave Ewart, Norman Farmer, Dennis Fielder, Stanley and Emily Galezowski, George Harris, C. Malcolm Hinds, John Hodge, Morris Jenkins, Robert Lindley, C. Frederick Matthews, Owen Maynard, Mario Pesando, Keith Richardson, Rod Rose, John Sandford, John Shoosmith, Leslie St. Leger, David Strangway, George Watts, and R. Lionel Whyte.

Owen Maynard spoke to me on several occasions, sometimes lasting several days, and provided me with a great deal of documentation that pertained not only to his role in the space program, but also to the contributions of many others. I also received a great deal of help from Owen's family, including his wife, Helen Maynard, son Ross Maynard, and daughters Merrill Marshall, Beth Devlin and Annette Maynard Franklin.

Rod Rose has been the unofficial historian of the Avro group. When I came to Texas, he and his wife Leila provided wonderful hospitality, and Rod provided me with a great deal of help in pointing me in the right directions. In Houston, Peter Armitage and his wife June put me up, and Peter was a great source of information and advice. Dennis Fielder provided all sorts of useful guidance when I visited Houston, and was one of the people I could call on for help by e-mail. I would also like to thank Bill and Willy Carpentier for their hospitality. John Shoosmith showed me around the Langley Research Center and the Hampton area, and he and his wife Carolin generously put me up. In Oregon, Norman and Helen Farmer were wonderful hosts for me and my wife Audrey. Fred Matthews provided many helpful documents. Bryan and Dona Erb provided a great deal of good advice and photos.

Many other family members helped me out. James A. Chamberlin's son Arthur was particularly helpful. I was also assisted by Chamberlin's brother William Falkner, daughter Shirley Ditloff, cousin Isabel Rowe, and nephew Peter Wilson. I also received important help from Frank Chalmers' wife June and son Ian, Joseph Farbridge's wife Joan and son Kevin, Eugene Duret's brother Maurice, Robert Vale's wife Marguerite and son Mark, and Carl Lindow's wife Jo and daughter Leslie.

Other people from NASA who provided me interviews include Christopher C. Kraft Jr., Charles W. Mathews, Paul Purser, James T. Rose, Bob Thompson and John Yardley. Thomas Kelly, who was Owen Maynard's counterpart at Grumman, also spoke to me, as did the University of Toronto experts who helped with Apollo 13: Bernard Etkin, J. Barry French and Peter Hughes.

James Floyd, who directed the technical efforts on the Jetliner, the CF-100 and the Arrow at Avro Canada was helpful to me throughout the project.

Peter Beck, Hal Smith and Frank Brame also helped me in my efforts. Mario Pesando, one of the top people at Avro Canada through its history, lives nearby and was always ready to provide sound advice about the story of Avro Canada.

Many other people made contributions to this book. First, the people from the Canadian Space Society who organized the 1994 International Space Development Conference, especially Keiran Carroll, Paul Swift and Greg Leach. In organizing the reunion of the "NASA Canadians," Keiran and his team assembled an invaluable list that gave me a big head start in the work of finding and getting in touch with the members of this group. Keiran has also been a source of advice and encouragement. Greg provided me access to the papers Owen Maynard left to the CSS.

Another person who worked at that meeting was my friend Andrew Yee, who gave me my first introduction to Owen Maynard and provided a great deal of assistance in the years that have followed.

At the same meeting I also met Randy Attwood and Paul Fjeld, who love space travel as much as I do and are world authorities on the workings of the lunar module. Both gave me invaluable help in writing this book, and I hope I've been able to help them with the book they are preparing on the LM.

James Oberg, whose books and articles provide the foundation of our knowledge of the Soviet space program, helped me whenever I asked. James R. Hansen, the Auburn University historian who has written the definitive account of the lunar orbit rendezvous debate, provided me with important information. Jerry Rosenberg of the NASA Alumni League helped me get in touch with former NASA employees. Marion Karasiuk interviewed members of the Avro-NASA group for an article she wrote for *Engineering Dimensions*, and generously shared her work with me.

Much of the material for this book comes from libraries and archives. These institutions and the people who work in them have suffered greatly in the past few years of budget cutting, and I salute them for persevering amidst such discouragements. Researching the story of Avro Canada in Ottawa, I made use of the Canada Aviation Museum research library and the National Archives of Canada. This work was eased by Glenn Wright's guide to the Avro Arrow papers in the National Archives. Diane Dupuis was able to search the National Archives and the National Library of Canada when I couldn't get to Ottawa myself. I also examined the papers of the Rt. Hon. John G. Diefenbaker at the Diefenbaker Canada Centre in Saskatoon, Saskatchewan, with the assistance of director Dr. Bruce Shepard and student archivist Johnson Kong.

In Houston, Texas, the Johnson Space Center's history office was closed down when I first visited in 1995. But many of the papers I required were available in the Johnson Space Center Archives at Rice University's Woodson Research Center in the Fondren Library. Lois Morris and Joan Ferry provided me with able assistance on my visits there. The papers at Rice have since been transferred to the University of Houston-Clear Lake.

When I returned to Houston in 1999, history was back at JSC in the form of the JSC Oral History Project. The staff there, especially my good friend Glen Swanson, provided me with a great deal of help in finding information and photos around JSC. The project provided me with fresh interviews that added to those I conducted. Glen, who has made a major contribution to the study of spaceflight history as the founder of *Quest*, also brought me together with my publisher.

Many of NASA's Mercury and Gemini papers reside at the U.S. National Archives and Records Administration's southwest office in Fort Worth, Texas. Kent Carter and his colleagues made my visit there productive.

At the NASA Headquarters History Office, I was assisted by Chief Historian Roger Launius, historian Stephen Garber and archivists Lee Saegesser and Mark Kahn. Bonita S. Smith at the History Office at John H. Glenn Research Center at Lewis Field answered my inquiries, and Jane Riddle of the Goddard Space Flight Center went the extra mile to find me information on NASA's tracking network.

Eric Ackermann of the special collections division of Virginia Tech's University Libraries gave me access to the papers of Chris Kraft and Bob Gilruth. Also helpful were Christy Vodden, communications officer with the Geological Survey of Canada in Ottawa, and Larry Merritt, an historian/archivist with The Boeing Company in St. Louis, where he works on the history of the former McDonnell Douglas Astronautics Company. I was also assisted by the staff of the Archives Division of the Smithsonian Air and Space Museum. Jennifer Romanko, curator of Canada's Aviation Hall of Fame, provided valuable assistance in my research on Jim Chamberlin. I also called on the resources of the British Columbia Legislative Library, the Vancouver Public Library, the Victoria Public Library, and the University of Victoria Library.

Most of the photos in this book are from NASA, and I obtained them with the assistance of G. Mike Gentry and Allen Bond of JSC's media center. Paul Cabot of the Toronto Aerospace Museum and Frank Harvey of the Aviation Heritage Foundation of Canada also helped me with photos, as did the staff of the Canadian Forces Photographic Unit, the National Archives of Canada and the Canada Aviation Museum.

R. Kyle Schmidt set up one of the earliest and best web pages on the Avro Arrow. He generously set up a web page for me, which helped connect me with members of the NASA-Avro group. My friends Ken Harman and Barry Shanko set up a subsequent web page on this book.

Others who helped me along the way include Rolf Maurer, Marcus Gee, Steve Pacholuk, Tom and Arlene Sheppard, Sue Birge, Lorne Slotnick, Ann Rauhala, Brian Lam, Howard White, Diane Kennedy, Susan Peterson, Nick Russell, Patrick Nagle, Harold A. Smith, Hal Doiron, Dr. Stephen Johnson, Geoff

Meggs, Lisa Fitterman, Julian Sher, Blair Redlin, Melva Forsberg, Steve Shallhorn, Christine Elwell, Sheryl and Kirk Salloum, Gwen Walter, Bob McCullough, Butch Head, Matt Nelson, Ron Caswell, "N-1 Ed" Cameron, Hugh S. Gregory, and Tony Beswick. Many others provided encouragement.

•

This book is based on interviews and documents that tell the story of the Canadian and British engineers and others who worked on the U.S. space program. These are listed in the sources section of this book.

The histories, chronologies, monographs and other works of the NASA History Office were invaluable resources in backing up, refuting or placing in context assertions made elsewhere. My copies of these works served time and time again when I needed to know what happened when, and who did what at NASA. They are listed in the sources section of this book and are recommended for anyone who wants to learn more about the history of space exploration. Another valuable source was David Baker's *The History of Manned Space Flight*. Three other books were also very useful to me: Charles Murray and Catherine Bly Cox's *Apollo: The Race to the Moon* tells the story of the people on the ground, including many of the engineers from Avro Canada, who got the astronauts to the Moon; Howard McCurdy's *The Space Station Decision: Incremental Politics and Technological Choice,* which was produced with assistance from NASA, outlines John Hodge's work on the space station; technical papers by many of the former Avro engineers are contained in *Manned Spacecraft: Engineering Design and Operation,* which is edited by Paul Purser, Maxime Faget and Norman Smith.

I first became aware of my publisher, Robert Godwin of CG Publishing, through his excellent series of NASA Mission Reports which I often referred to while writing this book.

This book would never have happened if my father hadn't pointed me to my interest in space exploration, or if my mother hadn't provided her indispensible support. My wife, Audrey McClellan, showed great patience and understanding as I spent time producing this book, and as a professional editor she took my manuscript and made it much better.

Chris Gainor
Victoria, B.C.
August 2001

Contents

Acronyms

AAP	Apollo Applications Program, later Skylab
ALSEP	Apollo Lunar Surface Experiments Package
AMU	Astronaut Maneuvering Unit
ASTP	Apollo Soyuz Test Program
ATDA	Augmented Target Docking Adapter
BIG	Biological Isolation Garment
CAE Ind.	Canadian Aviation Electronics
CAPCOM	Capsule Communicator
CARDE	Canadian Armament Research and Development Establishment
CM	Command Module
CSA	Canadian Space Agency
CSM	Command and Service Modules
CTS	Communications Technology Satellite, later known as Hermes
DDP	Department of Defence Production
DFI	Development Flight Instrumentation
EOR	Earth Orbit Rendezvous
ESA	European Space Agency
ESRO	European Space Research Organization
EVA	Extravehicular Activity, or space walk
FIDO	Flight Dynamics Officer
FOP	Flight Operations Plan
GE	General Electric
GSFC	Goddard Space Flight Center, Greenbelt, Maryland
HARP	High Altitude Research Program
IBM	International Business Machines
ICBC	Interagency Committee on Back Contamination
ICBM	Intercontinental Ballistic Missile
ILRV	Integral Launch and Reentry Vehicle
ISIS	International Satellites for Ionospheric Studies
ISS	International Space Station
IUS	Inertial Upper Stage
JPL	Jet Propulsion Laboratory, Pasadena, California
JSC	Johnson Space Center, Houston, Texas (from 1973)
LCM	Landing Craft, Medium
LCRU	Lunar Communications Relay Unit
LCU	Landing Craft, Utility
LLRV	Lunar Landing Research Vehicle
LLTV	Lunar Landing Training Vehicle
LM	Lunar Module (also known as LEM, Lunar Excursion Module)
LOR	Lunar Orbit Rendezvous
LRC	Langley Research Center, Hampton, Virginia
LRL	Lunar Receiving Laboratory
MOCR	Mission Operations Control Room
MOL	Manned Orbiting Laboratory
MQF	Mobile Quarantine Facility
MSC	Manned Spacecraft Center, Houston, Texas (1961 to 1973)
MSFC	Marshall Space Flight Centre, Huntsville, Alabama
MSS	Mobile Servicing System
NACA	National Advisory Committee for Aeronautics
NASA	National Aeronautics and Space Administration
NATO	North Atlantic Treaty Organization
NRC	National Research Council of Canada
OFC	Operational Flight Control
OMB	Office of Management and Budget
PLSS	Portable Life Support System
RAE	Royal Aircraft Establishment, England
RAF	Royal Air Force
RCA	Radio Corporation of America
RCAF	Royal Canadian Air Force
RETRO	Retrofire officer
RMS	Remote Manipulator System, the shuttle Canadarm
RPM	Revolutions Per Minute
RTG	Radioisotope Thermal Generator
SIG	Senior Interagency Group
SIM	Scientific Instrument Module
SM	Service Module
SNAP	Space Nuclear Auxiliary Power
SPAN	Spacecraft Analysis room
SPS	Service Propulsion System
SSRMS	Space Station Remote Manipulator System, or Canadarm2
STEM	Storable Tubular Extendible Member
STG	Space Task Group, forerunner to the Johnson Space Center
STS	Space Transportation System
SVS	Space Vision System
TCA	Trans Canada Airlines, now Air Canada
TWA	Trans World Airlines
TDRS	Tracking and Data Relay Satellite
UDT	Underwater Demolition Team
UHF/HF	Ultra High Frequency/High Frequency radio waves
USAF	United States Air Force
UTDC	Urban Transportation Development Corporation

Crowd gathers around Arrow 201 on rollout day, October 4, 1957.
Courtesy Aerospace Heritage Foundation of Canada.

October 4, 1957

The crowds that gathered outside a hangar on the outskirts of Toronto were expecting a glimpse of the future. For four years, hundreds of engineers and thousands of workers had been labouring on a new aircraft that would be seen in public for the first time that afternoon. It was Canada's first supersonic aircraft, the CF-105 Avro Arrow, and its manufacturer, Avro Aircraft Ltd., had pulled out all the stops to make this a memorable day for Canada's second largest city. At 2:00 p.m. on that bright and comfortably warm first Friday of October 1957, the band of the Royal Canadian Air Force (RCAF) formed up and began serenading those who were arriving at the Malton Airport from Toronto or the adjacent Avro plant for the ceremony that would begin a half hour later.

By the time the band finished with "God Save the Queen" and the "Knightsbridge March," more than 10,000 people stood in the sunshine before the decorated hangar and speakers' platform, which had seating for two dozen dignitaries representing Avro, the Canadian government and military, and the U.S. Air Force (USAF).

"The Avro Arrow is a twin-engine, long-range, day-and-night supersonic interceptor." The first speaker was Fred T. Smye, president and general manager of Avro Aircraft Ltd., a tall solid man wearing sunglasses and a well-tailored business suit. "It has a crew of two. It is a big, versatile aircraft. The primary armament of the aircraft is to be air-to-air guided missiles installed in a detachable armament bay in the fuselage. The versatility provided by this armament will enable the aircraft to perform other roles. The aircraft will be equipped with one of the most advanced integrated electronics systems, which will combine the navigation and operation of the aircraft with its fire control system."

Smye, who was the first employee hired by Avro 12 years earlier when the company was founded, spoke of the people and the work that had brought them to this day, and also of the event that most people who worked at the aircraft plant were waiting for. "We are hopeful," he told the crowd, "that the aircraft will make its first flight before the end of the year."

The next speaker was RCAF Air Marshal Hugh L. Campbell, Chief of the Air Staff, who with his close-cropped moustache and uniform resembled an officer of the Royal Air Force as much as its Canadian ally.

"Suffice to say the planned performance of this aircraft is such that it can effectively meet and deal with any likely bomber threat to this continent over the next decade. We in the air force look upon this aircraft as one component of a complex and elaborate air defence system covering in the first instance the whole of the North American continent, extending from Labrador to Hudson Bay to the Queen Charlotte Islands," Campbell said.

He noted that the Arrow would also play a part in the work of the North Atlantic Treaty Organization (NATO) by defending a perimeter that extended from Alaska to Turkey. Unmentioned but understood was the fact that NATO was defending its members against possible aggression by its Cold War adversary, the Soviet Union. Since Canada lay between the U.S. and Soviet Russia, the Arrow would play a key role in repelling any Soviet bomber attack on Canada and the U.S.

Campbell also spoke about the extent of the effort that brought about the Arrow. "There are some 38,000 parts in this aircraft and over 650 companies in Canada have been engaged in their manufacture." As well, government organizations in both Canada and the United States had contributed, including the USAF and the U.S. National Advisory Committee for Aeronautics.

"There are many difficult problems ahead — some can be foreseen, but some are hidden by the veil covering the unknown areas of aerodynamic science which have still to be explored."

Next the headline speaker was introduced. George R. Pearkes was the minister of national defence in the Progressive Conservative government that just three months before had taken the reins of the Canadian government, ending 22 years of rule by the Liberal Party. Pearkes had spent most of his adult life in the Canadian military, starting in the First World War where he won the

Victoria Cross, the highest decoration for bravery in what was then the British Empire. His military career ended after the Second World War with the rank of major-general.

"Much has been said of late about the coming missile age, and there have been suggestions from well intentioned people that the era of the manned aeroplane is over and that we should not be wasting our time and energy producing an aircraft of the performance, complexity and cost of the Avro Arrow," Pearkes said, his English accent reminding the audience of where he had been born and raised. "They suggest that we should put our faith in missiles and launch straight into an era of push-button war. I do not feel that missiles and manned aircraft have, as yet, reached the point where they should be considered competitive. They will, in fact, become complementary. However, the aircraft has this one great advantage over the missile. It can bring the judgment of a man into the battle and closer to the target where human judgment, combined with the technology of the aircraft, will provide the most sophisticated and effective defence that human ingenuity can devise."

"The aircraft now being produced in the various countries of our NATO alliance may or may not be the last of the manned interceptors. With the rapid strides being made in the fields of science and engineering, it would be unwise to attempt to forecast the future in this respect."

Pearkes ended his speech by gently pulling on a golden rope that hung next to the lectern on the speakers' platform. With that cue, a golden curtain in the door of the hangar bearing the words "Avro Arrow" parted, and a white tractor slowly pulled out a striking white jet aircraft with two engines and delta wings, a swept-back rudder and a black nose. The band struck up a fanfare and the jaunty airs of the "RCAF March Past" while a ceremonial flypast of CF-100 aircraft took place overhead. Those seated at the front stood up for a better glimpse of the single aircraft on the ground.

•

Pearkes' words about the future were more appropriate for the occasion than he or almost anyone else at the Malton ceremony could have imagined.

At a launching pad recently scratched out of the desert in Soviet Central Asia, another team of engineers and workers had spent that October 4, 1957, making final checks on a squat-looking rocket that had been stood up on the pad two days earlier. Fuelling had begun early that morning, and by late evening, the rocket was nearly ready for launch.

The launch would be the sixth of a rocket the Soviets called the R-7, their first intercontinental ballistic missile, which could lob a hydrogen bomb over the pole and onto a target in the United States within minutes. The first five flights of the R-7 were tests of this capability over a Soviet testing range. Three had failed, and two succeeded. This sixth R-7 carried in its nose cone a polished aluminized sphere, about the size of a medicine ball, with four trailing antennas. This launch was designed to carry out the Soviet government's public promise to launch a satellite of the Earth during the International Geophysical Year, a major international research effort going on at the time. The Soviet announcement had been made months before, complete with the satellite's planned radio frequencies, but it had been ignored in the west.

Inside the small launch control station at the secret launch site, operations were directed by Sergei Pavlovich Korolev, the hard-driving engineer and manager who had survived Stalin's gulags to become the preeminent leader of the Soviet missile and space programs. There had been some delays, but all the preparations were finally completed. Shortly after 10:28 p.m. Moscow Time, the firing command was given. The 32 engine nozzles at the base of the rocket roared to life, flames licked the exterior of the R-7 and smoke enveloped the rocket while the engines built up thrust. Finally, the rocket rose slowly on a brilliant pillar of flame and picked up speed as it arced eastward into the sky. Although all signals back to the launch control were positive, Korolev told the excited controllers to avoid any celebrations until success could be verified. That took place an hour and a half after launch with the first passage of the sphere, which was henceforth known as Sputnik, over the Soviet Union.

•

The R-7 rocket lifts off carrying the world's first artificial Earth-orbiting satellite - Sputnik
Courtesy Rocket & Space Corporation Energia

The historic launch took place just a few minutes after the ceremony at Malton ended. As the RCAF band serenaded them, the spectators surrounded the Avro Arrow for a close-up glimpse of an aircraft unlike any that had been seen before in Canada or most of the world.

"As many as could fit on the tarmac came out to see it," said Stanley Cohn, a computer engineer at Avro, remembering that day. "There was a tremendous sense of pride in the beauty of the aircraft at that stage of time."

Three hours later, after the crowd at Malton had returned home for the evening and after Sputnik had made its second orbit over the Soviet Union, the Soviet news agency TASS announced the launch to a world which was stunned by news that the Soviet Union, until then seen as a technological backwater, had made good on its audacious promise.

The Arrow's chief engineer, Robert N. Lindley, was driving two USAF generals back to the airport from a post-rollout celebration in Toronto when the news came over their car radio. Until they could confirm it, no one in the car believed that the Russians had done it. "I knew the Russians meant business," Lindley said. "They destroyed our rollout."

R. Bryan Erb, another Avro engineer, heard the news on the radio as he stood at the door of his infant son's bedroom. "It was a collective feeling that the aeronautical community had done it," Erb recalled thinking at the time. "The fact that it was the Russians didn't matter. It completely overwhelmed the Arrow rollout in significance for the day."

•

The disbelief inside Lindley's car was the most typical reaction that evening. The Soviet Union had given notice of its intention to launch satellites and had announced its first successful ICBM test. But no one, not even the Soviets themselves, were ready for the powerful reaction around the world that greeted the launch of Sputnik.

Soviet Premier Nikita Khrushchev had gone to bed calmly after Korolev told him that Sputnik was in orbit. The next day's edition of Pravda carried the news of Sputnik's launch as a routine report. But newspapers and other media outside the Soviet Union treated Sputnik's launch as a world-shaking event. Toronto's papers gave Sputnik the headlines that would otherwise have gone to the newly unveiled Arrow. Congratulations began pouring into Moscow. The following day's Pravda picked up on this and like every other newspaper, gave Sputnik banner headlines across page one. The strong worldwide reaction to Sputnik was a very pleasant surprise to the masters of the Kremlin, who lost no time in moving to exploit this propaganda triumph.

On November 3, under orders from Khrushchev, Korolev and his team launched Sputnik 2, which was much heavier than the first Sputnik and carried a live passenger, a mongrel dog named Laika. Although no attempt was made to recover Laika, her flight sent a clear signal that the Soviet Union was thinking of launching a human passenger.

•

In the United States, the Sputnik launches were seen as the opening shots of what would soon become known as the space race, a new field of rivalry between the two Cold War adversaries. Sputnik also had a major impact on the longer standing military rivalry between the U.S. and the Soviet Union because the ICBM that put both Sputniks into orbit could also deliver a deadly nuclear bomb to the U.S. or anywhere else in the world. Both sides were working feverishly to develop ICBMs and trying to determine what impact these missiles would have on their defence and warfighting strategies.

In Canada, Pearkes' statement about the need for both bombers and missiles came under question, both inside and outside government circles. Finally, 16 months and 16 days after Pearkes unveiled the Arrow on that October afternoon, the Arrow program was terminated. That was enough time for the Arrow to prove to its creators and to many others in Canada and elsewhere that it was the finest aircraft of its type at the time. More than 40 years later, the cancellation of the Arrow remains one of the most controversial events in Canada's history.

The demise of the Arrow meant the end of Avro Canada. Many of the engineers who lost their jobs as a result moved on to build other weapons for Canada's and America's Cold War arsenals. Some moved out of the aerospace business altogether. But a small number of them, including Cohn, Erb and Lindley, were snapped up by the U.S. government and private contractors to take leading roles on the American side in the space race. In 1969, the peaceful competition between the Soviet scientists who built and launched Sputnik and their American counterparts culminated in two U.S. astronauts standing on the surface of the Moon in one of the great technological feats of history.

This is the story of the select group of Canadian and British engineers who played a major part in putting humans on the Moon and opening the frontier of outer space to humankind. These newcomers to the U.S. worked alongside the best engineers America had to offer and held their own, in the process making history that will be remembered long after the other advances and the failures of the 20th century are forgotten.

1 The Avro Story

Among the proudest onlookers on the tarmac when the Arrow rolled out was the lanky, bespectacled, 42-year-old chief of technical design on the Avro Arrow. James A. Chamberlin had been one of the first hires at Avro Canada early in 1946, and he had played a key role in the three major aircraft projects that marked Avro's 12-year history up to that day. Those years had seen major changes that had permanently changed the aircraft industry and the people who worked in it. One program had led to the creation of a groundbreaking jet transport aircraft that never went into production. The second saw the design and production of Canada's first jet interceptor aircraft. The plane was a success, but it had a long and difficult development process that taught Chamberlin and his colleagues many hard lessons that they had put to use in their design for the third aircraft, the Arrow.

When Chamberlin and other engineers came to work at Malton in 1946, they had finished working for other firms on the propeller-driven aircraft of the Second World War. Just as the war was ending, the British and Germans had begun to fly aircraft powered by jet engines, and the Germans built the V-1, a pilotless jet aircraft also known as the buzz bomb, that delivered explosive charges to the enemy. But too few jets and V-1s had been produced to make a difference. At the same time, the Germans had launched liquid-fuelled V-2 rockets against their enemies, but the V-2 and the technical advances it represented came too late to affect the war's outcome.

Within months of war's end, the alliance that had defeated Germany and Japan split into western and Soviet camps. Both sides in the Cold War raced to defend themselves with the new technologies developed during the war. The experts and equipment behind the V-2 were swept up by the Soviet and American militaries to develop new weapons that would make their appearances in the late 1950s. But in 1946, the Jet Age had arrived, and the Avro plant at Malton was on its way to becoming the birthplace of jet aircraft in Canada.

The Malton plant had started in the 1930s as a factory for railcars, but the onset of the war in 1939 led to its conversion to aircraft production. Through most of the war, the plant was occupied by a nationalized company known as Victory Aircraft Ltd. that produced Lancaster bombers for use by the RAF. Victory Aircraft was one of the many wartime creations of Clarence Decatur (C.D.) Howe, the dynamic American-born minister of munitions and supply in Canada's wartime Liberal federal government. Known as the "Minister of Everything," Howe hired prominent business people and younger managers, including Fred Smye, to run the vast enterprises that made Canada one of the world's great industrial powers during the war. Canada's industrial output stood behind only the United States, Great Britain, and the Soviet Union among the Allied nations. The major contribution Canada made to the Allied victory in Europe gave Canadians a sense of pride and a belief that they could achieve almost anything.

Howe was eager to return the Malton plant to private hands at war's end, and he contacted Sir Roy H. Dobson, the managing director of Hawker-Siddeley Aircraft in England. By the end of the year, Dobson had arranged that the Malton plant would become part of Hawker-Siddeley. Its new name was A.V. Roe Canada, after the historic British aircraft company now part of Hawker-Siddeley where Dobson had got his start. Dobson immediately hired Smye to put together the new company. Avro drew on Canadian capital and on expertise that was already on site with Victory Aircraft, as well as bringing in talented engineers from elsewhere in Canada and from England to work on two new jet aircraft, one military and one civilian.

The Canadian government was anxious to patrol its own territory and defend against any Soviet bomber attack on the U.S., and as part of this effort the RCAF decided to develop a jet interceptor aircraft. The work of designing and building this aircraft, which became known as the CF-100, went to A.V. Roe Canada. At the same time, the team at Malton began to design a jet transport aircraft Avro hoped to sell to Trans-Canada Airlines (later Air Canada) and other companies, the C-102 Jetliner.

Among Avro's first engineering hires was Mario A. Pesando, a University of Toronto graduate who had started at Victory Aircraft and stayed on with Avro to work on the Jetliner. From Noorduyn Aircraft in Montreal, Chamberlin and Carl V. Lindow were hired for the engineering staff. As befits a firm with strong British ties, the Canadians at Malton worked side-by-side with a large contingent of British engineers and tradespeople, starting at the top with people like James C. Floyd, who headed up design of the Jetliner, and John Frost, who joined Avro in June 1947 to head up the effort on the CF-100. Many came to Canada to escape the postwar austerity still gripping Britain, whose aircraft industry suffered not only from stifling rules and customs, but also from postwar defence retrenchment. Canada's growing aircraft industry offered much more opportunity. And Canada was more attractive than the U.S. to these newcomers because it still had strong links and similarities to the United Kingdom.

•

Floyd and his team decided to build the Jetliner so that it could carry 40 passengers at a speed between 700 and 725 km per hour, with a range of about 2,400 km. These were above the specifications requested by TCA. Since the Jetliner was in a race to be the world's first jet transport aircraft, the design incorporated many new features. The original plan was to power the Jetliner with two Rolls Royce AJ65 engines, which were still under development. When it became clear in 1947 that the AJ65s wouldn't be ready for some time, the design was changed to four Rolls Royce Derwent engines.

C. Frederick Matthews was a young aeronautical engineering student at the U of T who worked summers in the stress office of the Avro plant while the Jetliner was being built. Matthews, a native of Guelph, Ontario, who had served in the RCAF during the war, spent one summer doing stress analysis on the center section of the Jetliner, where the wings joined the fuselage and where the two engines were attached to the wing. "The day I left, I presented my final report to the chief stress engineer, who said: 'This is great work. Too bad it's of no use.' I didn't understand what he meant," Matthews recalled. He was told that the change had just been made from two engines to four, giving Matthews a quick lesson in the vagaries of aircraft design that he would have occasion to recall in the future.

TCA withdrew its backing for the Jetliner at about the same time as the engines were changed, in part to avoid the financial and logistical risks it feared facing if it became the first airline to offer jet aircraft service. But with the financial backing of Howe, Avro continued to build the prototype Jetliner, which took to the air for the first time at Malton on August 10, 1949, and flew for an hour at altitudes up to 4,000 m. The Jetliner missed the title of the world's first jet transport, however, because two weeks before, the de Havilland Comet had earned the distinction when it took a short hop off an English runway. Despite the fact that TCA wasn't on board and the Jetliner could only claim a North American "first," flight testing went ahead. In April 1950, the Jetliner became the first jet transport in the skies of the United States when it flew to New York. It received heavy and favourable press coverage.

As the Jetliner took to the air, the Cold War was threatening to turn into a real war. In 1949, NATO was formed by the United States, Canada and Western European countries to face the Soviet Union. Communists under Mao Tse-tung took power in China. And the Soviet Union exploded its first atomic bomb. The speed with which the Soviets had obtained this technology stunned the West and spurred development of aircraft that could defend Europe and North America against the new Soviet nuclear threat. In, 1950, troops from Communist North Korea invaded South Korea. Under the flag of the United Nations, the United States, Canada and other countries went to war to save South Korea from North Korea and its Communist allies.

Back in Canada, these events caused Howe to agitate for acceleration of the CF-100 program. One of Howe's angry suggestions involved halting work on the Jetliner, and early in 1951, Smye reluctantly bowed to Howe's demand, though Avro continued to try to sell the Jetliner to customers in the United States.

Avro C-102 Jetliner in flight.
Courtesy Aerospace Heritage Foundation of Canada.

One such effort involved flying the Jetliner to California, where for several months it became the personal plaything of eccentric millionaire Howard Hughes, who owned, among other things, Trans World Airlines. Fred Matthews went to the Hughes airfield in the Los Angeles suburb of Culver City to prepare to run some flight tests as the Jetliner returned home to Malton, but his stay stretched into weeks as Hughes asked to continue to fly the Jetliner. Matthews and other Avro staff such as test pilot Don Rogers waited long hours for Hughes at the airfield or in the luxury Hollywood hotel suite Hughes had reserved for them. Matthews learned the hard way that Hughes, who was not yet the recluse that he later became, never carried money and was always asking for change for phone calls or to gas up his car. One night Hughes, Matthews and Rogers were talking about aircraft at Hughes' home on Sunset Boulevard when Hughes suddenly became silent. "Don Rogers looked at me, and I looked at him," Matthews recalled. "Finally, Don said to Hughes, 'Howard, what's wrong?' And he said, 'I can't remember where I left my Connie.' He couldn't remember where he'd left his four-engined Constellation that day. I heard of people who couldn't remember where they had parked their car, but that was the first time I'd met somebody who couldn't remember whey they'd parked their four-engined aeroplane."

Along with other airlines in the U.S., TWA was on the brink of ordering Jetliners, but the orders were stopped cold by Howe's insistence that the CF-100 program command all of Avro's resources. By the time the CF-100 was on track, the Jetliner's moment had passed. Avro continued to use it as a testbed, but after its last flight in November 1956, the Jetliner was dismantled. During its flying career, the only serious problem encountered was a landing gear failure on its second flight. To those at Avro who worked on the Jetliner, its loss was every bit as heartbreaking as the cancellation of the Avro Arrow. And in terms of the future of Canada's aviation industry in general and A.V. Roe Canada in particular, the loss of the Jetliner may have been more devastating than the demise of the Arrow. Canada had established a lead in the passenger jet industry, voluntarily surrendered it, and left the field for more than a generation.

•

For the CF-100, the RCAF asked in 1946 for a two-seater aircraft with two jet engines capable of flying at 900 km per hr or Mach 0.85 at an altitude of 12 km with a range of 1,200 km. The interceptor was designed to operate at any time of day in a full range of weather conditions. During the war, the Canadian government formed Turbo Research Ltd. to work on jet engines. When Dobson picked up Turbo in 1945, it became the Gas Turbine Division of Avro and built the Orenda engines that powered the CF-100.

Chamberlin, who was cutting his teeth on aircraft design at a time when jet aircraft were new and most available performance data on jets was what had been captured from the Germans, headed up design on the CF-100 until Frost came on board 18 months into the program. Frost and Chamberlin were in conflict because Frost had tried to change Chamberlin's design to make the CF-100 a higher performance aircraft, Pesando said. "There was no driving force behind the CF-100 like the Jetliner had with Jim Floyd."

Top designers of the Avro Arrow in January 1955: (l-r) Robert Lindley, then chief design engineer and later chief engineer; James Floyd, chief engineer, later vice-president and director of engineering; Guest Hake, Arrow project designer until 1957; and James Chamberlin, chief aerodynamicist and later chief of technical design. Courtesy James C. Floyd.

The first CF-100 flew on January 19, 1950. Although that inaugural flight was successful, subsequent test flights showed a serious defect in the aircraft: due to changes in the placement of the two jet engines, the structural supports for the engines could not withstand the loads and the flexing of the main wing spar. Other problems also cropped up, such as the aircraft's weight, fuel lines, canopy ejection failures and difficulty ejecting from the rear seat. There were troubles with the armament for the aircraft and, later on, with extended wings. Howe, by then minister of defence production, was angry about these problems when there was an urgent need for the aircraft. This caused a major shakeup at Avro in 1951. Crawford Gordon, an ambitious manager

The Avro CF-100

who had worked for Howe during the war, was named president of Avro, and Gordon gave Smye full authority over Avro's aircraft programs. Among the many changes Gordon and Smye made was moving Jim Floyd to the CF-100 project. To help him fix the troublesome spar and other problems dogging the aircraft, Floyd brought in Robert Lindley, who he had worked with in England. Lindley, a blunt, no-nonsense manager, took measures to fix the CF-100 spar problem. John Frost was moved to special projects.

As chief aerodynamicist, Jim Chamberlin was responsible for innovations in both the Jetliner and the CF-100, and along with his colleagues he also bore some responsibility for the problems that dogged the CF-100. Chamberlin sometimes locked horns with Frost, Lindley, Floyd, Pesando and Edgar Atkin, then Avro's chief engineer, but most of Chamberlin's colleagues agree that he was rarely wrong when it came to aerodynamics or physics. Pesando said Chamberlin was successful in shooting down bad ideas that came from others, and he persuaded Avro to procure a library for use by the designers and engineers.

To do the flight testing for the Jetliner and the CF-100, Avro formed a flight test department under Pesando, who had previously worked for Chamberlin on the aerodynamics of the CF-100 and the Jetliner, and then gone to A.V. Roe in England for training in flight testing. This department worked with Avro test pilots to assess aircraft performance, placing instruments in the aircraft and comparing actual performance of the aircraft to what had been estimated or projected from tests in the wind tunnel. The department had its work cut out for it ferreting out the problems and testing the fixes on the CF-100. Fred Matthews and Peter J. Armitage, a young engineer from Yorkshire who had just crossed the Atlantic to work at Avro in flight test, were both slated to fly in the back seat of the CF-100 as part of the flight test program. Both cheated death when they missed flights for medical reasons and the observers who took their places died in crashes.

Matthews, who missed his flight while recovering from eardrum injuries sustained in a diving aircraft that he rode in while suffering from a cold, was convinced that the observer who died in his place was killed by a canopy that would not come off. First, Matthews went looking for results of wind-tunnel tests on the CF-100 and found that the canopy covering the plane's back seat was not the same shape as the one on the model used in the wind tunnel. He ordered flight tests using a spare canopy drilled with holes for pressure gages. When that test flight was inconclusive, he put in more pressure gages, and on the next flight proved his theory that the canopy was not the right shape to come off the plane in an ejection. "All aeroplanes were grounded and the canopies were redesigned," said Matthews, who also worked on the tests that made sure that the redesigned canopies worked.

Once the canopies were redesigned, another problem revealed itself as an equivalent danger for those who flew in the back seat of the CF-100. After missing a fatal flight in which the pilot ejected safely but the observer died, still strapped in his seat, Armitage and others in flight test found out that if the aircraft was flying above a certain speed, the observer couldn't reach up and pull the over-the-head ejection handle because of the slipstream. The solution was to install a windshield in front of the observer.

To make sure the fixes worked, Avro held a series of tests of the ejection system from the rear seat of the CF-100. With Avro test pilot Jan Zurakowski at the controls, a test pilot with the ejection seat manufacturer Martin-Baker ejected safely from the back seat of the CF-100. The test was repeated twice, but with George, the Avro test dummy. As flight test engineer, Armitage observed the ejections from the Jetliner, which was flying alongside the CF-100.

The CF-100's placard speed was Mach 0.85, but Zurakowski and other test pilots managed to fly it at above Mach 1, which could be done safely in a dive under certain conditions. Many in the engineering department didn't believe that the CF-100 had broken the sound barrier, so Pesando and Zurakowski conspired to have a CF-100 reach supersonic speeds during a meeting held to discuss this controversy. Pesando remembers that the double sonic boom from the CF-100 quickly brought the meeting to an end.

After controversy over the CF-100 delays had spilled into the media and Parliament in Ottawa, the CF-100 entered regular production and full service in the RCAF just as the Korean War ended in 1953. By late 1958, when the last CF-100 rolled off the assembly line at Malton, 692 had been produced. Fifty-three were sold for use by the Belgian Air Force. The RCAF deployed CF-100s around Canada and in Europe until they were replaced starting in 1963. A small number of CF-100s remained in service until the Canadian Forces retired the aircraft in 1981. Its problems solved, the CF-100 had gone on to be a success, the only all-Canadian jet interceptor put into active service. And although the problems that plagued the CF-100 made Howe and others harsh critics of Avro, they also taught the Avro engineers and designers many important lessons that were put to good use on Avro's next military jet project and beyond.

•

The original version of the CF-100 was barely off the drawing board in the late 1940s when Avro designers and engineers began looking at swept-wing versions of the CF-100 that could go higher and faster. The first plane had flown faster than the speed of sound in 1947, and air forces around the world were eager to join the supersonic age. In 1953, the RCAF issued specifications for design studies on a supersonic all-weather fighter to succeed the CF-100 that were audacious even for British or American aircraft companies. But Avro Canada already had the inside track with its design studies on uprated CF-100s and on delta-winged aircraft. The design contract for the aircraft was given to Avro that summer, and work was soon underway at Malton on what became known as the CF-105 Avro Arrow.

Avro decided to fit the CF-105 with PS-13 Iroquois engines, then under development at Orenda. Until the Iroquois was ready, the first Arrows would fly with the less-powerful Pratt and Whitney J75 engine.

Designing the Arrow meant new challenges that almost no aircraft firm had ever faced. An aircraft that operated at twice the speed of sound would require much more than scores of engineers, aerodynamicists, designers and drafters. Among the team working under Floyd, chief

The "bullpen" at Avro Canada in Malton where engineers designed the Avro Arrow.
Courtesy Toronto Aerospace Museum.

engineer Lindley and chief of technical design Chamberlin were more than 2,000 talented engineers. An English engineer named John D. Hodge moved from the CF-100 to work on the Arrow's engine intakes and then took over the group of engineers dealing with airloads on the Arrow. His opposite number in the stress group was Owen E. Maynard, a native of Sarnia, Ontario, who worked his way through college at Avro and then concentrated on the CF-100 weapons pack after graduation from the U of T. Maynard recalled that Chamberlin frequently used him as a troubleshooter, dealing with issues such as the Arrow's landing gear, and fixing piping in the Arrow's hydraulic system that was failing in tests.

Wind-tunnel testing of the CF-105 began very early in the design process. During development of the aircraft, testing was done at the low-speed wind tunnel at the National Research Council (NRC) in Ottawa and in the U.S. at the Cornell Aeronautical Laboratories in Buffalo, New York, and the facilities of NACA, the U.S. National Advisory Committee for Aeronautics, at Langley Aeronautical Laboratory in Virginia and the Lewis Flight Propulsion Laboratory in Cleveland, Ohio. Floyd estimated that more than 4,000 hours of wind-tunnel work went into the Arrow.

Because of the high speeds the Arrow would fly at, rockets were required to test models of the aircraft. Nike rockets topped with models of the Arrow were launched at Point Petre, near Belleville, Ontario, and later in the U.S. at the NACA range at Wallops Island, Virginia, to take advantage of the telemetry facilities there. The tests examined the stability, flutter, spin and drag characteristics of the Arrow at subsonic, transonic and supersonic speeds. Data from these tests were assessed by officials from Avro, the RCAF and the NRC in Canada, the RAF, NACA and the U.S. Air Force. The positive impression the Avro engineers made during these tests would be remembered by the Americans when NACA's successor organization, the National Aeronautics and Space Administration (NASA), needed engineers in 1959.

The designers of the Arrow also required help from devices that up to that time were known only in the realm of science fiction: computers. Among the engineers hired by Avro were those who were knowledgeable in or could quickly learn the use of computers. At first, Avro had a CADAC computer, the first digital computer used in Canadian industry. Stanley H. Cohn, a mathematician educated at the U of T who had experience in the U.S. with the primitive punch-card computers of the time, was hired by Chamberlin. "They needed a few experienced mathematicians, and preferably with some computer exposure, but since there were no computers in Canada, it was difficult to get computer exposure," Cohn recalled. As second-in-command of Avro's computer program, Cohn hired a young engineering physics graduate from Queen's University in Kingston, Ontario, named John N. Shoosmith.

Soon Avro installed an IBM 704 computer, one of the first big mainframe computers IBM became famous for in the 1960s and 1970s. The IBM 704 had a memory of a few thousand words, a tiny fraction of the memory found in today's personal computers, but impressive in 1957. Because it had to be kept cool, the computer and the people who ran it were pulled out of the engineering "bullpen" and put in air-conditioned quarters, which Cohn, Shoosmith and others appreciated in Toronto's muggy summers. Once installed at Avro, the IBM 704 became a proud symbol of Avro's industrial preeminence and was used to do stress analysis and calculations of the Arrow's aerodynamic stability. "You used to do everything in wind tunnels before you had computers, but computers could simulate a lot of wind-tunnel work so that you would only have to do the refined wind-tunnel work," Cohn recalled. The computer was used to help build almost every part of the aircraft, including the Arrow's complicated landing gear, Shoosmith said. "We could simulate the motions and figure out the stresses the gear could withstand." Later on, the computer was used to analyze data obtained during Arrow test flights and recorded on equipment stored in the Arrow's weapons bay. As well, the Arrow flight simulator required computing time.

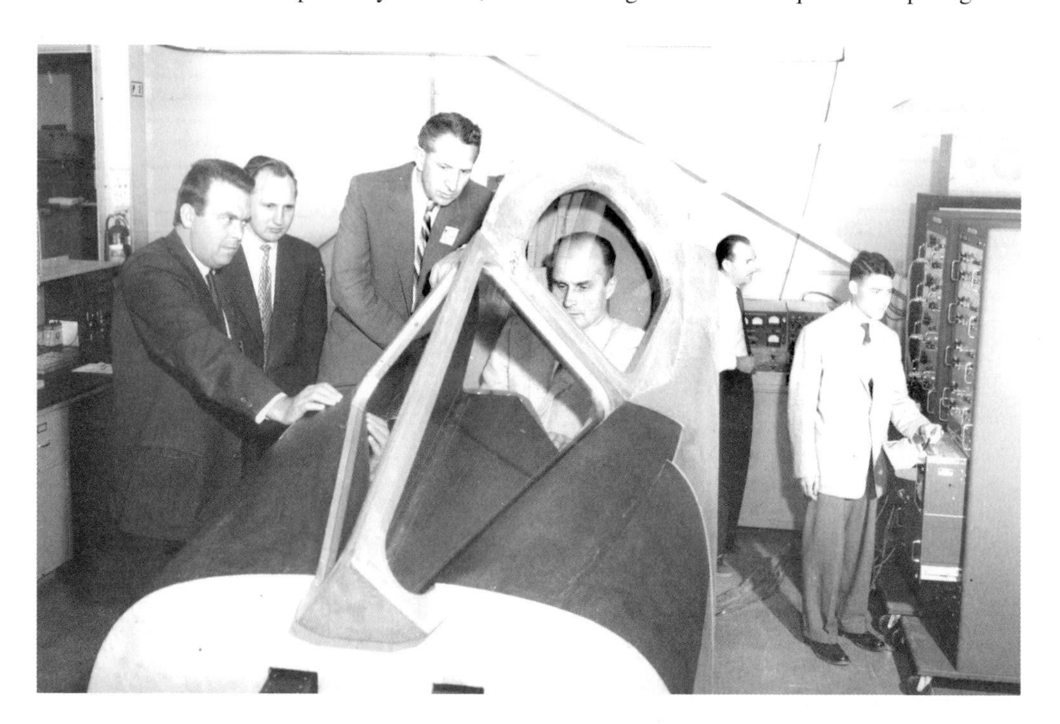

Avro test pilot Jan Zurakowski tries out the Arrow simulator at the Avro Canada plant in Malton. Dick Carley is second from left. Simulator supervisor Stan Kwiatkowski is at left, and Avro test pilot Spud Potocki leans on the canopy. Courtesy Toronto Aerospace Museum.

The design team at Avro did not invent any of the aerodynamic features of the Arrow, but they made use of the cutting edge concepts of the day. "We didn't invent anything," Pesando said. "We exploited it." Special materials on the Arrow's control surfaces, and powerful air conditioning inside the aircraft dealt with the effects of heating from high-speed flight. Sound effects from the engines and from the aerodynamics of supersonic flight put additional strain on the aircraft structure. Inside, the Arrow required strong hydraulic systems to move the control surfaces, and the Arrow's jet engines gulped huge amounts of fuel to move the aircraft at supersonic velocities.

Pesando's comment notwithstanding, the Avro Arrow did include a "first" in one area. It was the first aircraft to be equipped with a fly-by-wire control system, a system that used electronics and computers to move the aircraft control surfaces and control the aircraft. All previous aircraft relied on direct control inputs from their pilots. Fly-by-wire control systems, which used inputs both from the pilot's controls and from instruments, have found their way into all U.S. piloted spacecraft, today's high-performance combat aircraft and more recently into passenger aircraft.

Richard R. Carley, an engineer from Saskatchewan, helped develop the Arrow's new control system on the simulator built for the aircraft. Test data and estimates of the Arrow's flight characteristics were fed into the simulator by computer to show how, using a television screen with crude line drawings, the aircraft would react to control inputs by the pilot. "I learned to fly an airplane, because I thought, if you're going to do flight control systems, you ought to at least know how to fly an airplane," Carley said. "So I was pretty proficient on the simulator." Zurakowski flew the simulator before his first flight, although Carley remembers he was reluctant at first. But the test pilot soon became enthusiastic about the simulator, using it to determine the Arrow's stall characteristics before he took to the air in the real aircraft.

•

At the same time, John Frost's group was hard at work on the Avrocar, an air-cushioned, vertical takeoff vehicle that looked like a flying saucer. The initial plans for a saucer that would fly supersonic speeds fell by the wayside due to technical and aerodynamic problems, coupled with Howe's skepticism, but work continued with support from the U.S. military. Under the direction of Mario Pesando, engineers at Avro also took initial stabs at various projects such as a monorail, business jets, larger versions of the Jetliner, missiles, gyro copters, a supersonic transport aircraft, a satellite launch capability for the Arrow, and even a "space threshold vehicle."

As design work cranked up on the Arrow, the Avro plant was busy churning out CF-100s, which were then in full production. The years between the beginning of the Arrow program in 1954 and the winding down of the CF-100 production in 1958 were the height of Avro's existence. Housing developments sprouted in Malton, nearby communities such as Brampton, and what were then the northern suburbs of Toronto. When shifts changed at the Avro plant, the traffic jams were said to be the worst in Canada.

Crawford Gordon was reorganizing and expanding A.V. Roe Canada into a corporate colossus. The aircraft division of A.V. Roe Canada was split off into Avro Aircraft Limited, with Fred Smye at the helm, and the Gas Turbine Division became Orenda Engines Limited. Then Gordon acquired Canadian Car and Foundry, Canadian Applied Research, and control of the third largest steel producer in Canada, Dominion Steel and Coal Company (Dosco), which was based in Nova Scotia. Avro Canada, with 44 companies under its wing and 50,000 employees, had become one of the largest corporations in the world. In Canada, only Canadian Pacific and Alcan Aluminum were bigger.

The Arrow's mounting costs caused the Liberal government in Ottawa to look on with growing concern. When Avro requested additional money in 1955 for the Arrow, the government ordered a review and decided to put off any decision on the Arrow's future until after the next election, which was due to take place in 1957. Howe, always the disapproving father of Avro, said early in the program the costs of the Arrow "frightened" him and later told Parliament that the Arrow program "gives me the shudders." One reason was that the Arrow program involved more than the development of a new aircraft and a new engine. At the insistence of the RCAF, a new weapons control system called Astra was being developed by RCA in Montreal and Westinghouse in Hamilton, Ontario. On top of that, the Sparrow missile was being built at Canadair in Montreal. Similar equipment that was slightly less capable but much cheaper was available in the U.S. There was little coordination of all these contracts. An Arrow project office in government was not set up until October 1957, the same month the first Arrow rolled off the production line.

By then, Howe and the Liberals were gone from government. In an election the previous June, John Diefenbaker's Progressive Conservative Party won the most seats in Parliament, though not a majority. Among the seats won by the Conservatives were those in the Malton area. When Diefenbaker became Canada's 13th prime minister, one of the many difficult decisions awaiting him was the future of the Arrow program. Diefenbaker and his new cabinet shared Howe's deep concerns about the Arrow's mounting costs. And Diefenbaker, who was suspicious by nature due to a long political career spent in opposition to the Liberals, had little love for an A.V. Roe Canada that was led by Howe proteges such as Gordon and Smye.

Since it did not control a majority of the Parliament, the new government deferred its decision on the Arrow's future. In January 1958, Diefenbaker found the pretext he needed to call a new election, and on March 31, 1958, Diefenbaker won an unprecedented total of 208 seats in the 264-seat House of Commons. Diefenbaker campaigned in front of excited crowds on an undefined "vision" of Canada. The "vision" included "roads to resources" in Canada's north and aid to Canada's farmers, but little mention of the Avro Arrow.

•

Celebration of Arrow's first flight. Test pilot Jan Zurakowski shakes hands with Stan Brown surrounded by Avro employees and government officials including future NASA employees Fred Matthews (in parka, under engine intake), and Frank Chalmers (in jacket and tie, to right of Brown). Courtesy C. Fred Matthews.

Shortly before the 1958 election, on March 25, the Arrow took to the air for the first time. Although it had been delayed long beyond the time hoped for the previous October, Avro test pilot Jan Zurakowski reported only a few minor problems on that first flight. Everyone who could be spared came out to the tarmac to watch the flight, and afterwards, jubilant Avro workers hoisted Zurakowski on their shoulders in celebration. Dick Carley was anxious to hear from Zurakowski how the flight compared to the simulator. "The first thing he said to me was, 'It's a lot easier flying the airplane than it was a simulator.' Which I thought that's exactly what I wanted. If he could fly the simulator, we didn't have any worry about him flying the airplane. But I'm not sure how he really meant that."

Between the rollout and the first flight, the Avro team had been busy testing the engines, landing gear and brakes on the Arrow. The main landing gear had to fit inside the thin delta wings of the Arrow, and to do so it rotated and shortened as it folded up. The landing gear and the brakes were the source of some difficulty, and Peter Armitage was one of the people whose testing work helped remove the bugs. To test the anti-skid brakes, Armitage had a fire truck spread water on the tarmac one cold February day to make sure the brakes worked as advertised. "My most vivid memory is of the high-speed landing brake tests. I would dash out with a hand pyrometer to measure brake temperature. Only later on, with a more safety-conscious mind, did I realize how dangerous that was. The tire pressures were 400 pounds per square inch, and if one had blown - goodbye career!" On more than one occasion, the pressure and heat of landing caused the Arrow's tires to explode.

Once the Arrow was airborne, flight test needed to record huge amounts of data to determine how the systems were working. Weapons such as missiles were to be placed inside a special weapons pack in the belly of the Arrow that could be quickly changed out. When required, the weapons pack doors would open, the missiles would deploy and then be launched. During flight testing, the weapons pack carried recording and telemetry equipment instead of weapons.

Arrow 203 in flight over Malton in 1958. Courtesy Aerospace Heritage Foundation of Canada.

Arrow 201 flies over the Malton plant in 1958. *Courtesy Aerospace Heritage Foundation of Canada.*

This equipment was the province of the electronics group in flight testing, and people like George Harris Jr., an engineer from Britain, and Leonard E. Packham, a native of Saskatchewan, worked to make sure that data were recorded in a useful form. Harris recalls that one-inch magnetic tape, then a new medium, was used inside the aircraft to record data. The systems were tested in a CF-100 before being used in the Arrow, and the magnetic tape had the benefit of being compatible with the IBM 704 computer. But there was also a need for data to be provided live to the ground, so 11 strip chart recorders were set up in the control center to record the eight channels of data that came from the aircraft. In the center of the room sat the one person who could talk to the pilot. "For want of a better term, we called this the High Speed Flight Center," Harris said. "We could look at it with some kind of pride because to our knowledge, it was the first one of its kind that anybody had put together in North America."

The data obtained from these recorders were not only used to assess how the aircraft was standing up to the stresses it was under, or to compare flight data to wind-tunnel and rocket data, but to help pilots prepare for flight. Bruce A. Aikenhead, an Alberta-born engineer who had previously worked for the Canadian aircraft simulator firm CAE in Montreal, was now using actual flight data from the Arrow to make sure that the performance of the Arrow simulator matched the characteristics of the aircraft. As well, data from the flights were played back through the computer into the simulator to see what outputs appeared on the control panel.

In the High Speed Flight Center, Fred Matthews said he hired an ex-wing commander from the RAF to be the aircraft communicator "to keep the babble of engineers from talking to the pilot and confusing the issue." The center was also hooked up to a radar operator at an air defence radar station at Orillia, about 100 km north of Toronto.

One day in the fall of 1958, when Avro test pilot Spud Potocki was taking the second Arrow on a speed run to the edge of Mach 2, the radar operator, who Matthews believed had never seen anything on his screen faster than a CF-100, couldn't suppress himself. "Will you look at that sonofabitch go!" the operator yelled over the control center loudspeakers. "Will you look at that sonofabitch go!" Added Matthews: "We never did find out how fast the aeroplane could go. It was still accelerating and still climbing ... and that was with the J-75 engine and not the Iroquois engine. They never did find out how fast it could go."

•

By the time Potocki took the Arrow on its fastest flight, political storm clouds had gathered over the plane. The news that the Soviet Union possessed ICBMs that could loft atomic bombs to the United States in a matter of minutes threw U.S. and Canadian defence planning into confusion. The boasts of Soviet leader Nikita Khrushchev that the USSR was turning out ICBMs "like sausages" and that bombers were obsolete were widely believed until photos from U-2 spy planes and reconnaissance satellites disproved them. But in the interim, defence experts stated that the days of the nuclear bomber were numbered and that defence strategies should turn to dealing with the ICBM threat.

With the greatest mandate in Canadian history to that time, Prime Minister Diefenbaker had a free hand in 1958 to decide the future of the Arrow. He was still strongly concerned about the political fallout from an Arrow cancellation, and so the Arrow was on the agenda when U.S. President Dwight D. Eisenhower came to Ottawa in July. Later that summer, Defence Minister Pearkes went to Washington to meet U.S. defense secretary Neil McElroy, who offered to share the cost of installing new radars, control systems and Bomarc surface-to-air missiles in Canada. He told Pearkes the U.S. government would not buy any Arrows and offered to sell U.S. interceptors to Canada.

Following the meeting, the cabinet defence committee met and recommended that "consideration be given to" ending the CF-105 program. Starting in late August, a series of cabinet meetings pondered the fate of the Arrow. Pearkes kicked off the discussions by saying the aircraft should be abandoned. The Canadian chiefs of staff were divided on the Arrow, with the RCAF in favour and the other forces against because they feared losing funding for their own arms programs. But many cabinet members feared the economic consequences of cancelling the Arrow, particularly at a time when the economy was slowing down. Top officials from Avro came to Ottawa to try to save the Arrow.

Cabinet reached a series of decisions on September 21, and Diefenbaker announced them two days later in a five-page written statement. Cabinet had decided to build two Bomarc bases. As well, the Pinetree radar system would be strengthened and the semi-automatic ground environment (SAGE) control system would be installed to help direct the Bomarcs and aircraft toward any Soviet bombers. The statement obfuscated when it got to the difficult subject of the Arrow. It announced that the Arrow would not be put into production, but that the development program would continue until March 31, 1959, at which time a review of the program would be concluded. While the Arrow aircraft and Iroquois engine would continue to be developed, the government scrapped further work on the Astra weapons control system and the Sparrow missile. They would be replaced by off-the-shelf equipment from the U.S.

Many press reports suggested that the announcement was the death warrant for the Arrow, notably an article by veteran Ottawa correspondent Blair Fraser in *Maclean*'s magazine entitled "What Led Canada to Junk the Arrow?" Despite frantic lobbying efforts by Avro, sentiment against the Arrow grew in cabinet.

In mid-January 1959, cabinet began another long series of discussions on terminating the CF-105. The cabinet defence committee, fortified with doubts about the Arrow from the chiefs of staff, recommended in early February that the Arrow be scrapped. The consensus was virtually complete, and at a meeting on February 17, the Diefenbaker cabinet made the formal decision to cancel the Arrow program. The announcement of the Arrow cancellation was set for Friday, February 20, 1959.

•

What became known as "Black Friday" at the Avro plant began as a normal winter day for everyone except a few executives who noticed an ominous silence in Ottawa. Five Arrows with the J-75 engines had been constructed and were flying. The day before, Spud Potocki had taken the third Arrow, 25203, and then the first Arrow, 25201, up for short flights. The sixth Arrow - the first to be fitted with the more powerful Iroquois engine - stood at the head of the assembly line, just a few weeks away from its first flight and what everyone at Avro hoped would be a run at the world speed record. Avro test pilots were getting ready for further flight testing of an Iroquois engine which was attached to the back of the fuselage of a modified B-47 lent by the USAF.

At 9:30 a.m. Diefenbaker rose in the House of Commons and announced that the Arrow was cancelled. Noting that his September 23 statement had "foreshadowed" the cancellation, the prime minister then went into a lengthy set of justifications for the decision.

At the same time, top officials in the Department of Defence Production informed Avro of the cancellation in a series of phone calls and telexes. Soon, news reports of Diefenbaker's statements reached friends and relatives of Avro workers, and their telephone calls alerted workers inside the Malton plant. At 11:15 a.m., a loudspeaker announcement in the plant told the workers of the prime minister's announcement, but added that work would continue until further details were obtained. In another announcement that afternoon, Crawford Gordon ordered all workers except security personnel to put down their tools and go home. In one stroke, 14,528 employees of Avro Aircraft and Orenda Engines were laid off.

The events of Black Friday kicked off a debate that still hasn't ended: who was responsible for the demise of the Arrow and why? Although Diefenbaker's cabinet made the decision, there is strong evidence that the Liberals were also prepared to cancel the Arrow had they won the 1957 election. Armed with an opinion from C.D. Howe that "the CF-105 should be terminated - costs are completely out of hand," the Liberals did not directly attack the decision to cancel the Arrow but only the manner of the cancellation.

Many people blame the United States, pointing to its refusal to buy any Arrows and its offer to supply replacement aircraft, which was later taken up. This ignores the fact that both the USAF and NACA strongly assisted the development of the Arrow from beginning to end. The American decision not to buy the unproven Arrow was not unusual. The U.S. has always protected its aircraft industry, and aircraft plants are found in the districts of many powerful congressional leaders. Indeed, the only foreign-developed jet aircraft to make it into service in the U.S. armed forces in any significant numbers are the Canberra bomber and the Harrier Jump Jet, which were both developed and proven in Britain, and then manufactured in American plants under licence. Jim Floyd said it was almost impossible to sell an aircraft until it had been flown and proven. "If we had operated the Arrow, I'm sure the British would have bought them, and the Americans would have looked at it."

At bottom, the Diefenbaker cabinet feared a program which it was warned would have a price tag of $400 million a year for the next few years, an enormous sum of money at the time. These estimated costs covered what were expected to be huge maintenance and fuel costs for the Arrow. The figures were pumped up for public consumption by adding previously incurred development costs. Floyd pointed out that costs were driven in part by the RCAF's ambitious performance requirements and its demand for immediate production of the Arrow. "It was an expensive program because we had to get it right. You never get high technology for free."

There was a welter of confusing advice about whether or not the Arrow would soon be obsolete. The Diefenbaker government opted for what it thought was a quick and relatively inexpensive fix to deal with the Soviet bomber threat - the Bomarc missile. The Bomarc was obsolete by the time it was deployed four years later, and many believe that the Arrow was the best aircraft of its type in 1959 and would have long outlasted the planes that replaced it.

The controversy was deepened by the decision a few weeks later to break up for scrap the six aircraft that were built, and to destroy all the plans and other artifacts of the Arrow. Only the nose section of the sixth Arrow and some models, engines and other parts now remain. Many people believe that Diefenbaker ordered the destruction out of a vindictive streak or a desire to erase all traces of what he had ended. But there is no hard evidence to support this allegation. Attempts were made to persuade the Defence Research Board in Canada and the Royal Aeronautical Establishment in England to use the Arrows for aeronautical research, but both refused. The reasons behind the destruction of the Arrows will always remain open to argument.

A smaller controversy was generated by the government's refusal to assist further development and sales of the Iroquois engine, which had attracted interest from aerospace firms in the U.S. and overseas for use in other aircraft. Although its first installation in the Arrow was near on Black Friday, the Iroquois still faced serious production problems and was a long way from being ready for mass production. Orenda Engines survived the Avro cancellation, but Avro Canada itself and Canada's status as a manufacturer of military jets were dealt a fatal blow on Black Friday.

The CF-105 Avro Arrow has flown into Canadian myth, leaving a trail of controversy. Many see the Arrow cancellation as the end of Canada's period of wartime and postwar prosperity and power, and the beginning of Canada's conversion into a branch plant economy vulnerable to the whims of its southern neighbour.

But for the thousands of stunned Avro employees on the evening of Black Friday, those arguments could wait. The question on everyone's mind was, what comes next?

2 Moving South

Inside the Avro plant, a thunderous silence followed Crawford Gordon's announcement that everyone was laid off. Then conversation picked up as the thousands of shocked Avro employees downed their tools, gathered their clothes and headed out. A huge traffic jam developed as everyone drove away at once into a February snowstorm. John Hodge was part of a car pool of six Avro employees who lived in Oakville. "We went home, pulled everyone together and had a big party," Hodge recalled. "We didn't know what to do or what was going to happen." Engineer George A. Watts arrived home, told his wife Norma he'd been laid off, and was greeted with disbelief. "She looked at me and she laughed. It wasn't until the next day that she finally realized I was telling her the truth, when I was looking through the paper for jobs." Many of the laid-off employees could hardly believe it themselves. Diefenbaker's announcement was a surprise since the March 31 deadline for the Avro review was still weeks away. Was Crawford Gordon's layoff a gigantic bluff designed to force the government to reconsider, many wondered, and would it work?

Those hopes were reinforced over the weekend when many people were telephoned and told to return to work on Monday. But it became clear in the days following Black Friday that the government wouldn't reverse its decision. Most workers who were recalled to Avro were engaged in winding down the operation or helping others in their job searches and would soon be out of work. A small number came back to continue work on other projects, including those who were working on the Avrocar flying saucer until it, too, was cancelled in 1961.

The future arrived quickly for some of the laid-off people. George Harris Jr. awoke to the sound of the telephone on Saturday morning. He was feeling the effects of a night of drinking in his basement with his Avro colleagues. The voice on the phone invited him to breakfast at an airport hotel, and by noon that day, Harris had been hired by North American Aviation in Columbus, Ohio, to work on flight test instrumentation on the A2J Vigilante program. He was one of many Avro engineers who went to work in the U.S. aerospace industry.

But most weren't so lucky so fast. Although many people would get good offers of work elsewhere in Canada or in the U.S., these offers took time to materialize. Once the reality of the Arrow's cancellation had settled in, the need for new work became the overriding consideration. "Like everyone else, I had a mortgage and I had just had another child," Bryan Erb recalled. "So you've got to figure out how to put bread on the table. After the first week or two, you realize it's not going to get turned around. Yes, there's severance benefits, but they would run out before too long." George Watts remembered that he had voted for Diefenbaker to teach the arrogant Liberals a lesson: "We voted ourselves out of a job, out of a career, or in my case, out of the country. I haven't let my feelings influence my vote ever since. I think about it very carefully now."

Many of the newly unemployed never got a good offer and left aviation for good. Some of the British engineers looked homeward. Jim Floyd eventually returned to England where he began work on what would become the Anglo-French Concorde supersonic transport. But many decided against returning across the Atlantic to a country still emerging from postwar austerity and still afflicted by open class divisions. As engineer Burton Cour-Palais summed up an offer from his former company in England, Vickers Armstrong: "I knew what they would do to me. I would be put back right where I was when I started." Faced with a similar job offer back home, Peter Armitage looked into staying in Canada and enlisting with the RCAF. Because of the extremely high standards and high budgets of aircraft programs, he said, aeronautical engineers had trouble finding engineering jobs outside of the aircraft industry because of the fear that aeronautical engineers "might just call for too much sophistication."

Depending on their skills, experience and contacts, the Avro engineers dealt with many offers or a lack of them. Many journeyed to various parts of the U.S. and Canada for job interviews. Job offers included computer work with the oil industry in Calgary, Alberta, and work with a ball bearing company in Pittsburgh, Pennsylvania. Some prospective employers set up shop in Toronto hotels, and many ex-Avro workers saw each other repeatedly as they went from one job interview to another. Some weren't so lucky. "I put in 40 applications and never got an answer," Fred Matthews said. "Not only no acceptance, not even an answer. So things looked pretty bleak."

Among those most concerned with the fate of the engineering team at Avro were its leaders, chief engineer Bob Lindley and chief of technical design Jim Chamberlin. Both had developed strong contacts with their engineering colleagues at NACA in Langley Field in Virginia, including Robert R. Gilruth and Charles J. Donlan, who were now involved in a project to put an American into space. David D. Ewart, who was doing wind-tunnel tests of Arrow models, remembers meeting Donlan when Ewart went south to test the models in Langley's wind tunnels. "While I was down there, Charlie Donlan expressed an interest in the CF-105, and he asked me to set up a meeting with Jim Chamberlin, and he knew Jim was a mine of information on the 105. So I set it up, and Charlie Donlan and his staff pumped Jim Chamberlin out of information on the 105." Lindley and Chamberlin remembered the impression the Arrow had made and decided to see if the new U.S. space program was interested in taking on some engineers from Avro.

•

The launch of Sputnik and the flight of the dog Laika on Sputnik 2 in 1957 shocked the United States and impressed much of the rest of the world. The American scientific and political establishment was dealt a worse blow on December 6, 1957, when an attempt to launch a Vanguard satellite failed in a nationally televised fireball on a Cape Canaveral launch pad. The Vanguard launch, which was intended as a launch vehicle test, immediately became known to headline writers as "flopnik." President Dwight Eisenhower, who had tried to minimize the importance of the two Sputniks, came under fire for allowing America to start second in the space race. America's defences, technological capabilities and educational system were questioned. Senate majority leader Lyndon B. Johnson of Texas and other Democrats in the U.S. Congress attacked Eisenhower's space policy.

While the political heat increased, the U.S. government turned to Wernher von Braun's rocket team from Germany, now in the employ of the U.S. Army at the Redstone Arsenal in Huntsville, Alabama. On January 31, 1958, von Braun's team launched Explorer 1 atop a modified Redstone rocket. Explorer 1 was a smaller satellite than Sputnik, but its instruments led to the discovery of the Van Allen radiation belts girdling the Earth. Despite this success, it was clear that American space efforts were in disarray. The U.S. Army had launched a satellite, Vanguard was backed by the U.S. Navy, and the U.S. Air Force was busy planning satellite programs of its own, including human spacecraft.

Eisenhower, troubled by this infighting and the strong public reaction to the Soviet space successes, worked with Johnson and other Congressional leaders to design an administrative structure for the U.S. space program. Eisenhower chose to restrain military spending demands in the wake of Sputnik and place space research, including what was then called manned space flight, in civilian hands. Accordingly, he decided that the military would concentrate on building missiles and reconnaissance satellites. In April, Eisenhower proposed the creation of a civilian space agency, and after some hard bargaining the National Aeronautics and Space Act of 1958 passed through Congress and was signed into law on July 29. On October 1, the National Aeronautics and Space Administration (NASA) came into existence, absorbing the infrastructure of the National Advisory Committee for Aeronautics and some facilities that had previously been tied to the armed services. On NASA's first day, some of its engineers were already designing the new agency's first human spacecraft.

On October 7, NASA's new administrator, T. Keith Glennan, received the first reports on human satellites and approved the efforts by saying: "All Right. Let's get on with it." The next day the Space Task Group (STG) was unofficially set up under Bob Gilruth. It gained formal status on November 5, and in December the man-in-space project was named Project Mercury.

In January 1959, McDonnell Aircraft of St. Louis, Missouri, was selected to build the Mercury spacecraft. As the Avro Arrow was cancelled, NASA was gearing up for the challenges of Mercury, including hiring new engineers and selecting pilots for the Mercury spacecraft.

•

In the first week of March 1959, Bob Lindley and Jim Chamberlin flew from Malton to Washington, D.C. Lindley said that he and Chamberlin, armed with what Lindley called a "nifty brochure," went to the Canadian embassy in Washington and to NASA headquarters to sell the idea of the Canadian government supplying Avro engineers to the Mercury program in an arrangement similar to defence production sharing arrangements that were proliferating between the U.S. and Canada.

A March 5 meeting involving A.D. Belyea of the Canadian Department of Defence Production (DDP) and NASA's director of space flight development, Abe Silverstein, is recorded in a cable now in the National Archives of Canada.[1] Belyea cabled D.L. Thompson, the defence production coordinator at DDP headquarters in Ottawa, that he told Silverstein a "proposal for Canada to offer 30-45 senior engineers of Avro and Orenda was under active consideration in Ottawa and I expected yes or no today but in any event not later than Wednesday of next week. Meanwhile Chamberlin and senior project officer Mercury left for Langley AFB to discuss individuals concerned without prejudice. I advised [Canadian] government offer would be complete package deal for two years with renewal to be discussed at end of two-year program. Estimate of cost to DDP will largely depend on overhead we will agree today Avro [sic]. Naturally Glennan and Silverstein most anxious to close deal .... In my judgment this proposed deal represents the most significant move we could make in production sharing."

Belyea's cable went on to elaborate on the arguments in favour of the deal, including Canada's sharing in the prestige of space travel, and records Lindley and Chamberlin meeting with the Canadian ambassador to the U.S., Arnold Heeney. Belyea reported that Heeney planned to consult with his scientific attaché and then discuss the matter with David Golden, the deputy minister at DDP, something that "startled me some what [sic] to say the least." Like Belyea, Lindley recalled that the ambassador had responded positively to the idea. Lindley said the proposal went to Prime Minister Diefenbaker, who rejected the idea out of skepticism over the concept of space travel.

No other known document in U.S. or Canadian archives follows up on Belyea's cable, so the question of who in the Canadian government ultimately dealt with the proposal remains open. There is also the possibility that a direct role for the Canadian government in the U.S. space program was rejected by someone in the U.S. government. Perhaps Belyea and Lindley's views of the acceptance of their idea in Washington were affected by wishful thinking. What is clear is that the idea of official Canadian participation in Mercury died a quick death.

"So there we were," Lindley said. "We had NASA intrigued. But worst of all, we had all the good guys, who were our boys, all pumped up to go to work for NASA. So finally, in desperation, we said to NASA, 'How would you like to hire these guys?' They said, 'We'd love to hire them.' And they went and dug around and said, 'Yes, we'd do it.'"

There is no doubt that NASA was interested. Although accounts of what happened vary, it is clear that the top levels of NASA moved quickly to hire Avro engineers. At that time, NASA was in desperate need of engineers for Project Mercury. The Space Task Group began with 35 engineers from Langley and 10 others from the NASA Lewis Research Center in Cleveland, Ohio, when it was officially formed at Langley Research Center in November. By January 1959, STG had grown to 150 engineers, but many more would be needed to make Mercury fly, and qualified engineers were in limited supply. Available engineers were paid more in private industry. And many people were skeptical about the idea of shooting men into space on top of missiles that were blowing up with great regularity at the time. Even if it worked, Mercury might be a flash-in-the-pan project that would lead an ambitious engineer nowhere. By early 1959,

[1] First published in Campagna, Palmiro, *Storms of Controversy,* Stoddart Publishing, Toronto, 1992, pp. 156-7. The cable is located in National Archives of Canada RG 49, Vol. 781, File 207-6-3.

many NASA officials were resisting further recruitment from their established aeronautical research programs for Project Mercury. So the availability of the unemployed Avro engineers must have seemed like a godsend to those in charge of Mercury.

"Most Americans thought that Mercury was going to be a fly-by-night reaction to Sputnik that wouldn't last that long," Hodge said. "That was the general expectation. Most serious engineers in the States weren't the least bit interested in the Mercury program. So that left it wide open for the younger people."

Bob Gilruth, the head of Mercury, said he heard about the Avro engineers from an aviation acquaintance in Canada whose name he had since forgotten. "I got a call from a friend of mine in Canada who said that the Canadians had lost this big contract for building a fighter. They were going to have to get rid of some of their top people. Here we were in the Space Task Group just absolutely strapped for top people, and so he suggested I might want to come up and talk with some of these people and recruit. It was a big break for us."

Lindley and Chamberlin returned to Malton soon after their early March meetings in Washington with word that NASA would hire some of the Avro team. A meeting was called in the Avro cafeteria for interested engineers, and Lindley and Chamberlin explained what Project Mercury was about and what the work would involve. Those who were interested were invited to leave their resumes for consideration by NASA. A few days later, an aircraft landed at Malton bearing a group from NASA that included Gilruth, assistant project manager Charles Donlan, operations chief Charles W. Mathews, engineering chief Charles H. Zimmerman, Gilruth's assistant Paul E. Purser, and NASA Langley personnel man W. Kemble Johnson. After a night at a hotel, the NASA group came to the Avro plant and split up to interview about 100 of the applicants, who had been selected from the nearly 400 engineers who had left applications with Chamberlin and Lindley. The interviews took place on Saturday, March 14, 1959. It was a cold day at Malton, and the Avro plant had a forlorn, deserted look about it, according to Mathews. The interviews took place in anterooms off the silent engineering bullpens that less than a month before had hummed with activity.

Top NASA officials at Marshall Space Flight Center. (l to r) Delmar M. Morris, MSFC director for administration; Eberhard Rees, MSFC deputy director for research and development; Dr. Wernher von Braun, MSFC director, Dr. T. Keith Glennan, NASA administrator; and Maj. Gen. Don R. Ostrander, director, office of launch vehicle program, NASA. Courtesy NASA.

"We couldn't have asked for nicer treatment from Avro," Purser recalled. "We couldn't ask for better people than we got because they were all top notch. Avro said they were all top notch, and they wouldn't think of letting us see any of them, except that they lost the CF-105 contract."

•

For some of the applicants, it was a close thing. Peter Armitage recalls coming home at about 4 in the afternoon that day to find a message from Lindley. Armitage phoned back, and Lindley told him he had an hour to get to Malton for an interview with NASA. "I don't remember moving so fast in my life," Armitage said. "Malton was about 30 minutes from home, and I got there about 4:45 p.m. I was given a standard U.S. government form SF-57 [employment application] to fill in. There really wasn't time, so I just put my name on it and attached my resume. I remember meeting Bob Gilruth and Chuck Mathews for about five minutes, as well as Kemble Johnson. I was in and out in about 15 minutes - without much hope for a job offer."

John Shoosmith arrived home the day before the interviews from Calgary, where he had interviewed with an oil company that was seeking his expertise in computers. Shoosmith suffered on the long flight back to Toronto because he had caught the measles from a young cousin. Despite his pain and fever, he pulled himself out of bed the next day for his interview with NASA.

Dick Carley, who said he had recommended to Chamberlin that he check NASA as a possible source of employment for the Avro engineers, had gone skiing after offering that advice. He came back just in time to meet the NASA recruiters, in his case Donlan, for whom he had a question. "I had kind of an ideal situation at Avro. I had my own lab, I had computers, I had people who could build hardware. If you wanted to play with the hardware yourself or you didn't like the way the electronics was, you could go build it. And that's the only way you learn. So my question was, well, you know, are we going to be allowed to ever have hardware, or is the hardware just going to be designed by contractors or something like that. There was assurance, yes."

Rodney G. Rose was a British engineer who was reluctant to move to the U.S. because he had relatives near Toronto and entertained hopes of getting work with the National Research Council in Ottawa. He only applied to NASA after Chamberlin had spent two hours persuading him that there was little reason to expect new work at Avro and that helping put a man into space at NASA would be far more interesting than doing pure research at the NRC.

David Ewart was serving in an auxiliary RCAF squadron and was scheduled to fly a C-45 to Prince Edward Island from the nearby Downsview RCAF base the day of the interviews. Ewart showed up at Malton in RCAF uniform so he could head directly to Downsview for his flight. He was interviewed by Donlan, who he had met before during wind-tunnel tests of the Arrow at Langley. "By the time I got to Downsview, my co-pilot had the engine running and everyone was waiting for me," Ewart recalled. "But we did make it to Prince Edward Island in good time."

Fred Matthews says that the story of his experiences working on the fatal CF-100 canopy problem made an impression on the NASA officials who interviewed him. Owen Maynard, tipped off about the interviews just before they happened, said he was asked by Gilruth if he was interested in the man-in-space program. "I said I'd be interested in engineering or flying. Gilruth said, 'Well, we aren't recruiting flight crew.' I'm probably the first Canadian to volunteer to be an astronaut, and the first to be rejected." Maynard added that the fact that he had flown the Mosquito during the Second World War may have caused Gilruth to hire him.

After the interviews, the NASA group informed Chamberlin and Lindley of their choices for hiring. Before they boarded their aircraft to return to Langley, they paused to look at what their new recruits had done. "We went out to the hangar where the CF-105s were still in existence," Chuck Mathews said. "Very impressive airplane. In a way, much superior to anything the United States had at that time. Certainly, like everyone else, we were dismayed that they were going to break them up for scrap."

Starting the next day, Chamberlin and Lindley telephoned the successful applicants, including Armitage, Shoosmith, Ewart, Matthews, Rose, Erb, and Maynard, to let them know.

Erb said, "It didn't take very many microseconds to accept." The lack of another job offer made the decision easy in spite of the fact that he had United Empire Loyalist blood on both sides of his family. "My ancestors had left the United States at the time of the revolution." Two days later, Erb got another offer from the NRC in Ottawa. "I thought about it a bit, but I decided that no, [NASA] was too close to what I wanted to do, too interesting. I've got to move from where I was anyway." So Erb prepared to move to the U.S. But he had not heard the last of the NRC.

Armitage remembered that Chamberlin gave him an hour to decide whether he would accept or not. It was the first solid job offer he had got since Black Friday. "My wife wasn't too thrilled about the prospect of leaving Canada. Canada was a bit closer to being in England, and that's where she really wanted to be. But the bills were pressing, so I called Jim back and said yes." Others report similar understandable reactions from their wives.

George Watts waited at home a couple of days while others were being phoned. "I called up Jim and said, 'Jim, you didn't call me.' He said, 'Oh yeah, I forgot. Come on.' So that's the way I got hired."

The group hired by NASA included several people who had worked with Chamberlin designing and engineering the Arrow, but the fact that hiring choices were driven by NASA's needs was shown in the fact that many people associated with Chamberlin at Avro weren't selected or didn't apply. Many of the people chosen by NASA came from flight test, or were computer specialists.

•

On March 18, NASA administrator Glennan wrote a memo approving the hiring of the "research and development group of alien scientists having special qualifications in fields closely related to manned space flight." Listed on the memo were the names of 32 people from Avro, including Jim Chamberlin, but not Bob Lindley, who chose to remain at Avro for the time being. Seven of the 32 on the list didn't join NASA, presumably because they decided not to take the jobs.

Chamberlin, who according to Glennan's memo would be paid the then princely sum of $16,000 a year, twice the salary offered to many of the others, was among the first to go to Langley. John Hodge, the next most senior member of the group, stayed behind in Toronto to help organize matters on that end and was one of the last to go.

Anyone who has immigrated to the United States knows that obtaining the proper papers from the U.S. government is a long and difficult process. The 25 Avro engineers heading south also had to obtain security clearances because Project Mercury was a top priority effort involving missiles that were at the heart of the nation's defences. Most of them still sound mildly astonished when they recall what followed their hiring by NASA, which was confirmed in letters that went out March 25.

"NASA was as good as its word, and the consulate in Toronto was shut down for everyone except us, and we got the red carpet treatment," Bruce Aikenhead said. "Our papers were processed very quickly. But we still had to follow the rules." Everyone had to get medical exams, and police questioned old friends and acquaintances of the engineers as part of the security clearance process. Peter Armitage marvelled at the unusual speed and efficiency of their processing: "Someone high in the U.S. government had pulled a lot of strings - and the right ones!"

In spite of this, the processing did take time, and so did the work of preparing to move individuals and families from the Toronto suburbs to a peninsula in Virginia. This was complicated by the fact that hundreds of homes around Malton went on the market simultaneously as Avro workers moved to new jobs all over North America or back in England. The newly hired NASA engineers were told to report to work at the Langley Research Center no later than Monday, April 27, and in mid-April, most packed families and belongings into their cars and headed to the Canada-U.S. border at Niagara Falls for the journey south to new jobs.

As they arrived at the border in their 1953 Chevrolet, Rod Rose and his wife Leila got one more reminder of the priority they had in the eyes of the U.S. authorities: "The chief immigration guy there took a look at my papers and said, 'You guys must have somebody really pulling for you, coming down within two weeks of your application. It usually takes six months to get an interview with the consul in Toronto.' I said, 'Well, I guess they need us.'"

Peter Armitage drove to the border at Niagara Falls in his 1954 Chevrolet with his wife June, who was then pregnant, and son Mark. "I remember June was teary eyed as we passed the Canadian flag and drove by the U.S. flag in the middle of the Rainbow Bridge over the Niagara River. We had made many trips to Niagara, but mostly for the day and never to cross over the border to stay. Things soon cheered up as we began to see and think about the new adventure before us."

A friend gave George Watts a lift to Buffalo, where Watts bought a new Ford to continue the trip. Norman B. Farmer, a British engineer who had turned down an opportunity to work with Canadair in Montreal in order to join NASA, took the bus to Buffalo, where he picked up a 1957 Plymouth he had previously purchased there, and then he continued his trip. Like many other family members of the Avro group, his wife Helen didn't join him until later.

Most of the NASA recruits drove to their new home, a trip that took two days in most cases. Owen Maynard drove his family down in a new Plymouth, and Stanley H. Galezowski drove by himself in a Volkswagen Beetle. Wintry conditions, including ice and slush, still lingered in Canada and the northern parts of New York State and Pennsylvania. But as they passed through the warmer climes of Maryland and Washington, D.C., cherry trees were already in bloom. Crossing the Potomac, they took Virginia State Road 1 to Richmond, where they turned southeast. State Road 60 led through the forest and marshland of tidewater Virginia, past historic Williamsburg and out onto a peninsula that was tipped by Hampton and Newport News and surrounded by the York River to the north, the James River to the south, and Hampton Roads and Chesapeake Bay to the east.

"It was sort of a trap," Maynard said of his early impressions of the Hampton area. "First of all, you're in a foreign country. Second of all, you're in a desolate part of a foreign country."

Fred Matthews was also not impressed with what he found: "I drove down and arrived in the middle of the night. It was raining in downtown Newport News, which was just the most dismal greeting you could ever have. Newport News in those days was not very attractive. It was a shipbuilding center, and these days it's even worse because the small downtown has essentially disintegrated. And I was so discouraged, this dark, dismal night in the middle of nowhere. I almost turned around and went home. But I didn't."

Jim Chamberlin was already there, arranging cheap accommodation at the Empire Motel a block away from Buckroe Beach, a stretch of sand with an amusement park at the north end of Hampton. Chamberlin himself stayed there, but the two-storey motel was not to the taste of everyone, with "terrible" and "awful" being among the adjectives used to describe it. Some such as the Maynards and Roses, and Dave Ewart and his mother, chose to go a few miles south to Old Point Comfort. There they obtained accommodations in the Chamberlin Hotel, an elegant brick seven-storey facility with a commanding view of Hampton Roads.

"We headed straight to Buckroe Beach, perhaps thinking it might be like the picture we had seen of Miami Beach," said Armitage. "What a disappointment!"

Frank Chalmers, the lone Scot of the Avro group, wrote his wife June in Toronto promising to move to better quarters at the Chamberlin Hotel when she arrived: "I don't want you to move into the 'Empire Hotel' as the place is getting more and more like a dormitory every day, on top of which it is so poorly sound proofed that my next door neighbour's early morning pee comes through so loud and clear that I often start up half asleep thinking he is in my bathroom."

John Shoosmith had driven down in convoy with Morris V. Jenkins, a trajectory expert. "He was driving his car down at the same time I was. The first impressions of the Virginia peninsula were not that great. I remember driving in on a rather narrow two-lane road with lots of crooked shacks along the road. And the motel when we arrived was adequate, but it was not exactly a palace. We were enthusiastic when we arrived at the motel, and one of the first things we did when we arrived was to change into our bathing suits and jump into the bay. It was early for the natives, but we were there until we were able to find other accommodations."

Bruce Aikenhead and others flew down to Virginia: "Getting to the airport itself was a major operation. The dog didn't like being put in a crate. We [flew] to Long Island and then we had to change to get to Washington, and then we had to change again to get to Patrick Henry Airport outside Newport News, Virginia. We left the house about 9 in the morning and we didn't get to Virginia until well after dark. I remember Dick Carley and Tom Chambers coming to the airport to meet some of us. We had our hands full. Our baby was six months old, and the other three were there. We had tons of luggage, and somehow we managed to squeeze it into a car that took us to a motel in Hampton, where some of the other families were set up. We stayed in the motel for a few days until we had a chance to get houses or apartments. The guys went off to report to NASA in the morning, and the women looked after the kids."

George A. Watts in 1964. Courtesy NASA.

3 Building Mercury From the Ground Up

"Langley Field is a fantastic place, it covers a huge area and is occupied jointly by the Air Force and NASA," Frank Chalmers wrote home after his second day of work, April 21, 1959. "To give you some idea of the size, the personnel office ... is three miles away from the office in which I work. The whole area is very well kept up and has an established look about it. There are all kinds of research labs, wind-tunnels etc, built and rebuilt through the years where NASA have done and are doing the research work that has made them famous throughout the world .... To get down to everyday considerations, the working conditions appear good, and cheap, and the people are very friendly and helpful."

The newly arrived Avro engineers felt at home at the NASA center at Langley because it flanked a huge airfield, just like Avro did at Malton. Most of the NASA facilities were then located in Langley's East Area, a complex of buildings dating from the 1920s, sandwiched between the airfield and the Back River. The staff of Project Mercury worked in the East Area in two three-storey brick buildings with stone NACA logos set above the doorways. These office buildings sat near wind tunnels that from the outside looked like large steel plants. On the other side of the airfield, newer research facilities had been built during the Second World War in what was called the West Area, where Chalmers found the personnel office.

The first day, Chalmers was taken to the personnel office with another new recruit, and "there we spent almost the entire day filling out forms, being fingerprinted, photographed and instructed in civil service fringe benefits." The next day was spent in an orientation in the office on Project Mercury.

This was a typical induction for the new recruits at NASA. These formalities were followed almost immediately by the assumption of new responsibilities. According to Owen Maynard: "The first thing they did was to say, 'Here's our view of the spacecraft. Give us a critical review from your perspective. We understand you did studies on the escape system for the Avro Arrow.' I did, and pointed out a few things that hadn't been questioned. It wasn't a 'not invented here' defensive posture. They took criticism that was intended to be constructive. They invited it." In his early months at NASA, Maynard came up with a list of more than 40 possible problems worthy of investigation, and he was assigned other members of the team to help him look into these problems, which ranged from the design of the control panel to the capsule's aerodynamics during re-entry and landing.

Peter Armitage was initially assigned to help develop recovery devices and procedures for Mercury. A week into his new job, Armitage was summoned by his new boss, Jerome B. Hammack, to go to a meeting on an unspecified topic. When Armitage arrived, he found top Space Task Group engineers filling the conference room. He took a seat in the corner.

"Jerry started the meeting and announced that we were going to go over the test program plans for a series of air drops of full-scale Mercury boilerplate spacecraft. The object was to test parachute deployment, water landing dynamics, recovery aid deployment and shipboard recovery techniques. Next he introduced me. Jerry said, 'Peter is the project manager for these air drop tests and I'm sorry but I have to go to another meeting, so Peter why don't you take over and discuss the test program?' And then he left. I had no idea of the subject, let alone that I was to be in charge! This 'baptism under fire' happened a lot with NASA."

Armitage began talking, drawing on his experience in the RAF and at Avro, and soon others in the group joined in. "Caldwell Johnson said he had the boilerplate capsule design well in work. Bob Thompson said he could arrange for C-130 aircraft to do the air drops and the Navy surface vessels that would be needed. Everyone contributed, and soon we had the start of a test program. The meeting ended and I went off to write a test program."

"I'm not sure any other country could have absorbed as big a percentage of 'foreigners' as NASA did at that time and put them, basically, into middle management positions," Rod Rose said.

"At the Space Task Group we were given immense responsibility," Bryan Erb remembered. "It was really amazing. It was for a couple of reasons. On average, the people who went down from Avro were pretty experienced hands. Most had been through a number of aircraft projects including the Arrow. A lot of people at NASA were out of a research tradition. But few of them had been through a serious, large-scale aircraft project. Gilruth and [Mercury designer Maxime A.] Faget had flown rocket tests to gather high-speed aerodynamic data. They were the best in the world at their game. But they had not pushed hardware out the door on an industrial scale. The Avro people brought skills that were very complementary to the NASA skills, and in a way unique. NASA had extremely competent contractors, but as far as the customer side was concerned, the Avro people brought the industrial mindset and discipline to what had been to start off with a much more research-oriented organization. These people were given very substantial responsibilities to start off with."

The new recruits were also eager to show that they fit in with the NASA engineers. One was George Watts, a native of Trail, B.C., who studied at the Spartan School of Aeronautical Engineering in Tulsa, Oklahoma, and the University of Toronto, and had worked on loads for 10 years at Avro. Watts stepped in to help solve a problem with the Mercury parachute deployment. When the main parachute deployed, the steel cable that pulled the main chute and a drogue chute canister away from the capsule was snapping, resulting in crashed capsules. Watts calculated that the kinetic energy imparted by the canister hadn't been taken into account, and he proved his point to skeptical colleagues when a Mercury capsule went through a drop test with an extra cable. "So they dropped it out of the airplane. Sure enough, it came back with the first cable broken and the second one not broken. So they thought that was wonderful. Whenever you go to work for a new company as an engineer, you want to do one of those 'rabbit out of the hat' tricks for them just to get in well with people."

"Everybody had a job to do, and it was one of those cases where you sort of looked around and said, 'That needs to be done,' and you talked to somebody and they said, 'Well, go do it,' and that was it, and you went out and did it," John Hodge remembered. "Travel regulations and things like that, we broke every rule in the book. We really did."

The 25 Avro engineers reported for work on three successive Mondays, April 13, 20 and 27, 1959. Armitage, Maynard and Bruce Aikenhead were among those who started on April 27, along with seven other newcomers, the first persons to enter training to fly into space.

When Aikenhead reported for work, his new boss, Harold I. Johnson, declined to orient him right away. "He said, 'Instead of going into this right now, why don't you come along this afternoon when I have to brief these new people we've hired to be the astronauts? They arrived on station in the last couple of days and they're getting a series of lectures.' So the eight of us - the seven astronauts and myself - filed in to the lecture room and Harold Johnson explained what the training program was likely to be and the type of training devices he thought would be necessary. So that's when I met the astronauts for the first time. As it turned out, their office was immediately next to ours."

The seven astronauts were Lt.Cdr. Alan B. Shepard, Lt. Malcolm Scott Carpenter and Lt.Cdr. Walter M. Schirra of the Navy; Capt. Virgil I. "Gus" Grissom, Capt. Donald K. "Deke" Slayton and Capt. Leroy Gordon Cooper of the Air Force; and Lt. Col. John H. Glenn of the Marines. Based on the decision of President Eisenhower and the leadership of NASA, the new astronaut trainees came from the ranks of military test pilots and had undergone a rigorous selection process. Like the others at STG, the astronauts were trained in aeronautical engineering.

•

In their initial assignments at NASA, Jim Chamberlin was placed in the office of project director Bob Gilruth, with the task of providing technical assistance. Thirteen other members of the Avro group were assigned to the operations division of Mercury and the remaining eleven to the flight systems division.

In the operations division, Hodge was assigned to be assistant to the head of the division.

The seven Mercury astronauts: (l to r) Walter Schirra, Alan Shepard, Donald Slayton, Virgil Grissom, John Glenn, Gordon Cooper, Scott Carpenter. Courtesy NASA.

Aikenhead was project engineer on the procedures trainer and the flight simulator. Armitage was responsible for qualification tests of recovery systems. Chalmers worked on design of the control center at Cape Canaveral. Thomas V. Chambers was assigned to operational aspects of the capsule control system. Computer experts Jack Cohen, Stanley Cohn and John Shoosmith were assigned to mission analysis, with Cohen made assistant head of the branch. Dennis E. Fielder's first job was with the capsule and range instrumentation system. John K. Hughes, an English engineer, worked on the worldwide communications system. Fred Matthews' initial assignment was setting out Redstone blockhouse requirements. Len Packham, a communications expert, was assigned to capsule telemetry and instrumentation. Tecwyn Roberts began at NASA by drawing up procedures for range safety and the countdown.

In the flight systems division, Dick Carley was assigned to stabilization and control systems. Eugene L. Duret and Bryan Erb worked on aerodynamic heating analysis; David Ewart on reentry dynamics. Joseph E. Farbridge, one of the Englishmen, was assigned work on loads, pressure and aerodynamics. Norman Farmer's initial assignment was capsule instrumentation. Stan Galezowski and Morris Jenkins began work on stability and control. Owen Maynard started in aerodynamic heating and structures. Rod Rose was named project engineer on the Little Joe test vehicle. Watts had responsibility for aerodynamic and structural loads.

The growth of Project Mercury and the Space Task Group was reflected in a reorganization that took place in August. The new organization list for Project Mercury still fit on just a few pages, but Mercury was split into three divisions, which each had several branches. Many branches were split into sections, and the former Avro engineers, in common with their colleagues, got new assignments.

Chamberlin became acting head of the engineering and contract administration division, and the following January he was named head of the newly redesignated engineering division. He was also responsible for the capsule coordination office, which worked with the contractor that was building the Mercury capsules, McDonnell Aircraft Corporation, and was listed as Gilruth's technical assistant for the balance of 1959.[2]

In this period there were some comings and goings among the Avro group. Both Chalmers and Farbridge left NASA before the end of 1959. Late that year and in 1960, six other Avro engineers joined NASA and the Space Task Group: David Brown, Burton G. Cour-Palais, George Harris, John K. Meson, Leslie G. St. Leger, and Robert E. Vale. The most senior of these six engineers was Vale, who was named to head the structures branch and, on an acting basis, the structural analysis section. As well, Vale was assigned by Caldwell Johnson to review Mercury systems along with Maynard. St Leger worked on making sure that the Mercury capsule could withstand the stresses it would face in flight, and Cour-Palais did similar work with the escape rockets and the Atlas booster for Mercury. Harris and Meson found work developing the tracking and communications networks for Mercury. Brown was assigned to the engineering directorate.

Leslie G. St. Leger receives award in 1965. Courtesy NASA.

[2] Other changes: In the flight systems division, Maynard and Rose were assigned to the systems test branch; Ewart to the aerodynamics section; Watts and Farbridge to the loads section; Duret and Erb to the heat transfer section; Farmer to the electrical systems section; Carley, Chambers and Galezowski to the flight controls section; and Jenkins to the space mechanics section. Hodge remained assistant to the chief of the operations division, where Cohen, Cohn and Shoosmith worked in the mission analysis branch. Chalmers, Fielder, Hughes, Matthews, Packham and Roberts worked in the control central section; Aikenhead in training aids; and Armitage in the recovery operations branch.

David Brown in 1970. Courtesy NASA.

Harris recalls his move to NASA this way: One day a few months into his job at North American Aviation in Columbus, Ohio, he received a phone call. "A fellow said, 'Why don't you come over to the Holiday Inn in Columbus, have a drink with us. We bring you greetings from Tec Roberts and John Hodge, and they're all down in Langley Field, Virginia, and they want to know, why don't you come down and join them?' So I go down to meet this guy, and I think his name was Bill Gray, from NASA, Langley Field. Didn't know where he was from, actually. And I said, 'What are these guys doing down there?' And he said, 'Well, we've got this idea where we're going to put a guy in a capsule, going to fly him in space, and bring him back to Earth.' Sounded like a sci-fi story to me. So I asked him what kind of propellant, what kind of wings, what kind of engines, and there was none of that. The upshot of that was that I was asked to go ahead and join them and go down to Langley."

When Harris first told his bosses at North American that he planned to take a new job, they told him he couldn't go. Then he said he was going to NASA. "Immediately, they said: 'Sorry. You're allowed to go any time any place you want. Good luck.' That's how much pull NASA had in those days. I couldn't believe it. One hour, they're giving you a hard time. The next hour, they're just patting me on the head, 'Good boy, go.'" He reported for work at Langley on January 4, 1960.

•

By the time the Avro group began their new jobs at Langley, most of the major design features of Mercury had been set under the guidance of Max Faget and Caldwell Johnson at STG, and the contractor was already at work building the spacecraft. Flight schedules were being drawn up that included unmanned tests of the capsule and the boosters, followed by flights with chimpanzees aboard, and finally flights with astronauts. Suborbital flights into the Atlantic atop Redstone rockets would be followed by orbital flights using the more powerful Atlas missile as the launch vehicle.

The Mercury spacecraft was a bell-shaped vehicle that many called a capsule. It stood nearly 4 m high, including the retropack, with a diameter of nearly 1.9 m at the base. In orbit, Mercury weighed in at 1,355 kg. Except for an ablative heatshield on the blunt back end, the capsule was covered with dark gray metallic panels called shingles. The conical forward end faced forward during its ride into space, but contrary to the visions of most people, the capsule returned to Earth with the blunt heatshield facing forward. After an initial kick from retrorockets, Mercury would rely on the atmosphere to slow it down most of the way from its orbital speed of 28,000 km/hr. NACA researchers at the Ames Aeronautical Laboratory near San Francisco had discovered that a blunt shape would cause a shock wave to form in front of the spacecraft, carrying most of the heat and energy of re-entry away from the vehicle.

The Redstone rocket, designed and built by Wernher von Braun's group in Alabama, stood 25.3 m tall with its Mercury capsule, and provided 35,000 kg of thrust. The Atlas, by comparison, stood 29 m tall and packed 166,000 kg of thrust at launch.

As head of both the engineering division and the capsule coordination office, and later as the effective program manager, Jim Chamberlin had to make many decisions that touched on the booster rockets used in Mercury. But his major concentration was on the Mercury capsule. According to Erb: "Almost off the bat, [Chamberlin] introduced the whole notion of configuration management and control. Up until that time, whenever a NASA person had a better idea of what to do with the capsule, he would pick up a phone and call McDonnell and say, 'Do this instead of that.' Chamberlin instituted a configuration control board and said: 'We're only going to make changes when there's a good reason. We aren't going to whipsaw the contractor around with stray ideas. They may be good ideas, but that's not the way you run a program.' So he brought this kind of discipline to the program, and I think it was pivotal in the success of the Mercury program."

"The thing that's significant was that the NACA people didn't understand or recognize what a program office was," Chamberlin said in a 1966 interview, speaking of the former NACA researchers who had worked in large directorates set up according to research disciplines, not programs. "We did, in fact, during the Mercury program, establish such an operation. There was a lot of friction in doing it, a hell of a lot of friction. In any other organization, there wouldn't have been. The research people wanted to do everything themselves. And they had some peculiar manifestations of this."

NASA engineers who had been NACA researchers went to the McDonnell plant in St. Louis, Missouri, to monitor a particular issue in a capsule, and they would spend 20 hours observing work without taking a break, despite the fact that shifts had been set up. They wound up exhausted, Chamberlin said. "These people wouldn't act as a team and they wouldn't recognize that you have to organize and give people certain jobs and coordinate this, and all that. They wouldn't do that. That's why I got the job, because I had industrial experience and was familiar with this sort of stuff."

Working with the capsule coordination office, which supervised 17 groups making decisions on Mercury's systems and subsystems, Chamberlin and his counterpart at McDonnell, John F. Yardley, decided on the details of Mercury following Faget's basic design and flight concept. Faget said Chamberlin was "in charge of the day-to-day management of the Mercury Program."

Mercury never had an official program office until 1962, when Chamberlin officially moved to the Gemini program. But he had acted as a program manager from the time in 1960 when

Gilruth, Faget and others moved to Apollo or to supervise a human space program that was beginning to encompass more than Mercury. As Fred Matthews put it: "The father of Mercury was Max Faget. But Max Faget was off doing future things, so Chamberlin was put in charge of present things, Mercury. He became project manager of Mercury."

Chuck Mathews, one of the top people at STG, said Chamberlin was "the main contractual officer with regard to the McDonnell contract" to build Mercury. "But more than that, Jim took over the main job of the engineering responsibility for Mercury as it was being developed. Initially, that job was going to be undertaken by what was known as the flight systems division under Max Faget, but that group was not too experienced with production, design, manufacturing of airplanes and spacecraft, not that anybody was experienced with spacecraft. But Jim, even though he was an aerodynamicist, had that knowledge so he really took over that job. It goes without saying that he really dedicated himself strongly to accomplishing that. About as dedicated a person as you would find. Mercury had many design problems, and though far from a perfect vehicle ever, Jim did an excellent job in making it a vehicle that could fly and fly successfully."

Bob Gilruth called Chamberlin a "key figure" on Mercury, and explained: "He had a way of working with McDonnell that was very effective in that he put a lot of the burden on them. He really used McDonnell – if there was a problem he tasked it out to McDonnell and he, in this way, was able to give them a high feeling of responsibility, and at the same time use the few good people he had to maximum advantage."

Although Chamberlin is best known for what he did before and after Mercury, he played a crucial role in Mercury that made him arguably the biggest unsung hero of America's first human space project. Chamberlin provided the essential link between the designers and the people who actually built the Mercury capsules and made them work. Along the way, he added small design touches of his own.

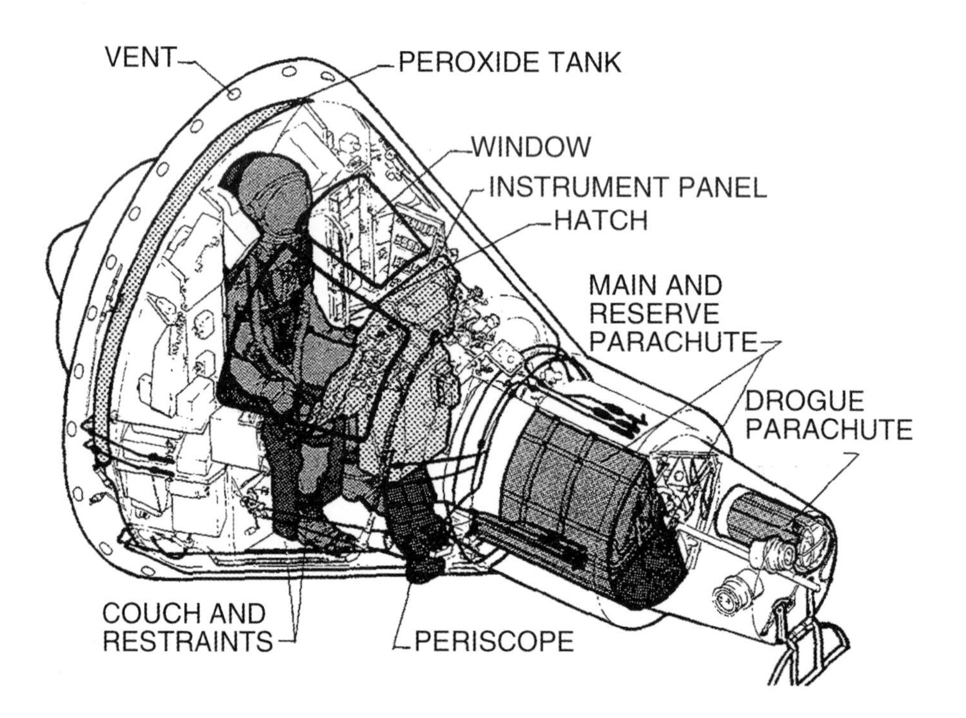

Schematic of Mercury capsule. Courtesy NASA.

Chamberlin and especially Dick Carley have been given credit for incorporating a fly-by-wire control system into the Mercury capsule. The Avro Arrow was the first aircraft that had been equipped with fly-by-wire, and the X-15 had followed soon after. Mercury management decided to equip the capsules with both manual and automatic control systems to control the capsule's attitude in space, but some people felt the two systems weren't adequate by themselves, so Carley, with his experience at Avro, proposed adding a fly-by-wire system. With Robert G. Chilton, who was in charge of Mercury's controls, Carley drew up plans to put together the mechanical and electrical interconnects between the manual and automatic control systems in Mercury for the fly-by-wire or "rate damping" control system. This third control system gave the pilot control over the capsule's movements and was more efficient than the manual system. Tom Chambers, who also came from Avro and worked on Mercury's control systems, stayed on to maintain these systems through the Mercury program when others left to work on Apollo and Gemini.

•

Chamberlin and his colleagues faced many serious problems with Mercury during 1960, not the least of which was a series of launch failures. The most serious failure was Mercury-Atlas 1, which blasted off from Cape Canaveral on July 29. MA-1 disappeared behind a thick bank of clouds and then, at 59 seconds into the flight, all contact was lost. The disintegrated capsule and parts of the rocket were later fished out of the Atlantic.

Maynard, who by then was moving to Apollo, was a guest at the launch, but before he could leave the Cape, he was assigned to help Chamberlin find the cause of the mishap. Maynard supervised the recovery of the rocket and capsule parts by Navy divers, and even took part in the diving himself. Determining the accident's cause without photos of the ascent proved to be difficult. At about the same time, NASA launches of lunar probes using Atlas-Able rockets had also failed. In the absence of hard proof, the failures were blamed on a structural failure of the upper part of the Atlas rocket. The remains of MA-1 didn't show conclusively what caused the MA-1 mishap, but they excluded other possible causes for the catastrophe. Atlas had a metallic skin as thin as a dime. If the tanks weren't pressurized, the rocket would crumple under its own weight. While this design was adequate for lobbing warheads at the Soviet Union, heavier payloads such as Mercury capsules were too much for the Atlas rockets then being made.

NASA ordered new "thick-skin" Atlases for Mercury, but with a space race on with the Soviet Union, there wasn't time to wait for one for the next shot, MA-2. Yardley and Chamberlin designed a "belly band" or metallic girdle wrapped around the base of the adapter that held the Mercury capsule to the Atlas for MA-2.

While preparations for MA-2 were going on, Mercury suffered more setbacks and criticism mounted. On November 8, a test of Mercury's escape system at Wallops Island failed when the escape rockets fired prematurely and the capsule didn't separate from the Little Joe booster rocket. That same day, American voters elected John F. Kennedy president of the United States, and no one knew what his attitude would be to the troubled Mercury program. On November 21, the most embarrassing failure of the Mercury program took place. Mercury-Redstone 1 lifted 10 cm off the pad, then settled back into place when the Redstone's engine cut off. The Mercury escape tower blasted away from the capsule, and then, like a cork on a champagne bottle, the capsule's antenna canister popped off, followed by the capsule's parachutes. The rocket stood on the pad for several hours before it was safe to approach. Those hours were filled with tension, because a gust of wind could have filled the parachutes and toppled the rocket.

MR-1 was repaired and launched successfully on December 19. On January 31, 1961, the chimpanzee Ham flew a suborbital flight aboard MR-2, a mission that was troubled by over-acceleration of the rocket and problems with the capsule's landing bag at splashdown. After the total loss of MA-1, the launch of MA-2 in February with the "belly band" and design changes to the adapter between the Atlas and the Mercury capsule was seen as a make-or-break event for Mercury. Another failure, and the newly installed Kennedy administration would be able to cancel Mercury and blame it on the Eisenhower administration. Kennedy's scientific advisors were known to be critical of Mercury, and the U.S. Air Force began intensely lobbying the incoming administration to give it the primary role in space flight that had been assigned to NASA. Bob Gilruth agonized over whether to fly MA-2 with the "belly band," but finally decided to go ahead. On February 21, MA-2 blasted off, and Mercury engineers were visibly relieved when the vehicle passed unscathed through the period of maximum dynamic pressure where MA-1 had failed. The suborbital flight ended successfully, and Mercury gained the confidence it needed to continue.

•

Rod Rose worked on tests of the Mercury capsule that were carried out using a jury-rigged vehicle made up of clustered solid fuel rockets called Little Joe. This low-cost rocket was designed to test Mercury and its escape systems under the severe conditions of launch. Little Joe launches took place at Wallops Island, Virginia, where models of the Avro Arrow had once flown atop Nike missiles.

Rose endured many hardships at Wallops Island, including nearly sinking in a boat that was ferrying him back from the island launch site to the mainland. Today he can laugh about the incident, along with the story of a careless camera operator. Little Joe leapt off the pad much like a fireworks rocket, quite unlike a liquid-fuelled Atlas, which lumbered off the pad before gaining speed. When Rose explained this to news media covering a Little Joe flight, he noticed that one camera operator wasn't paying attention. Rose asked his people to keep an eye on that man and was gratified when they returned after launch to report that the launch had left him asking: "Where'd it go? Where'd it go?"

Little Joe had more than its share of troubles. Thirty-one minutes before the first scheduled launch, the Mercury's escape rockets fired prematurely due to current passing through a "back door" circuit while batteries were being charged. Other flights were troubled by failures of the escape rocket or failure of the capsule to separate from the booster. Two small rhesus monkeys, Sam and Miss Sam, flew on successful Little Joe flights. The first monkey flight took place despite high seas in the recovery area, and Rose remembers that motion sickness medicine intended for the monkey wound up being used by a seasick medic assigned to the recovery ship.

Little Joe rocket. Courtesy NASA.

After the first six Little Joe flights, Rose joined others such as Chamberlin and Norman Farmer in helping to see Mercury capsules through the manufacturing process at McDonnell Aircraft in St. Louis. When the seventh and final Little Joe flew a successful test of the escape system just a week before the first astronaut flew a Mercury capsule, Fred Matthews was sent down to Wallops to help supervise. Even this test had its moment of drama: One of the solid motors in the Little Joe booster didn't fire at liftoff. Without this engine, the test would fail, but finally the heat from the other rockets caused the balky engine to fire, saving the test.

Bryan Erb once went out to Wallops Island to help determine why a bulkhead had failed on a Little Joe launch. As part of his work, he found himself applying paint to the rocket just a few feet away from the business end of the live rocket, mindful of the incident where the escape rockets had fired prematurely. After the launch, he watched on a recovery vessel with astronauts Scott Carpenter and John Glenn as the returned capsule bobbed in rough seas. When someone observed that being in the capsule would cause seasickness, Erb recalls Glenn responding, "If I get [through a flight] back to that stage, I'll be glad to be seasick."

Back at his office, Erb worked on determining how the Mercury heatshield would perform in flight. The heatshield, a fiberglass and phenolic resin composite about 3 cm thick, operated on the ablative or sacrificial principle, where the resin decomposed under the extreme heat of re-entry, generating a lot of gas and building up a char. The gas percolated out through this char layer and provided a relatively cool cushion against the extremely hot gas of the shock layer that formed as the capsule rammed into the atmosphere. Erb constructed an analytical model and computer program to predict how the ablator would perform. Flight data confirmed the predicted performance almost exactly.

When Erb ran some wind-tunnel tests at somewhat representative re-entry conditions, his tests indicated higher heating rates on the afterbody of the capsule than most experts had expected. Erb's findings were brushed aside as an "artifact of the test set up" and forgotten until the first flight of a full-size Mercury capsule, on a mission nicknamed Big Joe. When the capsule was recovered, part of the afterbody was found to have suffered buckling from severe heating. This finding sparked a crash redesign of the shingles on the cylindrical part of the afterbody, and the new design, employing beryllium plates, proved out well on a flight test of a production capsule on MA-2.

Peter Armitage was involved in developing the system that softened the capsule's parachute landing. Mercury's heatshield would detach from the capsule just before touchdown, but would remain hanging from metal straps inside a 1.3-metre-long landing bag that filled up with air. If Mercury landed on a hard surface instead of the ocean, the landing bag would reduce the shock of landing, where the high rate of deceleration otherwise would have endangered the astronaut's health. For water landings, the landing bag also acted as a sea anchor, keeping the capsule upright, at least in theory. Armitage's work helped show that the capsule would tip over after splashdown, so design changes were made to lower the capsule's center of gravity by moving ballast and instruments so that the capsule would remain upright after splashdown.

The landing bag system became the focus of attention after the troubled Atlantic recovery of Ham's MR-2 flight. Wave action had caused the metal straps inside the landing bag to snap. The landing bag tore and the heatshield fell off, but not before banging holes in the capsule that nearly caused it to sink. Rod Rose was told to find solutions, and he and his team decided to hang the heatshield from a set of stainless steel cables in addition to the straps when the landing bag was deployed. They found that the cables, and aircraft springs that connected the cables, took up the loads caused by wave action, preventing damage to the heatshield and landing bag. As well, a fiberglass shield and honeycomb panelling were added to the capsule bulkhead to protect the capsule. All this was tested in a series of trials in water tanks and air drops in the final weeks before the first manned suborbital Mercury flight.

Armitage helped develop and test other recovery aids such as a device to cut antennas that could interfere with lifting capsules out of the ocean. Armitage's group redesigned the cargo hook

Peter J. Armitage (left) and astronaut Virgil I. Grissom standing next to MR-2 aboard the USS Blandy *during recovery training exercises off the east coast of Virgina on October 26, 1961.*
Courtesy NASA.

used by military helicopters to pick up the capsule after splashdown after an investigation showed that the cargo hooks then in use on military helicopters were responsible for many dropped payloads. Similarly, he said, the shark repellent then in use in the military was found to attract rather than repel sharks, so a different repellent had to be found for the astronauts' survival kits.

•

Bruce Aikenhead was assigned to help prepare the seven astronauts for their flights. "Our first job in the training aids section was to analyze thoroughly all the pilot's tasks and what sort of training aids would be necessary to develop those skills." He was involved in the development of several different simulators used to ready the Mercury Seven for flight, notably the procedures trainer, a device where the fully-suited astronaut climbed into a cockpit that was an exact replica of his capsule to do a run-through of his entire flight. Aikenhead wrote the specifications for this simulator, and McDonnell built one for Langley and another for the launch site at Cape Canaveral. Aikenhead's work on this trainer brought him into discussions about cabin and switch layout, and the contentious question of whether to put a proper window into the capsule. Astronaut opinion eventually prevailed, and the window was in every manned Mercury capsule after the first one.

Mercury was the only U.S. human spacecraft that didn't carry a computer on board, but computers on the ground were essential for the success of its flights. Two of the ex-Avro engineers were hired for Mercury because of their knowledge of the IBM 704 digital computer, which Avro leased to help develop the Arrow. NASA handed over a 704 to the Space Task Group to do the computing necessary for Mercury. The 704, then one of the most advanced computers available, was a behemoth that used magnetic cores, magnetic tape drives and a card printer. It filled a room and had only a small fraction of the power found in today's average desktop computer, but Stanley Cohn, John Shoosmith and their colleagues used it to establish the Mercury capsules' paths in orbit and during re-entry.

Early in the program, Cohn and Shoosmith's group worked on preparing programs for flight and for testing various paths the capsules could take. Cohn pointed out that this task was not as simple as might be imagined. The atmosphere, which even at orbital altitudes exercises some influence over the paths of satellites, had to be factored in. Also complicating things was the fact that the Earth is not a perfect sphere. Early satellites showed that it was slightly pear-shaped, and this fact would have to be accounted for in calculating Mercury's orbits. One of the things computers were supposed to predict was when the capsule would appear over the various tracking stations in the Mercury network. "They didn't really know the exact latitude and longitude of some of the smaller islands [where tracking stations were located], so part of the development work in the tracking network was to make sure that those could be pinned down for accuracy, because we had to provide each ground station with information on how to pick up the satellite, when and where to look for it. So you had to know where the island was, to start with," Cohn said. The Earth's rotation also had to be worked into the calculations. Cohn, a native of Toronto, had moved into the fledgling area of computers after studying mathematics at the University of Toronto and Indiana University. At NASA he was head of his section, which included Shoosmith, the youngest member of the Avro group that went to NASA. Shoosmith, who was 24 at the time of the move to Langley, was born near London, England, but educated in Canada after his family moved to Ottawa when he was a teenager.

John N. Shoosmith in 1964. Courtesy NASA.

Another computer expert from Avro hired by STG was the British-born Jack Cohen. He headed the operational analysis section and was assistant head of the mission analysis branch, which included Cohn and Shoosmith's section. Both Cohen and Cohn were involved in collecting and tabulating statistical information on failures of Redstone and Atlas rockets to allow STG to anticipate and prevent the most common failures in these rockets.

Stan Galezowski, an engineer from Toronto who was almost as young as Shoosmith and had specialized in control systems at Avro, was assigned to the flight controls section at STG, where he worked with analog computers to develop simulations of Mercury's flight paths. "We simulated a rotating Earth with atmosphere, and used this to determine what sort of re-entry system to use, or theoretically test one that had been proposed," he said. At first Galezowski used analog computers in Princeton, New Jersey, to do this work, and before long he was working with Richard Carley and David Ewart, among others, performing simulations of re-entries with more advanced spacecraft such as Gemini and Apollo.

During Mercury flights, the 704 and later the IBM 7090 were used to plot emergency re-entry paths. "We had a whole set of contingency landing areas already programmed into the computer," Shoosmith said. "Once it was inserted into orbit, we knew its orbit pretty well. Then we would jump ahead and if something were to happen in the next several minutes, then we would fire the retros at such-and-such a time." To determine that time, Mercury's orbital parameters and the landing sites would be fed into the computer as the flight went along. Shoosmith remembers starting work at 2 a.m. on launch days to make sure the computer was ready to go. He was connected by telephone to Mercury control at the Cape. "The machine would just constantly recompute the orbit of the Mercury capsule, and for the particular retrofire points."

•

Mercury Control. *Courtesy NASA.*

After building the capsule itself, the biggest and most expensive job in Mercury was designing, building and operating a brand new control center and tracking network. Arguably, the Avro engineers' biggest contribution to Mercury was in this area.

The Mercury Control Center was being built at Cape Canaveral under the direction of NASA's original flight director, Christopher C. Kraft Jr., along with Chuck Mathews and Walter C. Williams. This center, which would control all Mercury flights and the first three Gemini launches, served as the model for space control centers that followed. It had three rows of desks with control consoles, monitors and switches, where the controllers sat. Behind them was a viewing room, and in front was a large world map showing the orbital path of the capsule, location of tracking stations, and the capsule's location above the Earth. Flanking the map were screens where information and television pictures could be projected. This basic design was refined when the control center moved to Houston in 1965, and was even adopted by the Russians starting in 1975.

"The detailed mechanical design was the responsibility of the Langley Research Center instrumentation group, who let the contracts out for the building of the network and the mission control center," Fred Matthews said. "The functionality requirements and so on were the strict responsibility of the Space Task Group. The key persons driving the operational concepts were Chuck Mathews and Chris Kraft. But in the detailed implementation of this functionality, a lot of the inputs to it were from ex-Avro people, in particular John Hodge and Tec Roberts and, to some extent, myself, although I was more involved with the downrange flight monitoring capabilities than I was with the mission control center." In addition, Dennis Fielder, John Hughes, George Harris and Len Packham from the Avro group were involved in this work.

Roberts was the ex-Avro engineer who probably had most impact on the design of the original mission control center. The balding Welshman became the first Flight Dynamics Officer or FIDO in the Mercury Control Center at the Cape, where he sat to the side along with the Retrofire Officer or RETRO facing trajectory plotboards. Later, in the Houston Mission Operations Control Center (MOCR), FIDO and RETRO were moved to the front row of the control room, which became known affectionately as "the trench" because the other rows were set on risers.

The job of Roberts and the FIDOs who followed in his footsteps has been to constantly watch the trajectory of the spacecraft and plan for the next needed maneuvers. In other words, FIDO is the person whose job it is to know where the spacecraft is and where it is going. This work is especially demanding during launch, when every second the booster fires changes the spacecraft's path. The RETRO sitting next to FIDO also watches the trajectory, preparing to abort the mission at any time should that become necessary.

In Mercury, most of FIDO's work took place during launch, because the Mercury capsule could not change its trajectory from the time the capsule's posigrade rockets separated the capsule from the launch vehicle, until the retrorockets fired to end the mission. If the booster was taking the capsule the wrong way, FIDO would be the first to know. And at the end of the boost phase, FIDO provided the crucial input into the Go/No Go decision that the flight director would have to make at that critical moment. If the trajectory had fallen short of orbit, FIDO could recommend using the retrorockets to change the impact point of the capsule. FIDO was also responsible for providing trajectory inputs for Mercury re-entry, whenever it happened. "He's what I called Mr. Trajectory," Fred Matthews said of Roberts. The constant orbit changes of Gemini, Apollo and the space shuttle meant even more challenging work for FIDO once orbit had been achieved in those programs.

Roberts was the lone Welshman of the Avro group at NASA. Born in Liverpool of Welsh parents in 1925, he studied and started his career in aviation on the Isle of Wight, specializing in hydrodynamics and flight testing with Saunders, Roe. He and his wife Doris came to Canada in 1952, and he continued flight testing with Avro Canada. During the Mercury program, Roberts

Robert R. Gilruth presents award to Tecwyn Roberts in 1964. His wife Doris Roberts looks on.
Courtesy NASA.

was one of the most prominent members of the Avro group due to his work developing the control center and his role as FIDO.

Chuck Mathews said Roberts "was almost Chris Kraft's right-hand man. As a matter of fact, when they went down for a launch one time, he and Chris always roomed together at whatever motel they were staying at. Chris really depended on him very heavily for his advice. Tec Roberts was not only very bright and very energetic, but at the same time he was a well-composed individual. He didn't get rattled easily."

Said Kraft: "Tec Roberts had a natural bent, it turned out, for things like radar, tracking and computer calculations of trajectories and how that all might play together, and how you would display data and pick apart data from a flight operations management point of view. I think he was a real diplomat. He knew how to get along with people. He didn't look like that kind of guy, but he was. Very sharp, not only technically, but with people. I think that the work he did in Project Mercury was outstanding."

During the early unmanned flight tests of Mercury, Roberts helped to coin a term that became one of the bywords of the early space age: "A-OK." While many people associate the term with the Mercury astronauts, it was actually used in public for the first time during the first manned Mercury flight by Col. John A. "Shorty" Powers, the astronauts' public relations man, who was known as the "Voice of Mercury Control." According to the official NASA history of Project Mercury, Powers picked up "A-OK" from reports written by Roberts and Air Force Capt. Henry E. Clements on unmanned Mercury tests. Roberts attributed the term to contractor employees involved in the construction of the tracking network who knew "A" has a strong sound. Others believe that "A-OK" was first used by railroad telegraphers to verify their equipment.

•

One of the U.S. space program's lasting traditions established in Mercury is the use of one person in the control center as the single point of contact with the astronaut crew. The capsule communicator or CAPCOM still maintains the Mercury-era name even in the time of the space shuttle. Except in unusual circumstances, the CAPCOM is the only person who speaks to the crew during flight. According to Harris and Matthews, they and other ex-Avro engineers were among those who urged the use of a CAPCOM rather than leaving communication with the crew to whoever wished to speak with them.

"When we were at Avro, we had set up a real-time flight monitoring capability which included real-time displays for engineering personnel to monitor the aeroplane's performance, etc," Matthews said. "Some of that experience proved to be of use in setting up the mission control center, and not only what it should do, but how it should do it."

Harris said similar ideas were being used in test flying aircraft in the U.S. and in England. "What we did, regardless of what flight pattern we were flying, we had this one guy talking. So I don't know, can we claim credit? Probably not. Did we bring the idea to a head? I think we did."

Chris Kraft said the idea for the CAPCOM comes from several sources, including the Avro people and the NACA flight test veterans including himself, Walter Williams and Howard C. Kyle. "So, we all did it. It is a natural thing for all of us to gravitate into that kind of thinking. You can't give the credit to anybody. You have to give the credit to all. We all came up with those ideas." Once the idea of a single CAPCOM had been established, Kraft said he and Williams took the idea a step further and decided that the CAPCOM should be an astronaut whenever possible. In Mercury, because each tracking station had to have its own CAPCOM, flight controllers from outside the astronaut corps had to be used in many stations.

Matthews, Hodge and Roberts were also involved in drawing up the mission rules for Mercury, the first in human space flight. These rules would, among other things, cover decisions on whether to continue a mission or abort it when a problem cropped up.

"Under Chris Kraft's direction and guidance, and with a lot of inputs from various sources, I generated the first set of mission rules defining what systems had to be working for launch and subsequent operations," Matthews said. "The task of updating these rules was subsequently taken over by Gene Kranz" when he joined the section late in 1960.[3]

Matthews said he was also involved in drawing up the flight controller handbook, which he described as a "12-colour multi-mode description of how every system in the spacecraft would operate. This became the bible of the flight controllers. It was no words, just diagrams of all the modes, and comparing what the astronaut could see and do with what the flight controllers could see and do, and every mode you could put the spacecraft in. We eventually came up with another handbook to supplement that, which was a procedural handbook as to what the flight controllers should do during countdowns and mission passes over their station, and so forth." The second handbook was put out under the supervision of Gene Duret.

Roberts said that he sat down with Chris Kraft and drew up the first countdown plan for a human spacecraft. They looked at how missile and satellite countdowns were handled at the Cape and used "a good imagination" to draw up the countdown checklist for Mercury Control.

Hodge said some of his own early work also concerned prelaunch checkout of the Mercury capsule.

It was standard procedure to equip all rockets with destruct charges that could be set off by the range safety officer in the launch control center if the rocket wandered off course and endangered populated areas, which they did often in the early days of the space program. Soon after they started at NASA, Matthews and Frank Chalmers were put to work on this problem, and

[3] Kranz didn't last long in the flight control section that Matthews headed. Kraft soon took the talented newcomer under his wing and Kranz was on his way to becoming the program's authority on flight rules and a legendary flight director.

they came up with a set of recommendations to assure both astronaut and public safety should a Mercury flight be aborted during the launch phase. Coordination was needed between the range safety officer, the flight director, the astronaut and any automatic systems that could cut short a launch. For example, they recommended a three-second pause between the destruction command and booster destruction to allow the astronaut time to fire the capsule escape system and safely clear the launch vehicle.

At the time the Mercury Control Center was built, Mercury had a very ambitious schedule that included a suborbital flight by each of the astronauts atop a Redstone rocket prior to the first orbital flight atop an Atlas. The Redstone flights were planned to take place in 1960, possibly before the control center would be ready. Matthews was assigned to prepare for this eventuality. "I came up with an evaluation of whether we could stuff some consoles into the Redstone launch blockhouse, and after deciding that wasn't a very good way of doing things, I came up with the idea of putting it in a trailer. So I schemed out a way of putting what we needed in a trailer, and it was a mini temporary mission control center. And rather than going out on bids, which we didn't have time to do, or to people who didn't have the detailed technical knowledge, I recommended we give the task of building it to McDonnell, who were building the capsule - this is what I call loose-rein management, because the specifications were only three 11 by 17 [28 cm by 43 cm] pieces of paper showing what it should look like and what it should do - and offered them a contract to build it, and they said yes, they could, and they would, and they did. They did a wonderful job," Matthews recalled.

•

On March 24, 1961, an extra test of the Mercury-Redstone vehicle was flown because of problems with Redstone in MR-1 and MR-2 that had concerned Wernher von Braun and other managers. This flight, designated MR-BD, for booster development, flew a near perfect mission. The Redstone rocket was now deemed to be safe to fly an astronaut. Meanwhile, the Soviets were launching and recovering dogs launched into orbit aboard what appeared to be test flights of their own human spacecraft. As the date neared for the first Mercury suborbital flight with an astronaut on board, everyone knew that the decision to add the MR-BD launch and delay the first manned flight would give the Soviets precious time to prepare their own vehicle. While the MR-BD flight was being prepared, two other items had to be dealt with before the first Mercury astronaut could safely fly. The first was the effort carried out by Rod Rose and his team to fix the landing system problems experienced in the recovery of MR-2. The second was the final Little Joe flight to verify that the launch escape system would work. Neither effort would be complete until the end of April. Meanwhile, the public had been told that Alan Shepard, Gus Grissom and John Glenn were training for the first flight. Only a few people knew that it was Shepard who was slated to fly atop Mercury-Redstone 3 in early May. But before Shepard had a chance to fly his capsule, there was another announcement from the Soviet Union.

4 How Do We Go to the Moon?

On April 11, 1961, President John F. Kennedy ended his day with a warning from his science advisor, Dr. Jerome Weisner, that the Soviet Union would likely launch a rocket with a man on board while he slept that night. Weisner's prediction was right: From the same launch pad where Sputnik had been launched more than three years before, a 27-year-old Soviet Air Force senior lieutenant named Yuri Alexeyevich Gagarin rose into space at 9:06 a.m. Moscow time on April 12. During his single orbit of the Earth aboard his Vostok spacecraft, the Soviet news agency TASS announced Gagarin's launch and his promotion to major. When Gagarin landed successfully after 108 minutes of flight, TASS trumpeted his success. But Gagarin's return to Earth nearly killed him, a fact that was covered up for 30 years. Cables tying an equipment module to the spherical Vostok capsule failed to disconnect, causing the capsule to spin wildly as it hit the atmosphere. The heat of re-entry eventually burned through the cables, and Gagarin was able to safely eject from his capsule and parachute to a landing in the Soviet Union. He flew to Moscow soon after for a delirious welcome as the Soviet government basked in the worldwide praise that followed his historic feat.

Kennedy, when reminded at a press conference the same day that a member of Congress had pronounced himself tired of the Soviet lead in space, said: "However tired anybody may be – and no one is more tired than I am – it is a fact that it's going to take some time. And I think we have to recognize it."

When John F. Kennedy took office in January 1961, the future of the U.S. space program was uncertain. Kennedy's attitude toward the space program was not well known. Mercury had not yet proved itself with a manned flight; the president's science advisor, Weisner, was a prominent critic of Mercury; and the U.S. Air Force was lobbying to get control of space back from NASA. By April, strong leadership was being felt in the U.S. space program. Kennedy's vice-president, Lyndon Johnson, who had championed the space program as the Senate majority leader, was put in charge of the National Aeronautics and Space Council. Under his influence, NASA was given a new administrator, James E. Webb, who was well known on Capitol Hill as a former member of the Truman Administration and as a protégé of powerful Oklahoma senator Robert Kerr. To complement Webb's strong administrative and political abilities, Hugh Dryden, the former head of NACA, remained as NASA's deputy administrator.

Two days after Gagarin's flight, Kennedy convened a meeting with Webb, Dryden and other advisors in the White House cabinet room to consider how to catch up with the Soviets in space. On April 19, Kennedy met with Vice-President Johnson and charged him with examining the space program and determining how the U.S. could gain the lead in space. The meeting was followed up with a written memorandum the next day calling for a report at the "earliest possible moment." One of the five questions in the memorandum asked: "Do we have a chance of beating the Soviets by putting a laboratory in space, or by a trip around the Moon, or by a rocket to land on the Moon, or by a rocket to go to the Moon and back with a man. Is there any other space program which promises dramatic results in which we could win?"

If Gagarin's flight was a setback for the Kennedy administration, an even worse one followed quickly. On April 17, Cuban exiles trained and backed by the U.S. Central Intelligence Agency landed on Cuban shores at the Bay of Pigs in an attempt to topple the Communist regime of Fidel Castro. When the troubled invasion began to fall apart two days later, Kennedy decided against committing American forces to back it. The failure at the Bay of Pigs marked the low point of the Kennedy presidency.

While Kennedy dealt with the fallout from this debacle, Johnson was meeting with NASA officials, scientists, politicians, and leaders from business, the military and Congress. Johnson had got congressional leaders to sign on for the expensive lunar effort and even won agreement from Defense Secretary Robert McNamara, who agreed to a massive funding increase for NASA because it would soften the blow of cuts he planned to make to defence programs.

On Friday, May 5, astronaut Alan Shepard boarded his Mercury capsule Freedom 7 at Cape Canaveral and blasted off atop a Redstone rocket on the 15-minute, 28-second flight of MR-3. He flew 187 km high and 488 km downrange into the Atlantic Ocean. Prior to the flight, government officials all the way up to the president debated whether to hold the launch in public. Prior to the original launch attempt on May 2, which was scrubbed due to bad weather, the name of the prime astronaut for MR-3 had been kept secret. But the first flight of an American into space took place live on television, and NASA public relations officer Shorty Powers reported Shepard was "A-OK," throughout. In a spontaneous move, Kennedy phoned Shepard soon after the astronaut was picked up by the recovery ship *Lake Champlain*.

Three days later, thousands of Americans came out to cheer their new hero when he came to Washington to address Congress and receive a medal from Kennedy at the White House that Monday. As Johnson departed on a trip to the Far East, he left the president with a memorandum endorsed by himself, McNamara and NASA administrator Webb calling for an aggressive expansion of the U.S. space program with a manned lunar landing as its central goal. On May 10, Kennedy was still basking in Shepard's reflected glory when he and a group of his officials considered Johnson's memorandum and decided to send astronauts to the Moon and support other space projects. To fund these ambitions, the president decided to propose an immediate boost in NASA's budget.

Kennedy appeared before a joint session of the U.S. Congress on May 25 for a speech on "Urgent National Needs," which covered a number of topics including the fight against Communism in developing countries, civil defence, social problems at home, and military and intelligence matters. Finally, he turned to space.

The launch of Mercury-Redstone 3 carrying Alan B. Shepard. Courtesy NASA.

After pointing out the importance of the space race in America's struggle with the Soviet Union for the hearts and minds of the world community, and warning of the Soviet lead in rocketry, Kennedy laid down a challenge to Congress and to Americans. "I believe that this nation should achieve the goal, before this decade is out, of landing a man on the Moon and returning him safely to the Earth. No single space project in this period will be more exciting, or more impressive to mankind, or more important for the long-range exploration of space; and none will be so difficult or expensive to accomplish. In a very real sense, it will not be one man going to the Moon – we make this judgement affirmatively – it will be an entire nation. For all of us must work to put him there." The president underlined the dimensions and monetary cost of the challenge and said there was no point deciding to attain leadership in space unless the country was prepared to bear these burdens. If Americans were not prepared to bear them, he said, "we should decide today."

Congressional support was quickly forthcoming due to wide public concern about the Soviet lead in space, and due to Johnson's work with his congressional colleagues. Although some people at NASA were nearly bowled over by the sheer audacity of the goal the president had set them, a group of people in the agency were well into their second year of planning how to send astronauts to the Moon.

•

Project Mercury was barely underway in 1959 when members of the Space Task Group decided to start thinking about what would follow it. In April, NASA set up a research steering committee for manned space flight under the chairmanship of Harry J. Goett, a top NASA manager. The committee quickly learned from Mercury designer Max Faget and others that flying to the Moon was high on the agenda of many at STG. That summer, Bob Gilruth formed a New Projects Panel at STG, which recommended that work begin on an advanced spacecraft capable of going to the Moon. On November 2, 1959, Gilruth assigned Robert O. Piland, Kurt Strass, John Hodge and Caldwell Johnson to begin work on a "preliminary design" of a spacecraft that could circle the Moon. For Hodge, it would be the first of many ventures into advanced planning, but he remembers his role at the time as simply representing the operations part of the organization in planning the new vehicle. The group did mission analyses, not only of flying around the Moon, but also of flying to the Moon's surface. NASA headquarters drew up a "Ten Year Plan," which included mechanical probes to the Moon and planets, and human circumlunar flight and space stations by the end of the 1960s, with a lunar landing after 1970.

Owen Maynard was assigned in September 1959 to a new projects group in STG's engineering branch under the direction of Caldwell Johnson, the draftsman who worked closely throughout his career with spacecraft designer Max Faget. The first few months there, Maynard concentrated on the Mercury capsule, and was assigned as project engineer on the MA-1 flight that would fly in July 1960. Bryan Erb had finished his heatshield work on Mercury and decided to try his hand at being a flight controller. But on May 25, 1960, Maynard was pulled away from MA-1. Erb was taken off flight controller training, and they were both placed in a newly formed group that would research and conduct studies on advanced spacecraft, including space stations and vehicles that would take crews to the Moon and Mars. The group reported directly to Gilruth.

"I got into Apollo very early on, before it had a name, in about May of 1960," Erb said. "We hadn't yet really got to the point where we could see the light at the end of the tunnel on Mercury, but we had gotten to the point where we said, 'What do we do next?' So they threw us into the advanced vehicles team, eight of us, and I was the heat transfer person. It was led by Bob Piland, and we were told, hey, you guys go and think about what we were going to do for a sequel to Mercury. And it was fairly clear, even at that time, that the only way we were going to really leapfrog the Russians was to go to the Moon. We started working on a lunar vehicle."

Soon NASA decided it was time to get private industry involved in the space program beyond Mercury, and meetings took place during the summer to inform the aerospace industry and prepare the way for initial contracts. At one such meeting, on July 28, 1960, NASA deputy administrator Hugh Dryden announced that the new spacecraft would be called Apollo, based on a suggestion by Abe Silverstein, who had also named Mercury. Apollo was designated to launch atop a Saturn rocket and be capable of flying three men for two weeks as either a space station in Earth orbit or as a vehicle orbiting the Moon.

"I or we asked Gilruth, what is Apollo?" Maynard recalled. "He said: 'That's up to you guys to figure out. That's why we hired you, to formulate guidelines and formulate programs.' I had never been in that position before and I kept asking for guidelines. He kept saying, 'That's your job.' After some time, he said, 'Well, I have one guideline for you.' I said, 'Oh, what's that?' He said: 'Formulate the program so that it can be performed by ordinary people.' I thought that was a ridiculous guideline. Everybody knows we're geniuses. I began to get religion as we got into it that we indeed were ordinary people. That Gilruth believed he was, believed I was, believed everybody was ordinary people. He didn't want to have to have the world's best astronaut to fly it, or the world's best program manager to manage it."

STG was reorganized in September with an Apollo Projects Office formed under Piland. In late October, three bidders won $250,000, six-month feasibility contracts to study Apollo. Members of STG who monitored the three contracts formed an informal fourth group. Maynard, Robert Vale, Stanley Cohn, Richard Carley and Bryan Erb were among the STG engineers who monitored the contract work of General Dynamics in San Diego, General Electric in Valley Forge, Pennsylvania, and Martin Marietta in Baltimore. These contract review trips turned out to be tests of endurance involving three very full days in a row at the three plants, with "red eye" flights from San Diego to the east coast to make sure no time was lost. At the same time, the group advanced work on STG's own design which, as it turned out, resembled Mercury more than the designs that came from the three contractors. Jack Cohen, a computer expert from Avro, was also moved to Apollo about this time.

•

With the government and the country behind Kennedy's goal in the spring of 1961, it was up to the people of NASA and American industry to carry it out. NASA faced innumerable decisions before the goal could be achieved. One question towered above all the others: How do we go to the Moon?

At first glance, the decision was obvious: Send a rocket up directly to the Moon, turn it around and land on the Moon. Then blast off from the Moon and fly home. This mode, known as direct ascent, was well known to aficionados of science fiction novels and movies, and it was the vision shared by most people at NASA.

But direct ascent is not as simple as it appears in the movies. For one thing, it requires a monstrous booster rocket. That's because every pound of spacecraft, fuel, passengers and gear for the voyage has to be boosted out of Earth's gravity, then slowed to land on the Moon, then boosted out of the Moon's gravity. Fuel burns off along the way, and stages fall off as well, but the fuel needed to escape lunar gravity and the heavy shielding to re-enter the Earth's atmosphere have to be taken to the lunar surface and lifted away, even though they are not needed on the Moon. That means an enormous burden of fuel must be taken away from Earth. Early plans for Apollo called for a behemoth rocket called Nova. As the engineers and designers started on Apollo, other problems with direct ascent appeared.

An idea to get around the need for Nova was to launch two or perhaps more smaller Saturn rockets, rendezvous in Earth orbit, then send the combined load on its way to the lunar surface. This mode was known as Earth orbit rendezvous or EOR.

Some engineers at the Jet Propulsion Laboratory in California had another idea. They proposed launching a vehicle that would land on the Moon near another already-landed vehicle that would contain the fuel needed to go home. The concept of lunar surface rendezvous never gained support in NASA outside of JPL.

A group of non-STG engineers at NASA's Langley Research Center, as well as a non-NASA group led by Thomas E. Dolan at Chance-Vought Astronautics, had been studying the problems of lunar flight and were both advancing another idea called lunar orbit rendezvous or LOR. The idea was to leave most of the spacecraft in lunar orbit, and descend to the Moon's surface in a small vehicle containing only what was needed to land on, explore and leave the Moon. The fuel savings were enormous, and it appeared that the whole package could be launched on a single Saturn rocket. But LOR had few backers in NASA when Kennedy issued his lunar challenge. NASA had a total of 15 minutes and 28 seconds of space flight experience, and the idea of astronauts' lives depending on a rendezvous in space 400,000 km from Earth was too frightening for most people in NASA to contemplate.

Lunar orbit rendezvous as an idea dated back to a 1916 paper by Russian theoretician Yuri Kondratyuk and a 1948 proposal by H.E. Ross of the British Interplanetary Society, but the researchers at Langley had no knowledge of the previous work.

Although he was not the first at Langley to think of LOR, John C. Houbolt, an aircraft structures expert and in 1959 the assistant chief of the dynamic loads division at Langley, had spent a lot of time thinking about the use of rendezvous in space, including the use of LOR to land on the Moon. As studies of lunar flight began to gear up in 1960, Houbolt became strongly convinced that LOR was the best way to go, and he began lobbying his colleagues all over NASA about his belief. On December 10, 1960, he met with leading members of STG, including Paul Purser, Bob Piland, Owen Maynard, Caldwell Johnson, Jim Chamberlin and Max Faget. Houbolt recalled his rendezvous ideas received a reception that was not hostile but not positive, either. Within days, Faget was criticizing Houbolt at another NASA meeting. The following January, Houbolt and some other Langley engineers met with three STG engineers, including Maynard, and again failed to persuade them of the virtues of LOR.

The story of how NASA came to accept lunar orbit rendezvous as the way to land on the Moon will always remain a subject of some controversy. Most histories of Apollo credit Houbolt's persistent lobbying, and the fact that he was unafraid to go over peoples' heads, for leading NASA to LOR. But Bob Gilruth later claimed that he came to favour LOR at an early date, an assertion that Houbolt strongly disputes. Others within STG say that they would have come to embrace LOR without Houbolt's high-pressure lobbying.

On April 19, 1961, with the wheels already in motion that would lead to Kennedy's commitment to the Moon, Houbolt held another briefing for STG engineers on rendezvous and LOR. Again, the response was negative.

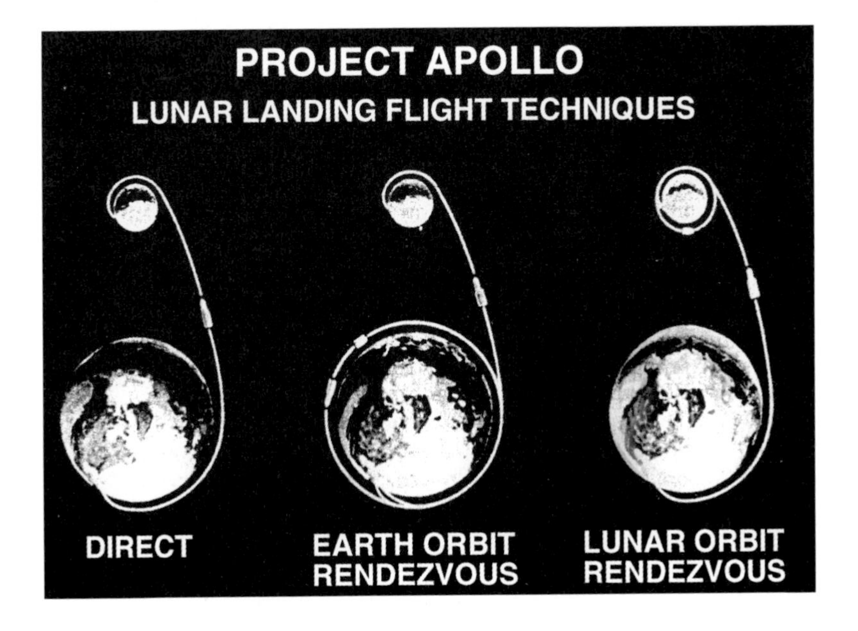

But after Houbolt finished his presentation, he was approached by Jim Chamberlin, who asked him for a copy of the handout he had prepared for his talk and any other relevant information. Both Houbolt and Chamberlin told NASA historians later that Chamberlin was aware of Langley's work on rendezvous, but was unfamiliar with the details of LOR. Chamberlin became the first member of STG to look seriously at LOR. As Auburn University historian James R. Hansen noted in his paper on Houbolt and LOR: "Perhaps it is significant that Chamberlin was not one of Gilruth's old associates from NACA. He was one of the relative newcomers – and a very talented one."

In spite of the fact that the end-of-decade deadline attached to Kennedy's challenge lent urgency to the question of how to land, Houbolt encountered great difficulty selling LOR around NASA through much of the rest of 1961. Two task forces looking into the issues around the lunar landing turned the idea of LOR down flat. That fall, a third committee, headed by Nicholas E. Golovin, began to take a serious look at LOR. While the Golovin committee did not call for LOR, it suggested that NASA use rendezvous in the lunar landing.

While the Golovin committee was deliberating in August and September 1961, Chamberlin gave Gilruth a proposal for a program to follow Mercury that included rendezvous in Earth orbit and then going for a lunar landing using LOR. Chamberlin's suggestions followed those given by Houbolt in his April presentation. As Hansen explained: "This was most significant. Never before had a member of the STG seriously offered any flight plan for a lunar landing involving any sort of rendezvous in lunar orbit. Although Gilruth was not convinced of the merits of such a scheme, he was open to further evaluation."

Jim Chamberlin circa 1962. Courtesy Arthur Chamberlin.

Chamberlin had been working much of the year on a successor spacecraft to Mercury, and his daring lunar plan involved using the two-man Mercury Mark II spacecraft he was developing, which would soon become known as Gemini. According to Chamberlin's proposal, the Mark II spacecraft launched by Titan rockets would fly test flights without and with astronauts, then try docking with Agena rockets in Earth orbit. Later flights would dock with the more powerful Centaur rocket, which would boost the two astronauts into deep-space orbit and then around the Moon. Then there would be tests in Earth orbit of astronaut transfers from the Mark II spacecraft to a lunar "bug," which could carry a single astronaut. Finally, the lunar landing would be achieved in 1966 using a Mark II spacecraft and a lunar "bug" flying an LOR mission after a launch atop a Saturn rocket. Initial plans for the "bug" called for a lander with no cabin. The astronaut would fly the entire landing dependent on his space suit.

Chamberlin later said that he was deliberately comparing his Mark II and the tiny 2,270 kg "bug" with the huge 68,000 kg Apollo vehicle then being developed for a direct ascent flight to the Moon. "The implied competition here, which was real, got [STG] changed around .... We did this because we thought we had a better mousetrap, and it was. We didn't get the cold shoulder on the idea [of LOR], but we got the cold shoulder with us doing it" using Mark II.

"Dr. Gilruth was very much impressed by my presentations, which emphasized the hardware aspects," Chamberlin wrote on another occasion. "I showed that there were great advantages in having a special vehicle designed only to land on the Moon. Making a vehicle that could also re-enter and land on the Earth was just too hard to do. Considerable design work was done on this [bug] and a display model made of it. Of course, we did not neglect the orbital mechanics advantages, which resulted in a great reduction of the size of booster required. As a result of these discussions, the STG started to seriously consider and work on lunar orbit rendezvous. The reason why the Langley work [of Houbolt and others] was largely ignored was because they were considered to be pure theorists with no practical experience.

"However, when we advocated [LOR] to also solve hardware design problems, we could not be ignored as being inexperienced in practical considerations. Although I was not encouraged to include this work in Gemini plans, I had frequent discussions with Gilruth on it, and I remember very well when we were riding together on an airplane, he told me that he had been thinking about lunar orbit rendezvous a lot, and that he was now convinced that it was the only logical way to go to the Moon."

"I think [Chamberlin's] purpose was really to shake them up to reality and take a real look" at LOR, recalled James T. Rose, who worked closely with Chamberlin in the early days of Gemini and on his proposal to land the "bug" on the Moon. Both Rose and Walter Williams remember a presentation that Chamberlin made in Washington on his Gemini "bug" proposal to NASA officials from STG and headquarters around Labour Day, 1961. Williams was on vacation and Gilruth called to summon him to Chamberlin's presentation. "I want you to hear Chamberlin's thoughts on a bug. It looks like he's got something pretty good here," Williams remembers Gilruth saying. Rose said this presentation led STG to meet again with Houbolt, and although the Langley engineer got a rough ride at the meeting, STG was finally looking seriously at LOR.

Rose said that NASA's LOR decision "might have never occurred if the bombshell hadn't been dropped" with the presentation Chamberlin did on the "bug" with Rose's help.

For his part, Gilruth said Chamberlin was the first to talk to him about LOR, but added that Chamberlin didn't convince him of LOR's virtues. "He came in and talked about it, and I was very much interested, but I didn't tell him I thought it was wonderful right then. But I did think about it a lot."

That fall, Williams recalled, NASA decided against using the bug in Gemini but to look at it for Apollo. "And of course, meanwhile, Max [Faget] came in with a proposal for a bug with Apollo, which I understand Owen Maynard had looked at some time before that, and Max didn't think much of and killed it. But when Jim came in with one for Gemini, then they dug up one quick for Apollo."

•

Maynard recalled not being impressed with his first exposure to Houbolt in January 1961, but after thinking about it over the weeks that followed, he began to find himself agreeing with Chamberlin that LOR was the way to go. In explaining his support for LOR, Maynard said much of it came from his work on Apollo before lunar landing was made part of the program. During this period, Apollo took shape with a command module, where the astronauts would ride during launch, re-entry and much of the time in space; a service module, which would contain rockets and other equipment needed to support the flight; and a mission module, which would change as the mission did. Many of Apollo's early designs came off the drawing board of Caldwell Johnson, who continued to work closely with Faget. Johnson credits several people with the design of the command module, including Faget, Maynard and Erb. As one of the first to work on Apollo's design, Maynard said he brought the idea of modularity to Apollo. "My associates had probably also thought about modularity, but the thing I recall that was quite convincing to them was, when someone asked me a question early in Mercury, how could we achieve artificial gravity most expeditiously, should it be decided that it is required. I immediately built on the modularity concept. Having two modules, say, one with men in it, and the other comparable to the mission module in Apollo that wound up being the lunar module."

Maynard said his belief in modularity came from Chamberlin, who had used the concept when he equipped the Arrow with an armaments bay in the underside of its fuselage. Different weapons could be placed inside standard modules and quickly changed out between flights. "One of the things I got from Avro - I learned it from Jim Chamberlin - was when I would design things, I would take all kinds of requirements that anybody threw at me, and I would figure out a way to put them into the design and integrate them. One day at Avro, Jim Chamberlin pointed out to me that I would be a lot better off if I took the functional requirements and I parceled them into separate modules. Instead of having 50 operational requirements integrated into one configuration, try putting 10 of the functions into each of five modules, and you'll find that you can get on with the design, development, manufacturing and testing a lot easier. If you have a problem with one function, it doesn't complicate all the others. With the concept of modularizing, you don't have all your eggs in one basket."

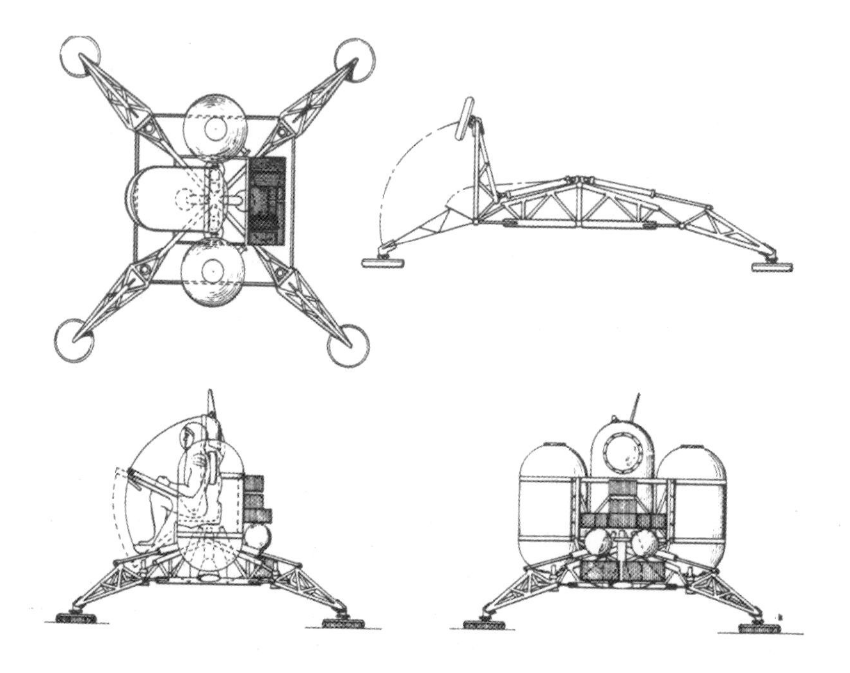

Chamberlin-Rose Lunar "Bug" as drawn by Harry C. Shoaf. Courtesy NASA.

Following Chamberlin's lead, Maynard saw great advantages in building separate vehicles to land on the Moon and return to Earth. Creating modules with different functional performance requirements considerably eased the problems of designing a spacecraft to land on the Moon. Maynard had been trying to design an Apollo command module in which the astronauts would lie on their backs for launch from Earth and then be able to sit up and look out the windows as they tried to land this same vehicle on the Moon. Then there was the matter of launching a fairly large and heavy vehicle off the Moon, and finally returning the crew to supine positions to re-enter the Earth's atmosphere. Maynard compared it to backing a Redstone onto the lunar surface and then launching it from the Moon without benefit of the launch crews and equipment of Cape Canaveral.

"As soon as you've drawn this picture of the command module sitting on top of this big long thing with the four little legs at the bottom, and you say, 'I'm going to land that thing, and I'm going to escape off that thing. I've got the guy sitting on his back and he can't see a damn thing. He's looking up.' You look at that picture, you say, 'That doesn't look right.'" The idea of a single spacecraft with so many different requirements placed upon it, including launching from the Moon without a launch crew, was what "torqued me around" to LOR, Maynard said.

He also had doubts about Earth orbit rendezvous, which involved two separate launches from Cape Canaveral. "At the time the evidence was overwhelming that the first vehicle could not have a long enough life [of its consumables and thermal systems] to survive long enough for the second vehicle to be launched." His concern was shown to be misplaced by the successful rendezvous work in the Gemini program in 1965 and 1966.

When Maynard was thinking in early 1961 about how to make a lunar landing, he played with the idea of leaving the fuel tanks for the return to Earth in lunar orbit. But this presented problems because the tanks would require equipment to assist in rendezvous and docking, along with thermal control equipment. As he thought further about LOR, Maynard made the first drawings of a new spacecraft that would eventually be known as the lunar module. A document dated September 27, 1961, by Maynard concerns a "preliminary evaluation" of a two-man "lunar excursion vehicle" that could remain on the Moon for 24 hours. Unlike Chamberlin's one-man unpressurized "bug" of a few weeks before, Maynard's vehicle had a pressurized cabin.

But Maynard was too junior to get involved in the arguments raging among people like Chamberlin, Faget and Gilruth, so he bided his time until others came to agree that LOR was the way to the Moon. A copy of a Maynard presentation made to a NASA committee in November 1961 still shows a direct ascent configuration for Apollo. As STG began to look seriously at LOR that fall, however, Maynard was ready when he was formally assigned to begin work on the design of the lunar module.

•

Another former Avro engineer at STG who quickly saw the superiority of LOR was Morris Jenkins, who early on was assigned to look at possible lunar trajectories. In November 1961, Jenkins was drawing up trajectories involving lunar orbit before and after landing. "There was a need for a precautionary orbit before they went down," he said later. LOR seemed to be the only plan that made sense to Jenkins, especially from a safety point of view.

Despite the changing attitudes at STG, Houbolt remained frustrated with the progress he was making in promoting LOR. In November, for the second time that year, he went well over his head in the NASA bureaucracy and wrote NASA associate administrator Robert Seamans to push for LOR. Seamans passed the letter around NASA headquarters. At the beginning of 1962, Joseph F. Shea, a sharp electrical engineer, had joined NASA at headquarters. After meeting Houbolt, Shea found himself in sympathy with the Langley researcher and also saw that STG was seriously considering LOR. In March, NASA awarded Chance-Vought Astronautics, one of the initial promoters of LOR, a contract to further study spacecraft rendezvous. Maynard was STG's representative on the contract. Soon, LOR officially became STG's preferred way to go to the Moon.

But there still remained others to convince, notably the group at NASA's George C. Marshall Space Center in Huntsville, Alabama, headed by Wernher von Braun. As a center concerned mainly with building booster rockets, it was clear that Marshall's work would be more important if the Nova rocket was required for direct ascent, or two Saturns for Earth orbit rendezvous. On April 16, 1962, a high-powered group from STG, by then known as the Manned Spacecraft Center, came to Marshall to talk about LOR. The group included Gilruth, astronaut Alan Shepard, Charles W. Frick, the project manager for Apollo, and many others, including Maynard.

It was not Maynard's first visit to Huntsville. He had been there before, under orders from Gilruth. When Maynard started work at NASA, he knew of von Braun and his group, but as a veteran of the Second World War who had lost many friends in the fight against Nazism, he didn't think much of von Braun. "He was tarred with the same brush the other Germans were from my experience in World War Two. All these guys were bad guys, and I had been trained to think that way. When I went to work for NASA I wasn't aware that we would have anything to do with Wernher von Braun or anybody else from Peenemunde. It never entered my mind."

Until 1960, Marshall was known as the Army Ballistic Missile Agency at the Redstone Arsenal. ABMA's core was von Braun's group of 118 German rocketeers who moved to the United States following the Second World War. Von Braun had overcome his controversial history at the German rocket development site at Peenemunde as the developer of the V-2 rocket that bombed London and other European cities, and had become a well-known and even popular salesman in the U.S. for space travel. After a period in New Mexico and the move to Huntsville in 1949, von Braun's team developed the Jupiter-C rocket that had launched America's first satellite, Explorer 1, and the Redstone rocket used in Mercury. By 1962, Marshall was hard at work developing a family of giant rockets called Saturn. If an even larger Nova rocket was required, it would come from Marshall.

"I never thought I would have to adjust my mental state to deal with these people," Maynard said. "But as time went on, Bob Gilruth said one day to me and some other people, that we should go to Huntsville and meet with Wernher von Braun and some of his people and begin a dialogue and begin to establish a rapport, because we were going to work with them. So we were ordered to go and establish a rapport.

"We dutifully went. Our hearts weren't in it. When we arrived in Huntsville, we were taken by car to the 'von Braun Hilton,' a multistory, air-conditioned office building. We got into an elevator. We were escorted into a room with long tables set with appropriate china and silverware." As he entered, Maynard thought of the more spartan facilities he occupied at Langley and earlier in Malton. Even though Maynard had not sat in the place assigned to him, von Braun arrived and greeted him, and the other STG engineers, by name. "We left liking them more than when we came in. I was sort of ashamed of myself. That personal encounter was definitely in a favourable direction."

The April 16 briefing at Marshall began with an introduction by Frick, an explanation of Apollo LOR flights by Max Faget and Chuck Mathews, and a briefing on the command and service modules (CSM) by Caldwell Johnson. The discussion on what was then called the lunar excursion module or LEM was kicked off by Maynard.

Illustrating his talk with slides, Maynard gave technical explanations of the features of the LEM, including weights, propulsion and other systems. The slides showed two variations in the design, both with four-legged LEMs carrying two astronauts. In both, the legs and tankage for descent were contained in a descent stage that would be left behind on the Moon, and the same engine would be used for both descent and ascent. The major difference between the two designs was in the structure of the astronauts' cabin in the ascent stage – the first design had a conical cabin with windows in the cone, and the second had astronauts seated behind wrap-around windows, much like the front of a helicopter. Neither design had a ladder for the astronauts to descend to the surface.

Maynard explained that the design with the helicopter bubble was the latest off the boards.

"The most important thing we have learned in the LEM configuration studies so far is that we must consider many more designs and develop the criteria in more detail before we fix the design."

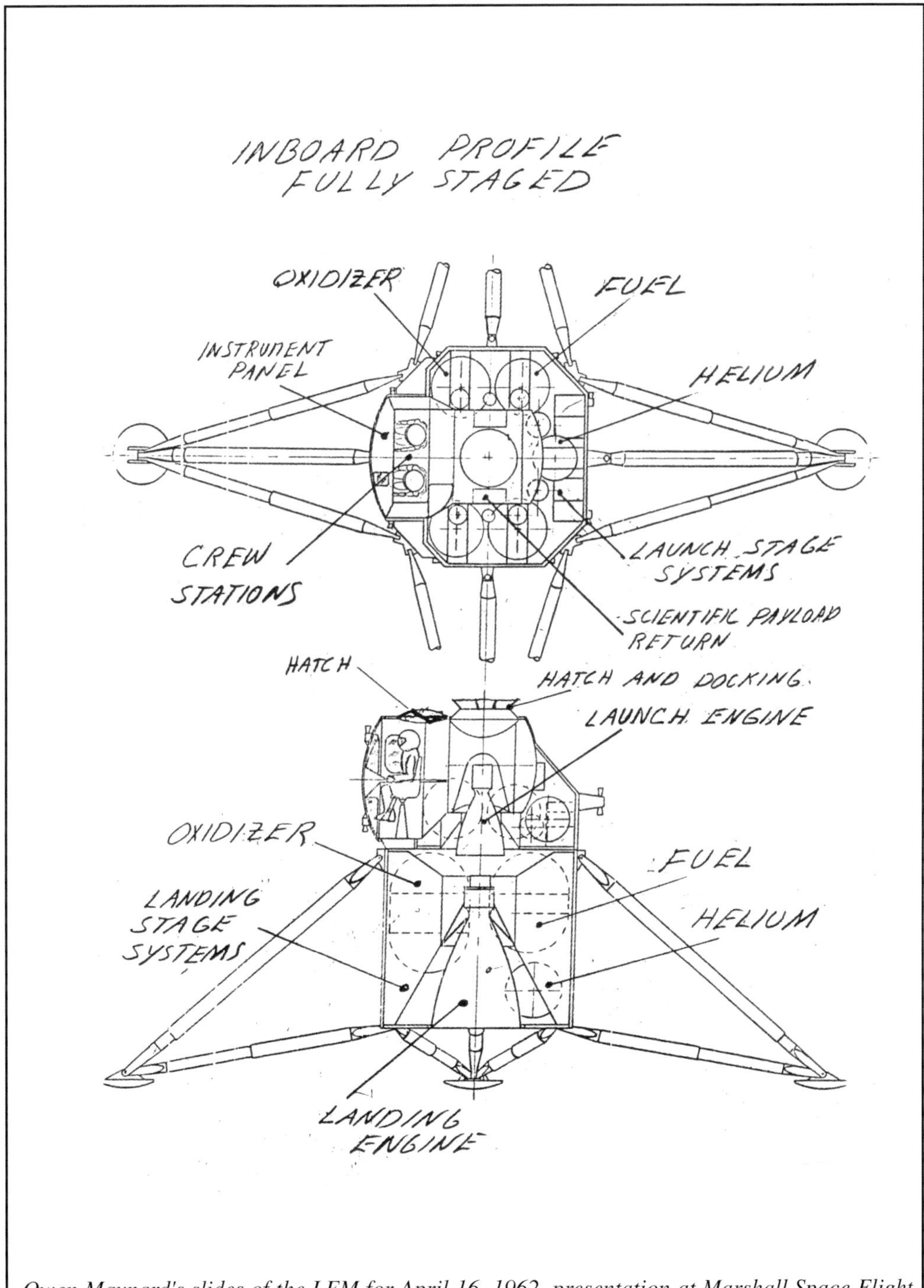

Owen Maynard's slides of the LEM for April 16, 1962, presentation at Marshall Space Flight Center. Courtesy Owen Maynard

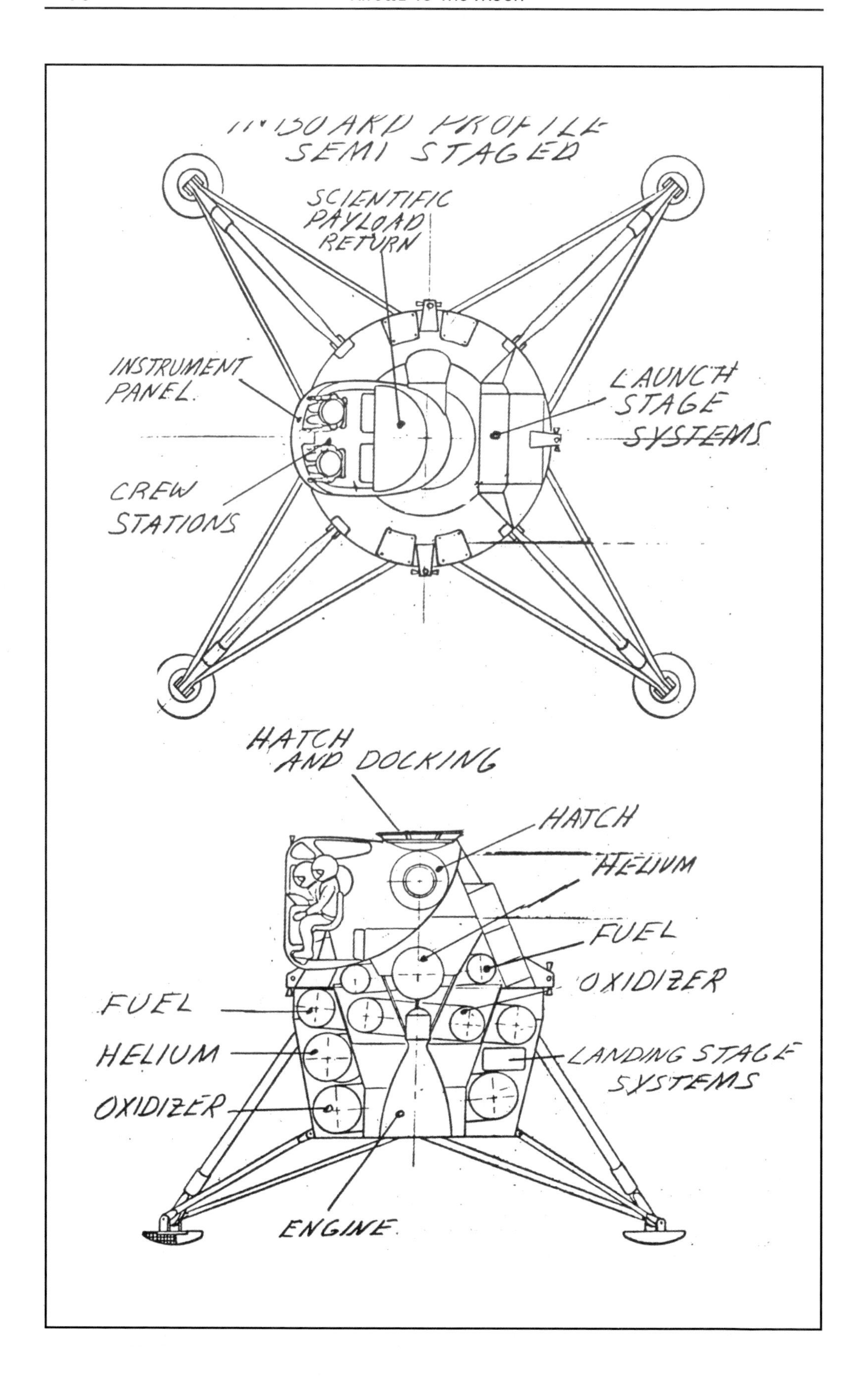
SEMI STAGED
SCIENTIFIC PAYLOAD RETURN
INSTRUMENT PANEL
LAUNCH STAGE SYSTEMS
CREW STATIONS
HATCH AND DOCKING
HATCH
HELIUM
FUEL
OXIDIZER
FUEL
HELIUM
OXIDIZER
LANDING STAGE SYSTEMS
ENGINE

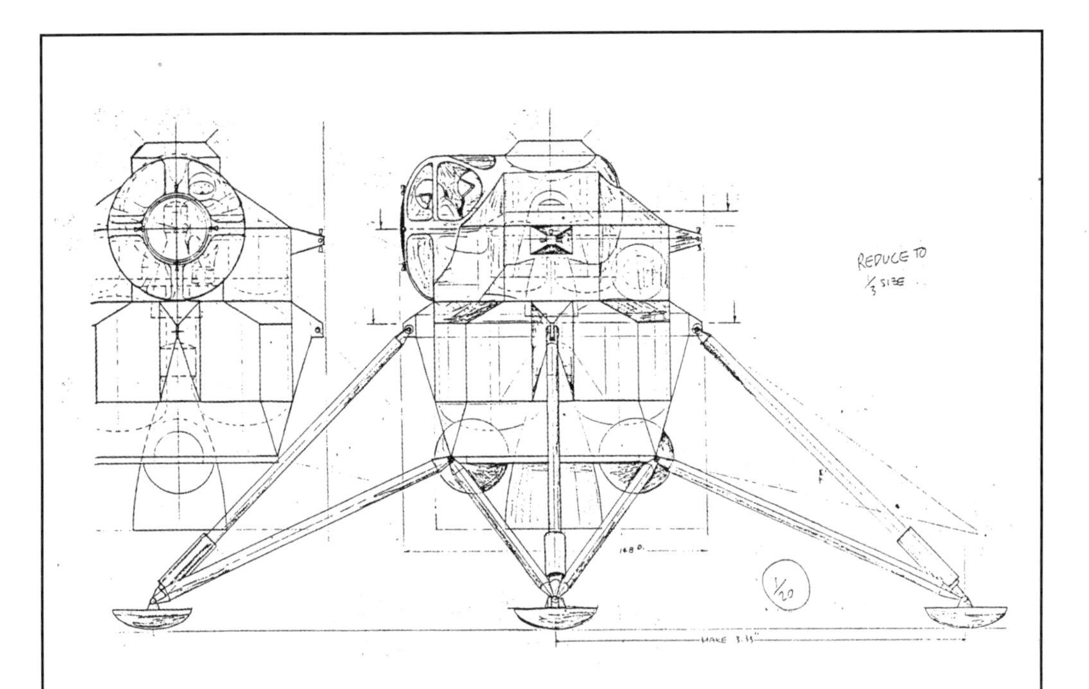

LUNAR EXCURSION MODULE
CONFIGURATION

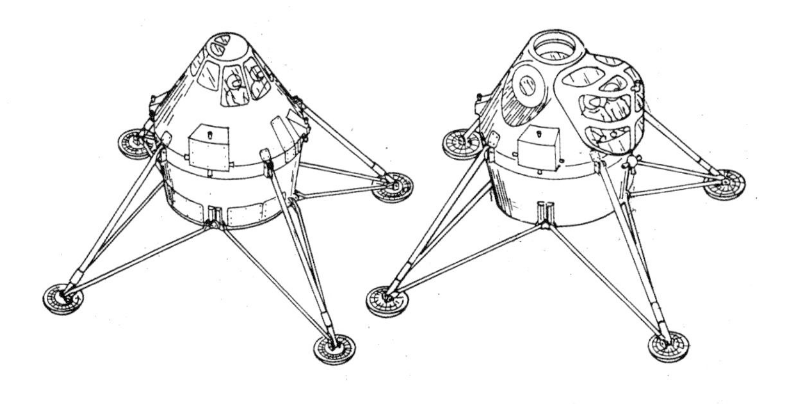

MSC LUNAR RENDEZVOUS · MSFC 16 APRIL 62 · MAYNARD S-137-4

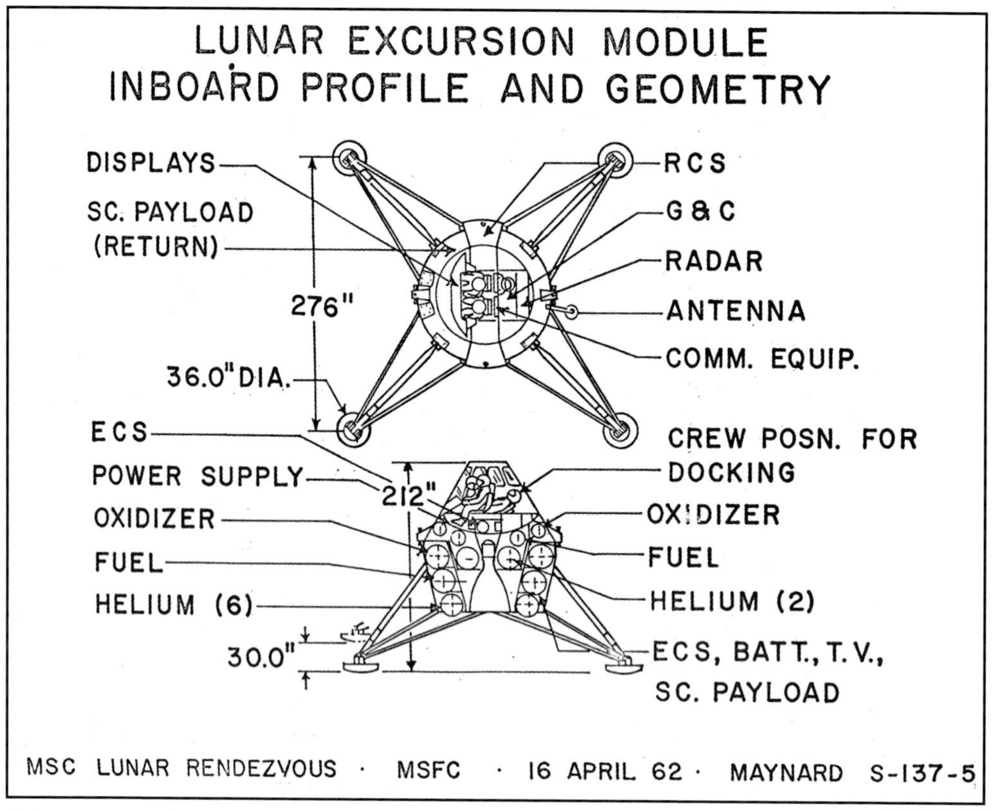
LUNAR EXCURSION MODULE
INBOARD PROFILE AND GEOMETRY
DISPLAYS
SC. PAYLOAD (RETURN)
276"
36.0" DIA.
RCS
G & C
RADAR
ANTENNA
COMM. EQUIP.
ECS
POWER SUPPLY
OXIDIZER
FUEL
HELIUM (6)
212"
30.0"
CREW POSN. FOR DOCKING
OXIDIZER
FUEL
HELIUM (2)
ECS, BATT., T.V., SC. PAYLOAD
MSC LUNAR RENDEZVOUS · MSFC · 16 APRIL 62 · MAYNARD S-137-5

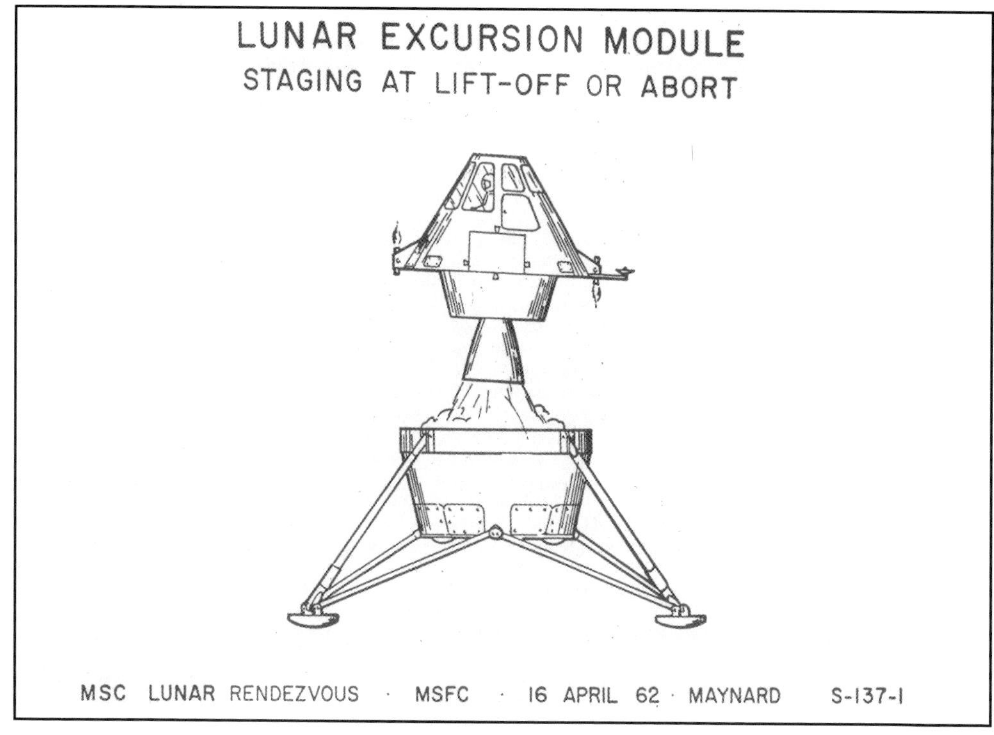
LUNAR EXCURSION MODULE
STAGING AT LIFT-OFF OR ABORT
MSC LUNAR RENDEZVOUS · MSFC · 16 APRIL 62 · MAYNARD S-137-1

But before the LEM design could be fixed, NASA as a whole would have to agree to LOR. STG now backed LOR, but von Braun and his center, which still favoured Earth orbit rendezvous, carried a lot of weight within NASA. Despite the high stakes involved in the decision, most of the discussion that followed the presentations was polite and technical in nature.

After the presentation at Huntsville, the MSC group took what Maynard and others called the "Charlie Frick road show" to Washington to sell LOR to officials at NASA headquarters.

Back at Huntsville on June 7, Marshall engineers held an all-day session on lunar landing, concentrating on Earth orbit rendezvous. Von Braun, who had spent the previous few weeks digesting the presentations from MSC and considering the offer of extra work on lunar surface operations to soften the blow of the loss of Nova, had a stunning finale for the day's discussions. Marshall, he told a silent audience that included Joe Shea from headquarters, would support LOR. With that, the meeting ended. "If Bob Gilruth had tried that at the Manned Spacecraft Center, we'd still have a half dozen independent guys saying, 'you may agree but I don't,'" Maynard commented later.

Most of Maynard's later encounters with von Braun were also favourable. "We got zero static from the Marshall guys. Von Braun showed us support, and he set the stage for many positive things that happened between us and the Marshall guys."

With von Braun behind LOR, the consensus built quickly within NASA, and on July 11, 1962, NASA administrator James Webb told a press conference that Apollo would use lunar orbit rendezvous to go to the Moon. Although President Kennedy's science advisor, Jerome Weisner, continued to oppose LOR, NASA's decision was final.

•

John F. Kennedy's Moon landing goal was the most important decision in the history of the U.S. space program. The decision to use lunar orbit rendezvous was probably the most important technical decision in the history of NASA. Without it, it is open to question whether NASA would have met Kennedy's goal to land on the Moon in the 1960s. Indeed, had NASA chosen another way to go to the Moon that took longer to build and prepare, the spending pressures from the Vietnam War and the ambitious social programs of the 1960s that eventually clipped NASA's wings in the 1970s might have killed Apollo before the first Moon landing. The first human footsteps on the Moon might have been postponed until the 21st century.

The LOR decision had other implications as well. The promoters of Earth orbit rendezvous argued that its use in Apollo would have led to the construction of a space station in Earth orbit as a base to launch astronauts to the Moon. But in the event, Apollo didn't need a space station and it ate up funds that might have been used for a space station. As a result, the only American experience with space stations in the 20th century was Skylab in 1973 and 1974, and later when Americans flew to the Russian space station Mir starting in 1995.

"The LOR concept was initially viewed as the product of pure theorists' deliberations with little practicality," Maynard wrote later. "The dedication and unswerving persistence of Dr. Houbolt in advancing these most meaningful theoretical and practical inputs (including the making of them heard by the right people at the right time), was most valuable to the program. The value might best be measured in terms of increased safety, straightforwardness in the way to proceed, time savings, and cost savings."

Maynard also said on another occasion that the use of modularity in Apollo was key in leading to the LOR decision. "What I'm saying is that LOR would have evolved with or without Dr. John Houbolt, and it would have been the mode used by NASA."

These statements encapsulate the historic debate about LOR. Few would debate the major contribution that Houbolt made to the decision, though they might argue about how important Houbolt's contribution was. Many others at NASA, including Gilruth, Faget, Caldwell Johnson, Shea, Seamans, von Braun and others were vital players in the LOR decision. Three engineers from Avro, Morris Jenkins, Owen Maynard and especially Jim Chamberlin, also played key roles in this historic turning point.

5 Mercury Orbits and NASA Moves

While some people at STG were embroiled in the arguments about how to get Apollo to the Moon, those who remained in the Mercury program still had to put an American astronaut into orbit around the Earth. Astronaut John Glenn had been selected to make the flight early in 1962. Some of the preparations for his launch were very unorthodox, as Fred Matthews remembers of a successful effort he and Tec Roberts made to keep the tracking network together.

"Just before John Glenn's flight, somebody in headquarters decided that it was time to renew the contract for the operations of the worldwide network and all the tracking, communications and telemetry stations around the world. This was going to save them money, they thought. They put out a request for proposal to industry and had received inputs from various contractors as to what they would do and how much their bid was. This upset Project Mercury people because we had spent a lot of time getting the people that were on stations trained. Here some beancounter was going to throw this experience away. Just before John Glenn's flight, we would have ended up with completely new people. Some of them would have transferred contractors, but essentially we would end up with a new organization, untrained, at a time when it was critical to the success of John Glenn's flight.

"So Tec Roberts and I were given the job at the last minute to see if there was something we could do to stop this proposed changeover. All the bids had been in, they'd been evaluated, and as far as I know, headquarters was ready to announce the winner and let the contract. Tec and I went up and reviewed all the bids. We took a day to go over them with a fine-toothed comb, hoping to find something in them that would allow us to put a stop to it. We'd almost given up at the end of the day when we suddenly concluded that there was something there that none of them had proposed. They all proposed how they were all going to train the operators so that they would be proficient for the mission. What none of them said was how they were going to train the trainers. This was just a big blank, and so we wrote a quick memo to the source selection board saying that if they went ahead with this, NASA would have been responsible for training the trainers, and we didn't have either the time or the personnel to do it, and in the end, it would jeopardize the potential success of the mission. The next day they cancelled the [request for proposals], much to the chagrin and disappointment of the contractors, who had spent a lot of time and money making the bids."

C. Frederick Matthews in 1960 at Langley Research Center. Courtesy NASA.

The worldwide tracking network NASA had ready for the Mercury orbital flights had been built almost from scratch. With the primitive communications tools available at the time, building this network had been a daunting task, as Roberts and Matthews knew well. It was made up of 18 tracking stations all over the world, including Cape Canaveral and five other stations in the United States, 10 ground stations in foreign territory, and two ships – one posted in the Atlantic Ocean and another in the Indian Ocean. To tie the whole network together across three oceans and three continents, NASA used 285,000 km of communications circuits and radio connections to bring information together through telephone, telemetry and teletype.[4] The network was tied together at NASA's new Goddard Space Flight Center in Greenbelt, Maryland, just north of Washington, D.C., where the information was processed and passed down to Mercury control at the Cape.

John Hodge, Dennis Fielder, George Harris and Fred Matthews were heavily involved in setting up this network under the direction of Chuck Mathews. The network was built under contract by Western Electric as the prime contractor. Hodge helped make some major decisions around the tracking network, including the crucial one concerning how much coverage of the flight there should be from the ground. In 1959, NASA physicians demanded that there be continuous and complete communication with the astronaut at every step of the flight, but Mercury engineers concluded that this would be almost impossible. Hodge said the doctors relented when they were asked what they could actually do to help an ill astronaut in orbit, and realized the answer was almost nothing. The doctors finally agreed that communication was most vital to ensure a safe re-entry in the atmosphere and a successful recovery of the astronaut.

For a variety of reasons, including tracking and recovery needs, NASA decided to fly Mercury in an orbit tilted at 32.5 degrees from the equator. This trajectory took it mainly over tropical and semi-tropical areas, and it required a launch in a northeastern direction from the Cape, a path that took it over the British-held island of Bermuda. "It became clear very quickly that we needed a backup control center at Bermuda, because the [booster] cutoff was at the very end of the Cape's capability. You only had a few seconds after cutoff to do the orbital calculations. So we put the Bermuda site in so that you had confirmation of orbit," Hodge said.

In a technical article at the time, Hodge and Roberts wrote that since there were no high-speed data lines available early in the program, an IBM 709 computer was installed at Bermuda to compute the capsule's trajectory at the moment it entered orbit. "In this respect, it acted as a backup control center," they wrote, in case the Cape was knocked out of action for any reason. For most Mercury orbital flights, Hodge was assigned to Bermuda, ready to take over as flight director. Only when communications links were upgraded, allowing Bermuda to operate directly under the control of Cape Canaveral, did Hodge return to the Mercury Control Center during flights.

Communications satellites were still in the future, so undersea cables provided telephone but not television links. These cables could only support a limited amount of data. Therefore, most tracking stations were staffed with a CAPCOM, a capsule systems monitor, and a medical monitor. "The most complex remote site consists of a UHF/HF transmitting and receiving system, an S-band and/or C-band radar, a telemetry receiving system, an acquisition system, and a command system," Hodge and Roberts wrote. The work on the control center and network began in the summer of 1959, Roberts said, and Western Electric, assisted by IBM and Bendix, developed the network. "It took only about 12 months to design and build that initial Mercury network including the control center. There were interesting problems with logistics. The original Mercury network and its remote site stations were designed and built on the concept of having flight controllers at each site. They would talk with the astronaut in the spacecraft as it came over and monitor the data that was being telemetered. All data transmission back to the control center was by teletype." By 1961, Roberts said, even most of the overseas stations could be connected to Mercury control by telephone. This meant that the CAPCOM at the Cape could talk to

[4] Teletype, which at the time was a major form of communications, had disappeared by the 1990s.

astronauts when the spacecraft passed over other stations, but the CAPCOMs in tracking stations didn't disappear until Apollo.

"I worked on ground support systems," said Dennis Fielder, who Hodge and others credit for his key role in setting up the network. "How much coverage we were going to get in real time, what instruments, what remote control command systems, and what voice communications could be expected in real time, and how could this time be most effectively used. Then we went out to look and see what the world looked like. It was joined together with cables across the oceans. The rest of it was radio. The reason there were flight controllers there was because we couldn't get direct voice contact with some of those sites and we couldn't relay to the mission control center. So we had controllers at that station." Fielder said his job was to supervise the contractors installing the equipment and to make decisions on such issues as monitoring of solar activity that would affect radio communications.

Dennis Fielder was born in Crouch End, London, in 1930 and took his apprenticeship in aeronautical engineering at the Royal Aircraft Establishment in Farnborough. After working on guided missiles at the RAE, he moved to Canada in 1954 and specialized in radar at General Electric, and on weapons systems and airborne flight test equipment at Avro Canada starting in 1956. Although this experience pointed him to working on the Mercury capsule when he joined NASA, most of those jobs were already taken, so he worked on ground facilities.

Chris Kraft said of Fielder: "He understood the teletype and the telephone, he understood the machines, he understood all the ramifications of dealing with communications to get the information back from around the world and in the control center. Dennis was a man of all seasons. He doesn't come over as being one of the more important 'Canadians,' but to me, he was."

Fielder also stood out in the group and in NASA due to the fact that he wore a beard and was fond of trendy clothing. Even when he was wearing the regulation jacket and tie, Fielder sported a pair of leather boots. Fielder was also noted for his love of cars. At Avro and in his early days at NASA, he drove a Jaguar.

Dennis Fielder receives a service award in 1969. Courtesy NASA.

Once in a simulation, Fielder was posted at Bermuda with John Hodge. As part of the exercise, the Mercury Control Center at the Cape cut all communications, and Hodge thought he would have to control the flight from Bermuda. Fielder took out his "black book" and, using regular commercial telephones, put a long distance call through to flight director Kraft at his console at the Cape, something that was not simple to do in the early 1960s. An annoyed Kraft refused the call, and Hodge and his Bermuda team wound up in charge.

As one of the designers of the tracking system, Fielder took part in checking out the various stations along with two other members of the Avro group, George Harris and Len Packham, who Fred Matthews described as a "whiz" at radar and communications. Harris used aircraft carrying Mercury communication equipment to test the tracking station setups before they were put into service. The first tests took place at Wallops Island, where NASA had built a complete replica of a tracking station.

"We had an old B-25, and we put a Mercury beacon and telemetry transmitter in the belly of it, and then flew up and down the coast, yelling over the radio, 'Wallops, this is King George. Wallops, this is King George.' It worked so well that when I got back, Barry Graves, who was the chief then of this whole network enterprise, said, 'Go pick out a better aircraft that can fly the seas and check out the tracking stations.' So we picked out two DC-4s. Fully equipped with a set of Mercury capsule gear, command, telemetry, beacons, and voice. The whole works." Harris and his colleagues also used two DC-3s - one posted in the U.S. and another in Australia – to test tracking stations.

"And don't forget, we had stations starting in Cape Canaveral, Bermuda, the Canary Islands, Kano in Nigeria, Zanzibar, Muchea in Australia, Carnarvon in Australia, Canton Island in the Phoenix group 1,000 miles south of Hawaii, Kauai, Point Arguello, Guaymas in Mexico, and Corpus Christi, Texas, and two tracking ships. It worked so well, we did all of the calibrations. One airplane going east, one airplane going west. We'd take a flight control team with us. We put the first global network in, and checked it out."

At the time he joined NASA, Harris was 30, a native of Willenhall, England. He attended Wolverhampton Technical College and then apprenticed at the Midlands Electricity Board, which was moving into communications and monitoring of power stations, a field where Harris learned his engineering. But he soon became restless, and decided to find work in the aircraft industry outside England. After narrowly missing an opportunity to go to New Zealand, Harris brought his family to Toronto, and after three months repairing radios he got a job at Avro Canada in 1954, where his work in flight test led him to NASA.

•

Bruce Aikenhead and Fred Matthews prepared the people who would work in these far-flung tracking stations. Matthews had been named as backup flight director to Kraft and Hodge, but his main concern was looking after the flight controllers in the stations outside of Cape Canaveral and Bermuda. There were only a few people in the flight control section at the time, responsible for staffing not only the Mercury control center and Bermuda, but all the other stations around the globe. STG engineers with other areas of responsibility, including Aikenhead himself and his colleague from Avro, Gene Duret, were pressed into service for the tracking stations during flights and simulations, and these people became known as "weekend warriors." At Matthews' behest, NASA contracted with Philco-Ford to supply an additional 19 engineers to fill out the work at the 16 tracking stations that operated during orbital flights.

"Fred Matthews eventually became the lead in training people and setting up the way in which we were going to train flight controllers," Kraft remembers. "He had tons of ideas that he developed. He had me direct what we called paper simulations for every operation. We built cubbyholes and we put people in them that were going to be in each one of the sites, and planned the use of how you were going, that sort of thing, from a simulation point of view. And then helped develop the procedures of how you were going to do it in real time in actual operations."

Bruce Aikenhead's training group was concerned not only with training astronauts, but also with training the people who would deal with them in Mercury control and the tracking stations. "We would have to train all these people going out to the tracking stations, so the logical way to do that was to build an adjunct to the [Mercury] procedures trainer. So in a corner of that room, we built a duplicate of the display console we would have in these tracking stations around the world. And then we put in an instructor's station where the instructor could put in problems according to a script. In this way, we could bring in a crew that was going to be trained for a tracking station. Meanwhile, the astronaut was being trained. We ran training exercises so that astronauts would get used to talking to these people and these people would get used to the astronauts," said Aikenhead, who helped write the document setting out the plan for the training equipment.

He said simulations involving every tracking station could be run using a set of cubicles, one cubicle representing each station. Reports duplicated on Gestetner machines stood in for teletype reports, and various scenarios were simulated, including one where Mercury Control at the Cape was shut down due to food poisoning.

"There was another fun simulator that involved the procedures trainer at the Cape," Aikenhead continued. "Someone down there realized that we had a beautiful way of training and preparing the remote sites. We recorded the data that was appropriate for Bermuda on one tape, and the data that was appropriate for the Canary Islands on another tape, etc. We prepared a whole set of tapes to simulate a three-orbit mission. This was a mission that was run by the simulation people only so that no one knew what was on that tape. These were scripted so that they developed in a realistic manner. When people went out to the tracking station, they took a tape with them, and these were given to the site manager who gave it to people in the back room. They would rack this up, and they would read appropriate responses on pieces of paper that came with the tape. This would simulate the astronaut. The guy in the back room would start the tape. The electrocardiogram readings would be data from an astronaut. This kind of training proved extremely effective. These were simulations that were happening in real time on a worldwide basis involving hundreds of people. This was a simulation that had gradually grown, and it paid off. The network worked extremely well."

MR-4 (Liberty Bell 7) Mission Review Conference at Cape Canaveral, July 1961. Left to right: astronaut Donald Slayton, astronaut Virgil Grissom, Kenneth N. Nagler, Warren J. North, William K. Douglas, astronaut John Glenn, astronaut Alan Shepard, Charles W. Mathews, John D. Hodge, Stanley C. White, and Christopher C. Kraft Jr. Courtesy NASA.

•

At the time Alan Shepard had flown in May 1961, NASA was still planning to fly most of the Mercury astronauts on suborbital flights boosted by Redstone rockets prior to going into orbital flights. As a follow-up to Shepard's success in May, Gus Grissom was tapped to fly the second Mercury flight aboard MR-4. Grissom boarded his Liberty Bell 7 capsule on July 21 and flew another 15-minute suborbital flight similar to Shepard's. But after splashdown in the Atlantic, the capsule's explosive hatch blew prematurely. Grissom nearly drowned before a helicopter pulled him from the water, and Liberty Bell 7 sank when it filled with water and a helicopter released the line that held it to the capsule. In spite of the loss of the capsule, NASA decided within a month to end the Mercury-Redstone flights and focus on matching the Soviet achievement of putting a man in orbit. The urgency of this task was underlined on August 6 and 7, when Gherman Titov flew for 17 orbits and a full day in space aboard Vostok 2.

No Mercury capsule had yet flown in Earth orbit. Earlier in the year the first orbital attempt, MA-3, had failed when the Atlas went off course. The capsule was carried away by the escape rocket and recovered. The next orbital attempt, MA-4, flew a nearly flawless circuit around the Earth on September 13 and splashed down near Bermuda. Mercury and Atlas had checked out for orbital flight, and so had the Mercury tracking network.

MA-5 was a bigger test of the system. Instead of instruments, a chimpanzee named Enos lifted off on November 29, and the flight got off to a nearly perfect start. "It had a greater degree of risk with a chimpanzee than with an astronaut because with the chimpanzee, you had to rely on the automated attitude control system, which was absolutely necessary to control re-entry," Matthews explained." We had hoped to have a three-orbit mission, and I had written a lengthy memo on the risks and problems and potential things that could happen, and on that basis they had made some decisions about what they would do if they couldn't bring the chimp back safely.

"During the second orbit pass over Australia, it was detected by the systems monitor that the automated control system was using up more fuel than it should have. So it was suspected that there was a leak. This was confirmed by the other Australian station, further confirmed by Canton Island, and further confirmed by Hawaii. And by that time, Chris Kraft had made up his mind to bring back the chimp at the end of the second orbit because he was afraid they would run out of attitude control fuel if they tried to make the third orbit.

"He made the decision to send the retrofire command to reset the clock from Point Arguello [California]. We got on the communications link to Point Arguello, called them, and got absolutely no response, dead line. If he set the clock at Point Arguello, it would have been set to fire the retros over Guaymas, Mexico, the next station in the chain. Chris gave up on Point Arguello, and decided to call Guaymas, and again there was no communications. Communications were dead. He kept trying, and the time for retrofire came closer and closer, and then suddenly Guaymas came on the line, and it was just in time to tell the capsule communicator at Guaymas, Arnie Aldrich, to fire the retros. He started the countdown, 5, 4, 3, 2, 1, fire. Arnie pushed the firing command and the retros fired and the capsule re-entered and was picked up in a contingency recovery area.

"This was an example of where flight controllers, both at the mission control center and the tracking stations, made a significant contribution to the success of the mission," Matthews said. An investigation later showed that the phone company had violated a rule not to use the same communication links between Mercury control and separate stations. When a farmer in Arizona ran a plough through the wire, communications were lost until the phone company switched Guaymas to another line.

Soon after Enos's safe recovery from the Atlantic, Bob Gilruth and Walter Williams decided to fly Glenn into orbit on the next flight, which was set for early in 1962.

•

While the engineers from Avro jumped into their work in Mercury and Apollo, they and their families also settled into their new homes in the unfamiliar environs of Hampton, Virginia.

NASA's new recruits found a variety of accommodations. Some moved into military-style apartments in Hampton, and many of those with families moved into ranch-style homes in a new subdivision in Newport News, Stoneybrook Estates, where three of the astronauts also settled in. Fred Matthews bought a split-level design called a Monterey. John Hodge, Tec Roberts, Len Packham and Stanley Cohn also moved into the neighbourhood, Cohn choosing a home that was located between the homes of astronauts Grissom and Slayton.

Frank Chalmers, who worked in the control central section of Mercury, wrote his wife June to say he was depressed by his early house hunting. "It compares very poorly with Toronto. There are some nice areas, but to live in these we would have to buy a house, or if we were lucky and got the opportunity, rent a home unfurnished. The apartments I mentioned [visiting] are nothing like city apartments and look like broken down barracks." But soon Chalmers found a furnished apartment he liked for $107 a month and paid his first month's rent and a $50 deposit with his first paycheck from NASA, which for two weeks' work totaled $266.95.

June Chalmers remembers that the Avro engineers got almost celebrity treatment when they arrived in Virginia in 1959. "I remember pulling into a gas station one day, and the young man asking me, 'Are you one of the scientists that's from Canada?' We felt we'd been elevated from engineers to scientists. It was a very welcoming atmosphere, so obviously these people knew that these Canadians were coming down, and they were always met with good feeling. You carried a NASA sticker on your car, you could go almost anywhere. Saluted and welcomed into almost any Air Force base, it seemed, if you had the inclination to go there. It was a good feeling."

Moving from Canada to Virginia was more of a change than a move to other parts of the U.S. might have been. Virginia was part of the South, and racial segregation was still very much a part of life there in 1959.

"There was no segregation in NASA, because it was a federal operation. The segregation was strictly in the State of Virginia. You couldn't fight it. You certainly didn't approve of it. It wasn't something we were ever used to, coming from Canada. But at that time, unless you moved to the northern states, you were going to run into segregation almost everywhere," June Chalmers said.

Matthews said he worked with a group of 64 doctors who took turns working as medical monitors at tracking stations, and one was a black Air Force colonel who was sent out to the tracking station in Kano, Nigeria. "The people in Nigeria were just ecstatic about us sending a black colonel medical doctor on the project out to their country. And so we got all sorts of accolades from them for that, and so after the mission, we brought everybody back for debriefing. I had to make arrangements for putting them up, because their home bases were scattered all over. Most of them stayed at the Chamberlin Hotel in Hampton, which was nearby. I got a call from this colonel saying he couldn't get into the Chamberlin. They wouldn't let him in because he was black. This kind of hit home to me that we were getting accolades from the other side of the world, and he couldn't even get into a hotel."

Within a decade, the overt forms of racial segregation were gone. Most of the credit for this change must go to the civil rights movements of the 1960s, but NASA played a modest role in creating the "New South" that arose from the ashes of the old. Many NASA facilities were put in the southern states, in part to bring new economic and social vitality to an area that had long depended on agriculture and had missed out on the industrial growth that the northern states had experienced earlier in the 20th century. Attitudes changed as people from outside the South moved there to take part in the space program, and as southerners went for higher education in growing numbers to take their own places in the space program.

•

STG was looking for a home when the new recruits arrived in Virginia from Canada in 1959. Officially, STG was a semi-autonomous unit of the Langley Research Center, and the STG personnel were shoehorned in among the wind tunnels and aeronautical researchers of Langley, but it was clear that they would have to move unless the idea of sending astronauts into space ended with Mercury. In May 1959, NASA announced that a new Goddard Space Flight Center would be set up north of Washington, D.C., near Greenbelt, Maryland. The core of Goddard was the former Vanguard satellite project team that had transferred to NASA from the U.S. Navy. Besides continued work on satellites, Goddard was handed responsibility for NASA tracking networks. STG officially became an administrative part of Goddard. When he first reported for work at STG, Bryan Erb recalls being told that the operation would be moving to Washington's Maryland suburbs in two years. When there was free time, the Erbs vacationed in the Carolinas, believing they would have ample opportunity to explore areas farther north once they were relocated near Washington. As the months went by, however, it became clear that the orientation of STG was quite different from that of Goddard. With the growing impetus to follow Mercury with Apollo and other projects, STG needed its own facility. Just before the Kennedy administration took office in January 1961, STG was separated from Goddard. Its future was left in the hands of the new administration.

By the time President Kennedy decided to go to the Moon in May of 1961, an organized effort was underway to decide where to put STG. Criteria for the new location were made public. They included an outlet for water transportation by barge, a moderate climate, the proximity of a large community with air services, an industrial complex, higher education institutions and utilities. NASA set up a site selection team in August, and it looked at 23 sites in Florida, Louisiana, Texas, California, and Missouri. Locations in Massachusetts, Rhode Island and Virginia also lobbied for consideration. On September 19, 1961, NASA announced in a press release that STG would be moved to a 1,000-acre site south of Houston, Texas. The site, a cattle pasture located near the road from Houston to Galveston, was donated to NASA by Rice University in Houston, although NASA quickly purchased an adjacent 600 acres. The new facility was called the Manned Spacecraft Center.

The decision proved to be a hard sell to the people at STG who would have to make the move from Virginia. Most of the year, Houston is very hot and humid. The air quality was worsened by emissions from petroleum and chemical plants. The new center was located near a murky and shallow body of water misleadingly named Clear Lake. And just in case anyone was unaware of the climate, Hurricane Carla struck the Houston area at the same time NASA announced that the new center was moving there. As well, Houston was not far from the Louisiana border, and in many ways it was as much a part of the South as Hampton, Virginia.

"We knew that that area had been considered a hardship posting for some people," Bruce Aikenhead said of Houston. "I remember being completely dismayed at this. It meant moving to Texas and raising our kids as Texans. It was farther away from home. Several of the Canadians and quite a few of the Americans decided to look for alternatives. Several of the Canadians looked for places elsewhere in the States, and some came back to Canada."

"There were a lot of people who were undecided about going, so Houston decided to send up a representative from the Houston Chamber of Commerce," Matthews recalled. "They invited all the members of the Space Task Group to attend and to bring their wives to a big auditorium. So the head of the Chamber of Commerce decided that since our wives were going to be there, he was going to bring his wife, too. So the two of them were sitting up on the stage and he was showing slides about how great Houston was and all the things that there were to see and do in the Houston area. And when he was all done, he asked if there were any questions, and one of the wives stood up in the back of the room and asked, 'How hot does it get?' And he hemmed and hawed around and said, 'Well, you've got air conditioning in the cars and you've got air conditioning in the houses and you've got air conditioning in the workplace, and so you don't really notice it at all.' He looked around again and asked, 'Are there any more questions?' And the woman in the back row got up and said, 'How hot does it get?' Again he started to hem and haw around and not answer the question, and finally his wife looked at him, as if to say, 'Dear, I'll answer the question, don't worry.' So he sat down, gave his wife the floor, she stood up, and he had a look of relief on his face that he wouldn't have to answer the question, and she said, 'It never feels more than 110 [43° C].' And the look on his face ..."

Adding to the STG engineers' unhappiness was the idea that Houston's selection was a matter of pure politics. Everyone knew that Vice-President Lyndon Johnson, who had been delegated heavy responsibilities in the space program by President Kennedy, came from Texas.

"The choice had already been made politically, and the whole thing had been a farce," Aikenhead said. "They were virtually instructed to choose Houston, because Rice University had made available some land. The impression was there that Lyndon Johnson and some of his political cronies had wanted Houston to be chosen. The announcement was made that it was to be Houston, and that very day we received complementary subscriptions to the Houston papers. So there were people in Houston who knew already. It's hearsay, but it's certainly common knowledge that this thing was a political choice."

According to the NASA history of the center, Houston was the second choice of the site selection committee, and moved to first place when the Air Force reversed a base closure that would have freed up the site of the original first choice, Tampa, Florida. But even the NASA history admitted there was enormous political pressure favouring Houston. There was not only Lyndon Johnson for NASA to contend with. Perhaps more importantly, the local congressman, Albert Thomas, was chair of the House independent offices appropriations committee, which decided NASA's funding. As well, Texas congressman Olin "Tiger" Teague chaired the House subcommittee on manned space flight. And Texan Sam Rayburn was speaker of the House of Representatives. The decision for Houston has always been shrouded in controversy, and the fact that the water access requirement was never used fuelled the arguments.

"I decided not to go down to Houston for several reasons," Matthews said. "One was that I had never been that thrilled by the South. I'd been down there on several occasions. In moving down, it would not have been a straight move into the Clear Lake area, because that was just a field. Nothing was built, so for two years they had to go down and work in office buildings scattered all over Houston until the Manned Spacecraft Center was built. This meant putting the kids into school in Houston and moving out [again] to some place like Clear Lake, or near it, and then changing schools all over again. The kids were that age where I didn't want to expose them to two moves when I could do it with one, because they'd already been exposed to one move. The other reason I didn't want to go was that both my parents were in Toronto and in not that good health. We wanted to be close enough to be able to go back and forth."

Enticed by his former boss at Avro, Mario Pesando, Matthews moved to Boston to work with RCA. Aikenhead, Jack Cohen, Stanley Cohn, and Stan Galezowski decided to return to Canada. Chalmers and Joe Farbridge had already left NASA in 1959.

"I wasn't too keen on going to Houston, because I had quite a bit of cultural adjustment just coming to Southern Virginia," said Cohn, who added that he had lived in the northern states earlier in his life without difficulty.

There were a number of reasons Galezowski gave for his return to Canada: "Four months of rain a year in Houston, hot and humid, 2,500 miles from anyone you'd want to know. The other thing was that I was getting technically obsolete. It was time to change careers."

Bryan Erb spoke for those who chose to move to Houston: "I just went where the job led." Climate didn't scare him off, said Erb, who had turned down a new job offer in Canada not long before the Houston move was announced. "By that time, I was involved in what became Apollo, and that was far too exciting to think of giving up." A survey conducted by NASA a few weeks after the move was announced showed that four out of five employees at STG agreed with Erb and intended to make the move to Houston.

Dona Erb shared her husband's feelings, albeit with one reservation. "The only concern that I really had was financial, because we had been told when we first got to NASA that the new facility for space projects was going to be at Goddard. Then that changed, and everybody, including Gilruth and all sorts of people all the way up the line, added to their house, bought a new house, or built extensions. We went out of our rental property and built a new house and got to live in that for seven months. When you have 400 or 500 families moving out of an area all at once, the real estate market kind of goes to blazes. That was more my concern. We were thinking that we were going to settle, and it turned out we weren't."

At least two of the Avro group made the move to Houston and decided that the climate didn't agree with them. Tec Roberts transferred to the Goddard Space Flight Center on doctor's orders in 1965, and George Watts came home one day and found his wife Norma passed out due to the industrial pollution that was a problem in Houston in the 1960s. This contributed to their decision to move to California in 1964.

For almost everyone at STG, the move to Houston or out of the space program didn't happen until 1962. STG's numbers continued to grow: at the time of the move there were 750 people working there, and by the time Gemini began flying in 1965, nearly 10 times as many people were working at the Manned Spacecraft Center.

•

On Friday, November 6, 1964, eight members of the Avro group and their families gathered at the United States District Court in the Houston Federal Building to be sworn in as citizens of the United States. Peter Armitage, David Brown, Thomas Chambers, John Hodge, Morris Jenkins, John Meson, Rod Rose and George Watts took the oath of citizenship and then celebrated at a reception.

Their decision to become U.S. citizens was a logical consequence of choosing to remain in the United States. Citizenship was a big issue because it sometimes affected security clearances. In 1960, all the engineers who had come from Avro had signed forms stating their intention to become U.S. citizens in an exercise that was clearly related to security clearances. By the end of the five-year citizenship waiting period, others including Jim Chamberlin also became U.S. citizens.

Avro engineers after becoming U.S. citizens, November 6, 1964, Houston Federal Building. Seated left to right: Peter J. Armitage, U.S. Circuit Court Judge John Brown, U.S. District Court Judge James Noel, John D. Hodge. Standing left to right: George A. Watts, Thomas V. Chambers, Rodney G. Rose, John K. Meson, David Brown, Morris V. Jenkins. Courtesy NASA.

Avro engineers after becoming U.S. citizens, November 6, 1964, Houston Federal Building. Seated left to right: U.S. Circuit Court Judge John Brown and U.S. District Court Judge James Noel. Standing left to right: John D. Hodge, Rodney G. Rose, George A. Watts, Peter J. Armitage, David Brown, Morris V. Jenkins, Thomas V. Chambers, John K. Meson.
Courtesy NASA.

Some who stayed, like Owen Maynard and Bryan Erb, did not immediately become U.S. citizens. Giving up one's Canadian citizenship "is not a decision you take lightly," Maynard said at the time. Eventually, Maynard became a U.S. citizen once the law changed so he could retain his Canadian citizenship. In the early 1980s, Bryan Erb applied for U.S. citizenship because his work in remote sensing would benefit from a Top Secret clearance. By the time the process was nearly complete, Erb was informed that his fingerprints were "obsolete" and a new set was required. Then a Top Secret report was leaked and the U.S. government, including NASA, responded by revoking many security clearances and making new ones virtually impossible to get. So Erb put his citizenship effort on hold and never resumed it.

With home so close, the Canadians were more likely to leave the U.S. after a short period of time. Only one of the British engineers is known to have returned home, and he eventually moved back to Canada. Most of the British engineers stayed in the U.S. and became Americans. In the 1964 citizenship ceremony, seven of the eight engineers who took the oath came from Great Britain.

The citizenship ceremony in 1964 was also notable because it was one of the very few gatherings of a large number of the Avro engineers since the early days at the Langley Research Center. Unlike Wernher von Braun's Germans at Huntsville, the Avro engineers at NASA did not form a cohesive group. While some knew each other at Avro, they had been drawn from different parts of the large Avro organization, so gatherings of more than a few of what NASA people called the "Canadians" were very unusual.

Aside from the citizenship ceremony, the only public event featuring the Avro engineers in the 1960s took place on the village green at Williamsburg, Virginia, on Sunday, October 29, 1961, not long before the move from nearby Hampton to Texas got underway. The event was the one and only game of the Space Task Group cricket team, which had accepted a challenge from a team at William and Mary University.

Cricket practice, Houston, Texas, 1964. Courtesy NASA.

Cricket practice, Houston, Texas, 1964. Thomas V. Chambers is at the bat (below).
Courtesy NASA.

Cricket practice, Houston, Texas, 1964. Burton G. Cour-Palais (right) is bowling. Courtesy NASA.

Cricket practice, Houston, Texas, 1964.
John D. Hodge (in dark pants) looks on as Peter J. Armitage bats. Courtesy NASA.

The STG team was made up of Dennis Fielder, Tom Chambers, Les St. Leger, John Hodge, Burt Cour-Palais, Rod Rose, Dave Brown, Peter Armitage, Dave Ewart, and two Americans - Bill Muhly, an original STG member who worked on scheduling and process flows, and Milton Windler, who went on to become a flight director in Apollo. The two Americans kept wicket. STG's three reserve players were Tec Roberts and astronauts Deke Slayton and Wally Schirra.

Armitage remembers the match this way: "We accepted the challenge since there were several of us who were from the U.K. and had played cricket in school. In fact we had some who were very good cricketers, although that didn't include me. Anyway, we decided to field a team and whip the living daylights out of William and Mary, who we assumed would not have as much cricket experience as we did. The British Embassy in Washington was contacted and they sent us a complete set of cricket gear for the whole NASA team. We were all decked out in whites and even got in a few practice sessions, although we did not think we needed it.

"The appointed day arrived and the local news media had picked it up so there was a sizable crowd assembled at the village green to see the NASA 'spacemen' beat up on William and Mary. When the William and Mary team came on the field, we got a bit of a shock. These were not the scruffy American college students that we had expected. They were English, Indian and Pakistan students, and to make matters worse, they were dressed in as fine cricket gear as we were. Well, you all can guess the rest of the story. NASA went down to a very humiliating defeat."

The final score was 84 runs for the William and Mary Colonials, and 58 runs for the NASA team. Rod Rose noted that one of the William and Mary players scored 28 runs on his own and then retired. The high-scoring William and Mary player was Dave McDougal, a Canadian according to the William and Mary college paper, the *Flat Hat*. The paper referred incorrectly to the STG team as "a Royal Air Force team from Langley Air Force Base" and noted that the game closed out the college team's fall schedule.

The NASA side reformed at least once more after the move to Houston. A Royal Navy ship was coming to town in 1964, and the NASA cricketers practised with a group of cricket players from the University of Houston in an effort to form a team to play the sailors. But as Armitage recalls, "The game never took place since a hurricane warning appeared, and the British ship put out to sea."

•

On January 27, 1962, John Glenn boarded his Friendship 7 capsule atop its Atlas booster on Pad 14 at Cape Canaveral, but the countdown was halted at T-minus 29 minutes because of bad weather. Three days later, another postponement was announced, this time due to problems with the Atlas. Weather conditions intervened when the next launch attempt was due to be made two weeks later, and another attempt was set for February 20. This time the weather and minor technical problems held up the launch only 2 hours and 17 minutes. The final hold, for two minutes at T-minus 6 minutes and 30 seconds, was called by John Hodge at Bermuda while he made sure his computer was ready for its work of verifying the orbit of Glenn's capsule. "We've just lost the main power on the computer and we want to make sure that nothing else has gone wrong," Hodge explained to Mercury Control. Under pressure from Chris Kraft, Hodge's team quickly decided that the computer was ready for its important job.

At 9:47 a.m. eastern time, the flight of MA-6 began and Glenn became the first American to rocket into orbit. Kraft confirmed Glenn's orbit with Hodge and Tec Roberts, and gave Glenn a "go" to continue the flight. Friendship 7 performed well during the first orbit. Gene Duret, the Saskatchewan native who had worked on heatshields, the flight controllers' handbook and the tracking network, was working the flight as CAPCOM at the Kano, Nigeria, tracking station. "What is your status?" Duret asked as Glenn came into contact for his first pass over Africa 22 minutes after launch. "My status is excellent. I feel fine," Glenn replied. During his pass over Kano, Glenn made the first test of his attitude control system, and Duret read Glenn information he would need if he had to cut his flight short. "Roger, this is Friendship Seven," Glenn said. "Out the window, [I] can see some fires down on the ground, long smoke trails right on the edge of the

The launch of Mercury-Atlas 6 carrying John Glenn. Courtesy NASA.

desert. Over."

"Roger," said Duret. "We've had dusty weather here, and as far as we can see, this part of Africa is covered with dust."

"That's just the way it looks from here," Glenn replied.

Glenn's attitude control thrusters were working fine as the astronaut made his first pass at Kano, but by the end of his first orbit, thruster problems similar to those that cut short Enos's flight appeared. During his short conversation with Duret on the second pass, Glenn reported that one of his yaw thrusters was not working properly. Unlike Enos, Glenn could compensate for this thruster problem, and he controlled his spacecraft for the rest of the flight.

Controllers also noticed a light that indicated that the landing bag had prematurely deployed. The landing bag was not supposed to extend until after re-entry to cushion the splashdown. If it had deployed early, it meant that the heatshield could come loose during re-entry after the three straps holding the retrorockets to the back of the heatshield were cut, and if the heatshield were loose, Glenn would be incinerated during re-entry.

As Friendship 7 flew through its second and third orbits, worried controllers put out the call to anyone who knew about the heatshield. Max Faget tried and failed to get hold of Erb, who had worked on Mercury's heatshield and was already working on the heatshield for Apollo. On his own, Faget recommended to Kraft that he keep the retrorockets attached. "Even if the heatshield had come loose, it could be kept in place by keeping the retropack on until the pressure on the heatshield had built up enough during re-entry to keep it in place. Faget also assessed correctly that the retropack straps would soon burn through without serious effects," Erb said. "Faget was brilliant, one of the most brilliant intuitive engineers that I have ever known. Absolutely incredible guy. He was no heat transfer expert, but he sort of knew how the physical world behaved."

Following Kraft's decision to keep the retrorockets attached, John Glenn splashed down safely after 4 hours 55 minutes and 23 seconds in orbit. NASA later found out that the indicator was out of order, not the landing bag. John Glenn, the first American to break the Soviet monopoly on orbital flights, returned to America and the biggest welcome for a returning hero since Charles Lindbergh flew from New York to Paris 35 years earlier.

Mercury's greatest triumph took place on the third anniversary of Avro's Black Friday.

The reason Faget couldn't reach Erb during Glenn's flight was because Erb was driving from Virginia to Houston as he moved his family to their new home. And Erb wasn't the only person on the move. The next day, Fred Matthews ended his career at NASA and moved to Boston.

Those who moved to Houston in the early months of 1962 went to work in temporary workplaces that would serve until the buildings on the new campus of the Manned Spacecraft Center were ready in 1964. A string of 13 buildings along the Gulf Freeway from downtown Houston south to Ellington Air Force Base, including a former television studio, an ex-bottling plant, and a converted bank, served as temporary homes for MSC.

•

Despite the disruption caused by the move, Mercury flights continued. On May 24, Scott Carpenter flew MA-7, the second Mercury orbital flight. Carpenter had replaced Deke Slayton as pilot when doctors became concerned that an erratic heartbeat could endanger Slayton's health in space. Carpenter flew his Aurora 7 capsule for the three scheduled orbits, but due to a manual retrofire that took place late and in an incorrect attitude, Aurora 7 splashed down 250 miles beyond the planned impact point. Carpenter spent nearly three hours in the water after splashdown before he was picked up by recovery helicopters.

The next Mercury flight took place on October 3, when Walter Schirra flew his MA-8 Sigma 7 capsule on a six-orbit "engineering flight." Shortly after launch, the temperature in Schirra's space suit rose, and Mercury control began to think about ending the flight early. But each astronaut had been assigned a different area of expertise during the development phase of Mercury, and Schirra's area was the space suit. He began making adjustments that brought the temperature down. For this flight, Gene Duret was assigned as CAPCOM at the Muchea tracking station in Australia. Duret asked Schirra on his first pass, "Do you feel too hot or anything?" "No, I have beads of perspiration on my lip. That's about all," Schirra replied, despite a temperature of 82 degrees. "The systems are all green. I'm practically using no auto fuel. My only problem area is my suit circuit, which I'm monitoring very carefully."

They also conversed about a flare test in the Australian desert, but Schirra could not see the flare due to clouds. He did see some lightning nearby.

On his second orbit, Schirra had reported to Duret that his suit temperature had fallen to 70 degrees. During this pass, Schirra conducted an experiment in celestial navigation, using the Moon and other celestial bodies to point his capsule. "Ooh! I've got some lights down there. How about that?" Schirra exclaimed as he spotted the lights of nearby Perth. "Good. That's us," Duret replied.

During his flight, John Glenn had captivated Mercury control and the millions following his flight on radio and television with his reports of luminous "fireflies" circling his capsule at sunrise. Scott Carpenter had accidentally tapped the wall of his capsule at sunrise and sparked a flurry of "fireflies." On the next orbit of Sigma 7, Duret asked Schirra for a repeat of Carpenter's test on behalf of Glenn, who was a CAPCOM in California that day.

"John Glenn suggests that if you have time, when the luminous particles first appear at sunrise, to tap the side of the capsule and test his favourite theory," Duret said.

"Roger, I have done that, Gene, and they do come from the capsule," Schirra replied.

"Roger, and he suggests later on, when they appear like white particles, that you do the same, and this might prove that they are the same particles."

"I tried that too."

"And got the same results?"

"That's affirm."

"Roger."

The mystery of the "fireflies" was solved: They were tiny blobs of thruster fuel that had followed the capsule.

As Schirra continued, his later orbits took him out of range of most of the tracking stations he had overflown on his first three orbits, including Muchea. After six orbits, and with most of his maneuvering fuel still unused after a successful conservation effort, Schirra fired his retrorockets and splashed down in the Pacific Ocean within sight of the recovery ship *Kearsarge* after 9 hours and 13 minutes of flight. Sigma 7 had been the most successful Mercury flight so far.

By this time, MSC personnel were shifting over to new programs, and some suggested that Mercury end with Schirra's flight. But one major goal remained – to fly Mercury in space for a full day. While original plans spoke of an 18-orbit flight, Mercury management finally settled on flying for up to 22 orbits. Gordon Cooper was selected as the pilot for MA-9, and modifications were made to his capsule to prepare it for the long flight.

Previous flights had been short enough that only one group of flight controllers was needed in Mercury control. But with a flight lasting nearly 34 hours, shifts would have to be introduced. For Cooper's flight, NASA decided to have 12-hour shifts with two teams of flight controllers. Chris Kraft had been flight director for every Mercury flight, but for MA-9, Kraft was flight director on the day shift and John Hodge, who had been backup flight director posted at the Bermuda tracking station for the early Mercury orbital flights, was flight director on the night shift. Hodge would become the first – and only – person other than Kraft to take the reins of Mercury control. This assignment added to Hodge's growing responsibilities in the flight control area. In 1961, Hodge had been named chief of the flight control branch with responsibility for preparing controllers for flights. The next year he moved up to become Kraft's deputy as chief of the flight control division.

On May 15, 1963, Cooper blasted off from Cape Canaveral aboard a spacecraft he called Faith 7. "The thing you have to remember was that he was so damn cool," Hodge remembered. "He went to sleep during the countdown." His capsule worked well during the early hours of the mission, which followed the path laid down by Glenn, Carpenter and Schirra. As Cooper moved into the latter part of his first full day in space and Hodge took over, the astronaut rested and took pictures as he floated over parts of the Earth that an American had never seen from space, such as the Himalayas. For much of this part of the flight, Cooper was out of contact with the ground because the tracking network was set up to provide nearly continuous coverage of a three-orbit mission but not longer flights. The rotation of the Earth beneath him shifted his orbit until the start of the second day, when he began to retrace his initial orbits.

"I can't remember anything spectacular during that period," Hodge said of his first shift as flight director. "Most of the time I was on, we were planning for the next phase and he was asleep. Our job was really a planning function. Which is a job I sort of took over in later missions, when we started going to Gemini and Apollo. My reputation was for planning for future events, which was a job I liked doing anyway." One lesson learned during this flight was that two shifts of flight controllers probably wouldn't be enough for longer duration flights, so most Gemini flights had three daily shifts of flight directors.

As Cooper's mission continued into its second day, problems began to crop up. Cooper lost attitude readings, an inverter tied into his stabilization and control system failed due to a short circuit, and the carbon dioxide level began to rise in the capsule. Cooper was forced to fire his retrorockets manually, which he did perfectly, and Faith 7 splashed down in the Pacific within four miles of the recovery ship, again the *Kearsarge*. Cooper's 34-hour flight resulted in a hero's welcome, including a reception by Congress, a medal presentation at the White House, and a ticker tape parade in New York.

Although some people lobbied for an even longer Mercury mission, NASA administrator Webb decided to end the program. A month later, the Soviet Union, which the previous August had launched Andrian Nikolayev aboard Vostok 3 and Pavel Popovich aboard Vostok 4 for simultaneous flights, repeated the feat, with a sensational twist. This time, Valery Bykovsky was launched aboard Vostok 5. Two days later, Vostok 6 left the pad at the Baikonur Cosmodrome with a woman aboard. The Soviets trumpeted Valentina Tereshkova's flight as a symbol of equality for Soviet women, although it would be nearly two decades before another woman would fly into orbit. Tereshkova returned to Earth after three days in space with more flight time than all the Mercury astronauts combined, and Bykovsky set a record for the longest space flight at five days. The Soviets still maintained a strong lead in the space race.

•

A few months later, the celebrations of Mercury's great success gave way to criticism when the press picked up some disturbing information contained in a paper delivered at the Mercury summary conference. The authors found that inspections of the MA-9 capsule at Cape Canaveral had turned up 720 discrepancies in systems or components, and 536 of them were due to faulty workmanship.

At least one of the former Avro engineers endorsed the concerns raised in the paper. Norman Farmer, who worked as an instrumentation subsystems engineer inspecting capsules at the McDonnell Aircraft plant in St. Louis, remembers: "I spent a lot of time out there ripping into capsules. My conclusion looking at those vehicles and some of the other stuff I saw going on there, was that they weren't doing a very good job at all. Their testing was very slipshod in many areas. They weren't checking it out. They were only concerned with one thing: Getting that thing out the back door and down to the Cape. That was their main object." Farmer had been born in London, educated in electronics and worked at EMI, English Electric and in the aircraft industry in England before coming to Avro Canada in 1957. He was never afraid to criticize the space program: "I was quite concerned. I was proven wrong, but I didn't think that there was going to be much success on those Mercury vehicles. They did work, but only just."

Not long after the summary conference, on November 16, 1963, President Kennedy flew down to Cape Canaveral to inspect the work on Apollo facilities. The next week, he went to Texas on a political trip, and during his stop at Houston on November 21 he spoke about the effort to meet his lunar landing goal. John Shoosmith, who was working in a NASA computer laboratory converted from a television studio, remembered stepping out from his workplace to watch the president pass by in a motorcade. The next day, Shoosmith was interrupted at his work by the news that Kennedy had been assassinated while riding in a similar motorcade in Dallas.

A few days later, the new president, Lyndon Johnson, designated the NASA launching facilities at Cape Canaveral the John F. Kennedy Space Center. The Cape itself was renamed Cape Kennedy.[5] Johnson was, if anything, a stronger believer in the space program than Kennedy had been. Using his unique powers as a legislator and the nation's desire to honour the memory of President Kennedy, Johnson was able to maintain public and political support for NASA, and with Mercury over, a new program would now take center stage in the American effort to fly to the Moon.

[5] The name reverted to Cape Canaveral in 1973.

Andre J. Meyer Jr., Paul M. Sturtevant and James A. Chamberlin in Gemini Project Manager's office, Houston, 1962. Courtesy NASA.

6 Jim Chamberlin and the Creation of Gemini

On January 20, 1961, the day John F. Kennedy was sworn in as president, Space Task Group managers led by Bob Gilruth and Max Faget held a capsule review board meeting at their office at Langley Research Center. The first item on the agenda was a follow-on program to Mercury, and ideas raised included longer duration flights, and missions investigating rendezvous in space and artificial gravity. The group also talked about flights using more powerful boosters than Atlas, but the only conclusion reached was that more study was needed before decisions could be made.

This meeting was far from the first time that talk at STG had turned to Mercury's successor craft. In the early days of Mercury, its designers believed that major changes would be needed to make the Mercury capsule last longer than the three orbits it was originally designed to fly. At the same time STG engineers first pondered flying to the Moon in 1959 and 1960, they also began talking about how to extend Mercury's capabilities, even though it had yet to carry an astronaut. Dick Carley, who specialized in controls in Mercury, was working with an associate, Robert G. Chilton, on the idea of allowing astronauts to control Mercury during re-entry by offsetting its center of gravity so that the capsule could be maneuvered toward different landing spots simply by rolling the spacecraft. Such a spacecraft would need a much more advanced control system than Mercury had. Carley said he took the idea of the vehicle, which he and Chilton called "Lifting Mercury" after its re-entry control characteristics, to the management of STG.

Shortly after the January 20 meeting, the ideas were transformed into action. On February 1, Gilruth assigned Jim Chamberlin, the head of engineering for STG and de facto Mercury project manager, to begin work on an upgraded Mercury capsule for more ambitious flights.

This assignment would have major consequences both for Chamberlin's career and the U.S. space program. Before he was done, Chamberlin helped lead Apollo on a new path to the Moon, and he joined the exclusive club of spacecraft designers. Even today, only a very small number of people have designed spacecraft that have carried astronauts and cosmonauts. The Soviet Vostok, Voskhod and Soyuz spacecraft, and later the Buran shuttle, came out of the design bureau of Sergei Korolev, Vasily Mishin and Valentin Glushko, spurred by the work of the rival design bureau of Vladimir Chelomei. In the United States, Max Faget and Caldwell Johnson were the primary designers of Mercury, the Apollo command and service modules, and the space shuttle. In the United States, only the Apollo lunar module and the new spacecraft that Chamberlin began work on that February, were designed by others.

•

James Arthur Chamberlin was then 45, at the height of his career. He had come to Mercury and the brand-new field of spacecraft design two years before, when Mercury's major features had already been decided on. Responsible for nursing the Mercury capsules through production and through many small but vital design problems, Chamberlin was ready to apply the lessons he had learned in Mercury to the new upgraded Mercury, just as the sometimes chastening experience of the CF-100 a decade before prepared him to do much better with the Avro Arrow.

Chamberlin was 193 cm tall and in 1962 weighed 88 kg, and the effect of his height was increased by a slim frame that caused people to call him gangly. At one time, an increase in weight gave him what some called a slight pear shape. His curly brown hair was kept closely cropped. Behind the rimless glasses he favoured, he usually looked the serious engineer he was. When he did smile, his face lit up in an almost boyish way. Most people remember Chamberlin as being focused on the task at hand.

The colourful Canadian journalist Tom Alderman drew this portrait of Chamberlin when they met in 1969: "He sits primly at his desk, precise, correct, blue-gray eyes peering out behind the bifocals .... I ask a question. He considers. 'Yes!' he says crisply. I ask another question. He

James A. Chamberlin in early 1930s and unknown person as winners of Toronto Mail and Empire *Model Aircraft Contest.*

considers once more. 'No!' he says more crisply .... 'Jim's a strange guy,' they'd warned me. 'He doesn't talk much. Doesn't like wasting words. But he's one helluva genius of an engineer' .... 'He's one of the best,' they say of Chamberlin. 'He'd be even bigger around here if he could only talk to people.'" Alderman noted that Chamberlin carried a slide rule in his pocket, which Chamberlin explained by saying, "I'm not sure that I can always trust computers. They make mistakes occasionally."

Another writer described Chamberlin in a 1962 profile: "He is 47 but looks 27, with a youthful, almost diamond shaped face, wide at the cheekbones. His wavy brown hair shows no trace of gray." He was described as a "hulking man" but having "the same quiet manner he had as a boy."

Chamberlin's manner of speaking made him sound simpler than he was. Combined with his sometimes abrupt and taciturn nature, this quality did not impress some of those he worked with. But Peter Armitage said Chamberlin made full use of this apparent disability. In meetings, Chamberlin's manner would cause other people to try and take over. "Then when the end of the meeting came, Jim would put on what must be thought of as another mask and sum up the meeting, and make a decision. That's what I thought was an interesting management method. What he was doing was forcing everybody to talk, and not talking himself. And he forced people to talk intelligently and foolishly. Jim's mind was working on time and was able to do a lot during these discussions. He was in a position to make an intelligent summary, and then he would shift gears and say, 'Here's what we're going to do.'"

Born in Kamloops, B.C., on May 23, 1915, Chamberlin and his mother moved around B.C. to Victoria and then Summerland after his father, Arthur Chamberlin, died in the First World War at the battle of Vimy Ridge in April 1917. Chamberlin's mother, born Theresa Goldie in 1889,

James Arthur Chamberlin
Toronto, Ont.
Mechanical Engineering;
J. A. Findlay Scholarship III.

Chamberlin's yearbook photo from University of Toronto, mid-1930s.

met and married John William Falkner in 1921 in Vernon, B.C. The young Chamberlin had started schooling nearby in Summerland, but he, his mother and stepfather soon moved to Toronto, where he got most of his education.

"Almost as soon as he could walk, Jim was building airplanes," his mother wrote in a 1962 newspaper article. "These were a great novelty in those days. He built them out of blocks and later balsa models that would fly. He was never interested in flying but he loved designing aircraft." She recalled taking him to a small airbase in north Toronto, and would let him sit in the aircraft as long as they stayed on the ground. "Everyone tried to discourage his interest in aircraft design, but I defended him because that's what he wanted to do. He kept at it and later took mechanical engineering because there was no course in aeronautics. He specialized in hydraulics. When he does anything, he is confident about it. He says: 'I knew it would work!'"

Chamberlin earned a degree in mechanical engineering at the University of Toronto, graduating in 1936. He then went to London, England, where he earned a master's degree from the Imperial College of Science and Technology in 1939. Before returning to Canada, Chamberlin worked briefly at Martin-Baker, the ejection seat manufacturers. Moving to Montreal, he worked for Federal Aircraft Ltd. on the Canadian version of the British Avro Anson aircraft. During the war, he also worked for Clark Ruse Aircraft in Dartmouth, Nova Scotia, and Noorduyn Aviation in Montreal, before he joined Avro Canada in Malton in 1945 as one of its first hires.

While in Halifax, Chamberlin met Ella Scrapek, a nurse from Saskatchewan whom he married in February 1942. The following year they had a daughter, Shirley, and in 1946, a son,

Arthur. Ella Chamberlin became a homemaker who her son Arthur remembers was more socially inclined than her husband.

Bryan Erb described Chamberlin this way: "He was a very bright guy, really good engineer, very resolute to do what he set his mind to. He just sort of plowed on and did what he thought was technically the right thing to do. He was a blunt engineer who spoke his mind. He was always very fair and had a high level of integrity."

Dave Ewart said: "It sometimes took a little while for him to express himself. But when he did, he always had some very good ideas. And he was a good boss to work for. He was a heck of a nice guy. I remember when my dad died in England, he didn't blink an eye about letting me go for a couple of weeks."

Mario Pesando said Chamberlin often had strong differences of opinion with those he worked with at Avro. Chamberlin was also a mentor to many younger engineers there. Rod Rose, Owen Maynard, Pesando and others remember Chamberlin helping them along. James T. Rose, a North Carolinian engineer (no relation to Rod Rose), calls Chamberlin his mentor and said he left NASA when he could no longer work for Chamberlin. "Chamberlin was a magnificent designer," Rose said. "He was very quick, and he knew what he wanted to do. I was just very fortunate to have the opportunity to learn from him. He gave you all the responsibility. I mean, if he was confident in you, boy, you had responsibility." Pesando, who later parted ways with Chamberlin in internal office disputes at Avro, retains his admiration and gratitude for the help Chamberlin gave him in his early days at Avro.

Maynard, who credits Chamberlin with bringing the concept of modules to Avro and NASA, said Chamberlin had a hard driving side. When Chamberlin had Maynard change some piping in the Arrow, Maynard's immediate boss, Hank Shoji, complained that Chamberlin's demands on Maynard were excessive and unfair. "Jim looked at Hank and said, 'What are you talking about? I have no requirement to be fair.' This is a very good expression of the tone and the level of intensity that we were working under," Maynard said.

Another example of Chamberlin's style at Avro came after his boss Jim Floyd wrote a memo in late 1956 outlining a new project research group. Chamberlin's reply began: "I have felt for a long time that the function of studying new projects was seriously hampered by lack of interest, encouragement, and policy direction from a management point of view." Floyd's reply showed that he did not appreciate the criticism. Floyd, who said such disagreements were rare, is one of Chamberlin's biggest boosters, calling him "absolutely a genius."

"He was very smart but didn't have the appearance of being very smart," John Hodge remembers. "He was totally and absolutely dedicated to his job. 100 per cent. I never heard him talk of his family or anything like that. And very good. Because of the way we were able to operate in Avro, which was very independently, he was his own boss and really didn't have to answer to anybody. And when he came down [to NASA] he tried that and got mixed up in how much independence he was allowed."

"I just couldn't figure out where he was coming from," Chris Kraft said. "But on the other hand, his contribution to the development of the hardware, both in Mercury and in Gemini, was of great importance."

One secretary at NASA, Margaret Marshall, said Chamberlin regularly started work at 8 a.m. and remained at his desk hours after everyone else had left. In the workaholic world of NASA in the 1960s, that was saying something.

"I've been in this rocket business ever since it started, and I've never seen anyone with his capability," said Paul Sturtevant, an American colleague. "Problems that confound other engineers are solved by him sometimes in a matter of minutes."

"He did seem to be much on the technical wavelength all the time," Rod Rose said. "He had a prodigious memory. That's something that struck me about him early on. One of the first times he came to my desk to give me a task, he said, 'Well, I think you'll find something about that on page so-and-so of such-and-such an issue of this technical journal.' And I looked it up. And sure enough, there's the article. And I found out later that that was just Jim. He had a prodigious memory. I learned later through experience with Jim that he had terrific retention and recall."

Despite Chamberlin's no-nonsense image, he did have another side. His son Arthur remembers that Chamberlin did not talk much about his work at home. A favourite pastime was sailing. And he was fond of swimming. When the Avro group arrived at Buckroe Beach in 1959, Chamberlin quickly took to the water. Bob Lindley remembered the day Chamberlin joined him and other McDonnell workers for some water-skiing on the inland waterways that surround Cape Canaveral. At one point, Chamberlin fell while the boat that had been pulling him carried on. Lindley recalled the feeling of horror when someone looked back and discovered that Chamberlin was nowhere to be seen. John Yardley, another McDonnell manager who was in the boat, said Chamberlin, who swam back to shore and caught a ride back to his hotel, wasn't found until after dark, when NASA officials were already starting to ponder how to tell the press that one of their top engineers had been lost in a water-skiing accident.

"Funny things could get to him, and he could laugh as long and hard as anybody," James Rose said. "People never saw that." Once, Rose joined Chamberlin and two McDonnell managers at a St. Louis nightclub where two scantily clad dancers joined the group for drinks, much to Rose's discomfiture. As they left the club, Chamberlin turned to Rose and said, "The food wasn't any good, and the show wasn't any good, but I could have stayed all night and watched you!"

"He would do the unexpected, I would guess," Chamberlin's colleague George Watts said. Once Chamberlin was going up the steps when someone noticed an undone seam in his pants. "I guess I noticed it too. This guy said, 'Jim, Jim, you've got a great big six-inch hole in your pants.' Jim reaches down and feels the hole and says, 'Nah, it's only three inches.' That's the way Jim was. That was pure Jim. He put on no airs. None whatsoever."

•

Chamberlin began working with McDonnell Aircraft Corporation, the builders of Mercury, on the improved Mercury spacecraft on February 13, 1961. His routine included spending the middle three days of each week at the McDonnell plant at St. Louis. Due to his work shepherding the Mercury capsules through production, Chamberlin was already a familiar sight at the St. Louis plant and had a strong working relationship with the people of McDonnell. "I had an office there," Chamberlin recalled in a 1966 interview. "I had my own office, and I did the work from my own office, more or less as if I was part of McDonnell or they a part of me."

"He disappeared for a month and was buried behind the doors of the preliminary design office of McDonnell," said Fred Matthews, who was still with NASA at the time. "It's an unusual arrangement to have a customer working in the contract organization, doing what a contractor is supposed to be doing."

Although most people at NASA believed that only moderate changes to Mercury were needed, Chamberlin disagreed. He made his views plain a month after starting work on his new assignment when he joined other STG managers, including Gilruth, at a retreat at Wallops Island. The retreat, which NASA historians place on the weekend of March 17, 1961, was also attended by Abe Silverstein, NASA's head of space flight programs. It dealt with some immediate issues facing Mercury, then just weeks away from its first manned flight, but focused on Apollo flights to the Moon, which NASA was preparing to sell to the Kennedy administration. When Silverstein raised the idea of future missions for Mercury, Chamberlin began to talk.

"My role in this meeting was intended to be a discussion of the progress on Mercury," Chamberlin wrote later. "During this, I naturally discussed the enormous difficulties being experienced in checking out the spacecraft. It was the prevailing NASA view that only by extremely dedicated 'tender loving care' in checkout, especially at the Cape, could successful manned flights be made. Since we all felt that the prestige of the country and the future of NASA manned space flight depended on this, there was a reluctance to accept the risk of tempting fortune too far by extending Mercury missions beyond the program commitment [of three orbits]. However, it was apparent that there were many more missions required for the development of manned spaceflight to reach the maturity required for more ambitious programs such as the lunar program. I expressed the view that the difficulties in checkout were not fundamental to manned spacecraft but were due to inherent deficiencies in the conceptual design of the Mercury vehicle, and could only be fixed by altering the design of the equipment installations to something similar to that used for fire control and weapons delivery systems on fighter aircraft with which I was familiar. I then proceeded to make more sketches on a blackboard which was set up on the left-hand side of the room – showing how the equipment could be arranged outside the cockpit to be readily accessible for installation and checkout, in the same way as fighter aircraft. Because these features were commonplace in aircraft, we could not afford the unnecessary risks involved in prolonging the life of the Mercury design, and should follow up with a repackaged design as rapidly as possible. This presentation was largely impromptu, but got the immediate support of both Silverstein and Gilruth, who authorized me to pursue these ideas with McDonnell in detail in order to make specific proposals."

According to others at the meeting later interviewed by NASA historians, Chamberlin's presentation wasn't entirely impromptu. He produced a "brochure" outlining a flight around the Moon using the improved Mercury. The proposal went nowhere, but it wouldn't be the last time Chamberlin raised the idea.

The meeting at Wallops Island, Chamberlin said, marked the beginning of what became known as Project Gemini. The Gemini spacecraft, which Chamberlin went on to design in cooperation with McDonnell, owed a great deal to Chamberlin's experience developing the Avro Arrow. The "modularized" approach of Gemini differed from Mercury, which Chamberlin described this way: "In Project Mercury, most system components were in the [pressurized] pilot's cabin; and often, to pack them in this very confined space, they had to be stacked like a layer cake and components of one system had to be scattered about the craft to use all available space. This arrangement generated a maze of interconnecting wires, tubing, and mechanical linkages. To repair one malfunctioning system, other systems had to be disturbed; and then, after the trouble had been corrected, the systems that had been disturbed as well as the malfunctioning systems had to be checked out again. Only one technician could work inside the Mercury cabin at any one time." Chamberlin was proposing to turn the guts of Mercury inside out to deal with the welter of problems that dogged Mercury capsules to the end, causing concern to Chamberlin, Norman Farmer and others.

After the Wallops Island meeting, Chamberlin went back to work developing the advanced Mercury spacecraft. He was joined by another STG engineer, James Rose, and by a McDonnell team headed by William J. Blatz. NASA signed an agreement with McDonnell extending the Mercury contract in anticipation of the advanced Mercury spacecraft. Chamberlin and Blatz gave their first report on their plans for Mercury to another capsule coordination meeting in June, shortly after President Kennedy had given NASA the task of flying a man to the Moon. Rod Rose remembers joining James Rose and Chamberlin in Chamberlin's basement to work out design ideas for the new spacecraft.

Most STG engineers expected to see designs for the minimal changes required to keep Mercury flying for 18 orbits, a one-day mission. Instead, Chamberlin and Blatz produced plans for a radically changed spacecraft, with modular systems placed outside the pressure vessel and easier to service.

There was more. The new spacecraft used an ejection seat rather than the escape tower that was used in Mercury to pull the capsule away in case of emergency. Chamberlin disliked the escape capsule concept, in part because of the complicated wiring, relays and sequencing it required, and also due to the fact that the Mercury escape tower was unreliable. Escape towers had fired without need or warning on several test shots, and each Mercury flight that followed required elaborate preparation to make sure that the escape tower worked properly. Carley noted that the Mercury escape tower worsened the vehicle's vibrations during the launch phase. Furthermore, replacing the escape tower with an ejection seat meant that Gemini wouldn't need an expensive test program for the escape tower, as Mercury and Apollo required with Little Joe rockets. The Gemini ejection seats could be tested on their own on the ground, without launching boilerplate capsules.

The ejection seat meant that the new spacecraft would have to be equipped with a larger hatch that would be easier to get in and out of. Finally, the proposal included the idea that the spacecraft would return to Earth following a controlled re-entry using a paraglider, a flexible wing that would unfurl like a parachute and allow a controlled landing on the ground, instead of Mercury's parachute and splashdown. STG engineers were already looking at using the paraglider to land Apollo.

The meeting on June 9, and a follow-up meeting on the 12th did not lead to an endorsement of the Chamberlin-Blatz proposal, but they were allowed to continue working on advanced Mercury ideas. At the same time, McDonnell was also asked to work on determining the minimum changes needed to keep Mercury flying for 18 orbits. With the involvement of McDonnell's Mercury manager, Walter Burke, the contractor began to look seriously at various types of advanced Mercury spacecraft, including a minimum-change Mercury Mark I for the 18-orbit mission, and two variations of Mercury Mark II along the lines of Chamberlin and Blatz's proposal. The first was a one-man spacecraft as proposed, and another was a two-man spacecraft following a suggestion by Mercury's designer, Max Faget, who was thinking already of space walks.

On July 27 and 28, a number of astronauts and senior NASA officials, including Silverstein, gathered at McDonnell's plant at St. Louis, where they were shown quarter-scale models of Mercury Mark I and both versions of Mercury Mark II. There was also a full-scale model of the cabin for the two-man version of Mark II. Silverstein directed McDonnell to continue working on the two-man Mark II. Although the Mark I designation disappeared, the concept was incorporated into Gordon Cooper's Faith 7 capsule, which flew the final Mercury mission of 22 orbits in 1963.

•

Chamberlin, now with a larger team from both McDonnell and NASA behind him, continued working on the new spacecraft called Mercury Mark II or just Mark II. By August 14, 1961, he was ready with a preliminary project plan for Mark II that set out six ambitious objectives in 10 flights during 1963 and 1964. The first objective was long-duration flights with astronauts and animals, the second was an unmanned flight through the Van Allen radiation belts, third was controlled landings with the paraglider on all manned flights, fourth was rendezvous and docking with an Agena rocket in orbit, and fifth was astronaut training. The sixth objective was audacious and in direct competition with Apollo: Dock Mark II with a Centaur rocket and use Centaur to shoot the spacecraft and its crew around the Moon.

A few days later, the sixth objective disappeared from the project plan, but Chamberlin's provision for rendezvous in Mark II reflected the growing sentiment within NASA that rendezvous would be an important part of space operations, possibly including the lunar landing. Early in September, Chamberlin came back to STG with plans to use Mark II and a lunar "bug" as the means to land on the Moon. Although this proposal was soon rejected, it marked the first serious look within STG at lunar orbit rendezvous as a way to meet President Kennedy's goal.

During the fall of 1961, discussion of Mercury Mark II, minus the lunar proposals, moved higher up the management chain of NASA. At the end of October, Chamberlin's latest project development plan had eliminated both the animal flights and the radiation tests and put more emphasis on rendezvous and docking. Chamberlin returned to St Louis to gear up for the approval he expected for Mark II, and McDonnell began to pour more resources into the program. NASA approved Mercury Mark II on December 7. The same day, Bob Gilruth, now the director of the Manned Spacecraft Center that was moving to Houston, announced in a speech to the Houston Chamber of Commerce that a "two-man Mercury" would fly after Mercury and before Apollo. McDonnell had a contract by Christmas to build the vehicle, and on January 3, 1962,

NASA released the first public drawings of the new spacecraft and announced that it would be called Gemini. Chamberlin was named project manager on January 15.

Chamberlin and Kenneth S. Kleinknecht, the new Mercury project manager, split Chamberlin's engineering division in two, and the staff that joined Chamberlin's Gemini Project Office were among the first to move from Virginia to temporary quarters in Houston. Gemini enjoyed more autonomy from the rest of MSC than Apollo and Mercury, in part because NASA's focus at the time was on Mercury and its preparations for John Glenn's orbital flight, and on Apollo, which was still facing a decision on how to go to the Moon. Most of the remaining ex-Avro engineers who weren't still in Mercury were moving to Apollo. Only Dick Carley, who worked in guidance and computers, and Rod Rose, who was assigned to the paraglider program, followed Chamberlin to Gemini at the time.

Carley said the "Lifting Mercury" concept he and Robert Chilton had advanced formed the basis of Gemini. Indeed, when the two control experts presented their idea to STG, Carley said his old boss Chamberlin was one of the skeptics. But the two often had dinner together, and Chamberlin was won over, according to Carley's account.

"Dick Carley worked on Gemini a great deal," said George Watts, another Avro engineer who only worked briefly on Gemini. "He did the controls and horizon scanner and the instruments. He followed the development of the actual instruments themselves. He really was responsible for the success of the Gemini missions. He worked very, very closely with Jim Chamberlin."

"The Gemini Program Office had Dick Carley overlooking the [guidance and control] development," Chilton said. "He was pretty much a one-man show as far as the government monitoring of the McDonnell activities," at least until the program was well advanced. One of the small group of people helping Carley out was John Shoosmith, the youngest member of the Avro group.

Chamberlin's succesor as Gemini Program Manager, Charles W. Mathews (left), presents award to Richard R. Carley in 1965. Courtesy NASA.

"Dick was just a very outstanding scientist and engineer in his field, which was guidance and navigation," said Chuck Mathews, who worked closely with Carley during his career at NASA. "And so outstanding that other people could hardly live with him, because they had the impression that he was almost talking down to them. He was the guy who was in charge of that particular subsystem in the Gemini program, and he stayed in that particular capacity all the way through the program. He had a lot of trouble getting along with people, but as long as you used him mostly by himself, he could do the work and come up with the ideas of 10 people."

When Gemini spacecraft made the first rendezvous in space, Carley and his staff were given a NASA Group Achievement Award for their work.

Born in 1927 in Saskatoon, Saskatchewan, Carley grew up with an interest in airplanes. His father owned bakeries, and one of Carley's summer jobs while in university was as a management trainee, but his interests lay in how things were made, and this led him to engineering. Carley studied engineering at the University of Saskatchewan. At first he took mechanical engineering and later electrical engineering. During this time he became interested in cybernetics, the science that was leading to development of computers. After graduation, Carley joined Canadian National Telegraph, where he worked on installation of telephone systems. Carley's interest in control systems grew, and this led him to join Avro Canada in 1954. There he worked on the Arrow's control system. His pioneering work on fly-by-wire control systems at Avro continued when he joined NASA and the Mercury program in 1959.

•

More than Mercury, Gemini was developed by the contractor, the McDonnell Aircraft Corporation. On McDonnell's end, Gemini engineering manager Robert Lindley was given the responsibility of overseeing the design and production of Gemini.

Lindley and Chamberlin had worked together before on the Avro Arrow, Lindley as chief engineer. Born in Manchester, England, Lindley had worked at A.V. Roe in England as a protégé of famed aircraft designer Roy Chadwick and had helped design the Vulcan bomber. Lindley came to Avro Canada in 1949 after a stint at Canadair in Montreal. At Avro, he put the CF-100 back on track and then worked on the Arrow. During that time, the no-nonsense Lindley had differed on more than one occasion with Chamberlin. Although he joined Chamberlin in trying to arrange the loan of Avro engineers to NASA after the Arrow cancellation, Lindley chose to remain at Avro rather than join NASA with Chamberlin and the others. But when Sir Roy Dobson put Harvey Smith in charge of what remained of Avro Canada, Lindley left Avro. His job search ended with a call from McDonnell Aircraft's president James S. McDonnell, who was known as "Mr. Mac." Lindley's personality is illustrated by his account of his first meeting with McDonnell: "He had just built a new engineering building in St. Louis. A beautiful building. He was a great stutterer: 'W-w-what do you th-think of my engineering b-building?' I said, 'You don't agree with Parkinson's law.' I was reading the book. I had it with me. 'W-what is Parkinson's law?' he said. I said, 'What he shows is that the virility of the organization and the splendour of the surroundings are in inverse ratio. The British colonies all disappeared when the colonial office got a beautiful building.' 'You don't believe that.' I said, 'Here's the book.' Well, anyhow, he went off. And then I got called to come back down again and talk to his people, which I did. Anyhow, I went down there and talked to people, and everyone had a copy of *Parkinson's Law*."

While he waited for U.S. security clearances, Lindley worked on a proposal for a passenger aircraft, and once cleared, he worked on McDonnell's proposal for the swing-wing TFX, Tactical Fighter Experimental, which later became the F-111. When McDonnell lost the bidding for the F-111 and won the Gemini contract, Lindley was assigned to Gemini, where he remained until the end of the program.

While many features of Gemini were already in place, Lindley worked on the detailed design and development of Gemini under the watchful eyes of Chamberlin and his successor, Chuck Mathews. "He really ran the engineering people in the project office, and then he was the chief

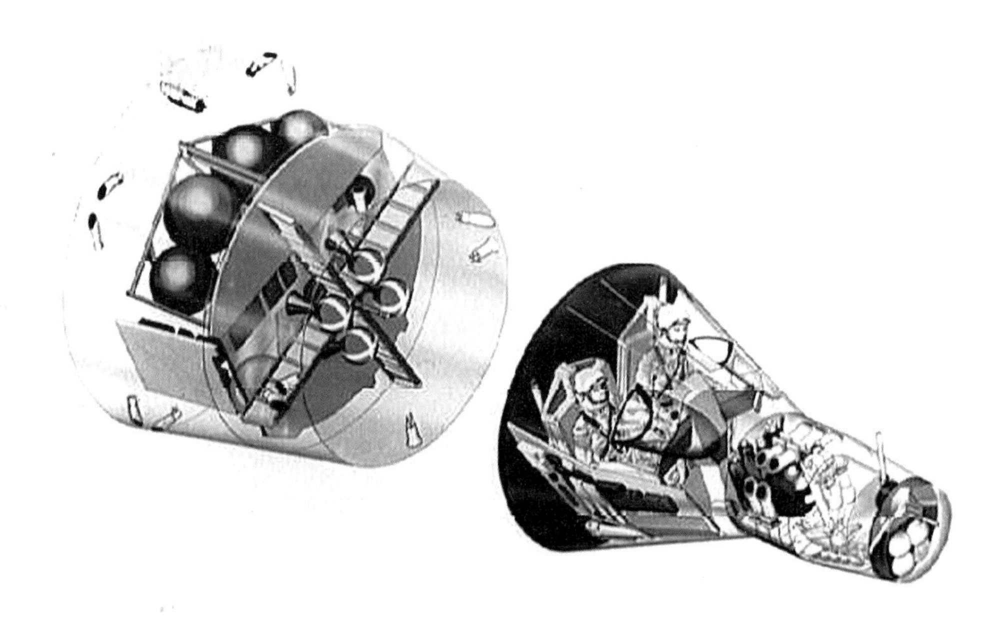

Cutaway of Gemini spacecraft. Courtesy NASA.

contact with other engineering groups at McDonnell in getting support for the technical activity involved in the Gemini program," Mathews said of Lindley.

"Actually, Lindley did an excellent job of pulling together the design and the production and all that stuff for Gemini," said John Yardley, McDonnell's Gemini program director and Lindley's boss. "Of course Chamberlin was involved, but he was involved at the top – let's do this instead of that – but somebody has got to do a lot of detail beyond that. And I would say that Lindley did a very good job of managing the technical activities of getting Gemini designed." Lindley, for his part, praised the Gemini team at McDonnell as a motivated, "close-knit" group that "really put their backs into it."

•

During 1962, the first year that Gemini existed as a stand-alone program, Lindley said the early design work took an unusual course. "Jim Chamberlin didn't really go too much, you might say, for formal systems in getting this design nailed down, but he recognized very clearly the need for close monitoring of our efforts and came up with what I think is a rather novel approach. He started a series of meetings, or you might say he started a meeting which got going about real early in '62, and that meeting just kept on going all year as far as I could make out. We just sent platoons in and the meeting just ran. Blatz ran that operation."

While the designers of Mercury concentrated on meeting their goal of getting astronauts into space and back alive, the serviceability of the capsule wasn't given much thought. Chamberlin decided that Gemini must be serviceable so that the act of putting astronauts into orbit would be routine.

A new twist in Gemini's design was an adapter section that stayed with the spacecraft rather than falling away with the booster as Mercury's did after serving as the interface between the booster and spacecraft. This idea had been under consideration for some time at STG. The Gemini adapter contained orbital maneuvering thrusters, retrorockets, heat radiators, fuel for orbital operations, and Gemini's power source. It separated just before re-entry, and Gemini returned to Earth using a set of re-entry thrusters, backup fuel and power supplies. The use of the adapter vastly simplified the design, building and checkout of Gemini. The modular concept was extended to the adapter, and on more than one occasion, troubled systems were quickly changed on the launch pad, preventing or shortening launch delays.

Gemini's booster rocket was the Titan II, built in Baltimore, Maryland, by the Martin Company. The two-stage booster was similar to the spacecraft in many ways. Chamberlin called Gemini and Titan "an excellent marriage," because "there was a second generation missile with a little bit bigger payload, and it had all these design improvements that go with second generations, second thoughts." Titan II's design concepts were "almost identical [to] what we were thinking about on Gemini." First-generation missiles such as Redstone and Atlas, and the R-7 used by the Soviets, relied on traditional fuels and oxidizers that required igniters and chilling for the liquid oxygen. In the early 1960s, both the Soviets and Americans were switching their nuclear forces to missiles like the Titan, which used hypergolic fuels, fuels that ignite on contact and can be stored at room temperature. Once engineers had dealt with the corrosion problems inherent in these fuels, these new missiles were better fitted for their roles in the nuclear arsenal than the earlier missiles. Chamberlin saw that the Titan II would be an excellent fit for Gemini, allowing easier setup and checkout.

Since the fuels used on Titan II weren't as explosive as traditional liquid rocket fuels and there would be a smaller fireball if the booster exploded, Chamberlin could equip Gemini with ejection seats rather than an escape tower. The ejection seats remained unpopular with people such as Max Faget and Walter Williams and even some astronauts, but Chamberlin, Lindley and their team pressed on. In case of a mishap on the launch pad, the seats would have to lift the astronauts away from an explosion from a height of only 30 m. To make sure the parachute came out, the first thing out of the astronaut's parachute pack in the Gemini ejection sequence was not a small drogue chute, but a combination drogue chute and balloon that was called a ballute.

The ejection seats meant that Gemini had large hatches that could be used by astronauts leaving the spacecraft in orbit for a "space walk." Both Chamberlin and Max Faget said they thought early on of attempting space walks in Gemini, and Chamberlin pressed on with the idea.

As a champion of space rendezvous, Chamberlin made rendezvous and docking a central feature of Gemini, even before NASA decided to use lunar orbit rendezvous in Apollo. He chose an Agena rocket fitted with a docking device as the rendezvous and docking target for Gemini, because Agena already had a strong track record as a second-stage booster vehicle. The Agena was boosted into orbit by Atlas rockets; then, when it was docked to Gemini, the Agena's large engine could be used to boost Gemini into higher orbits. Adding another booster and another orbital vehicle to Gemini made the program much bigger and more complicated than just producing a new generation of spacecraft.

To make the rendezvous and docking possible, the astronauts would have to be able to change Gemini's orbit, so thrusters were built into Gemini's adapter to achieve this, and Gemini became the first spacecraft that could be maneuvered in orbit. Neither Mercury nor the Soviet Vostok and Voskhod spacecraft could change their orbits until the retrorockets fired to take them out of orbit. That is why Gemini and the vehicles that followed were called "spacecraft," and the word "capsule" disappeared from NASA jargon with the end of Mercury. The only exception was the retention of the CAPCOM, "capsule communicator," title after an attempt to rename the position to reflect the change of jargon failed.

Rendezvous and Gemini's other capabilities also required computing, and so Gemini became the first spacecraft with a computer on board. In those days before computer chips, when virtually every computer filled a room, Gemini flew with what was then a tiny 27 kg computer built by IBM. The computer had a core memory of 4,096 words, which was supplemented with a small magnetic tape memory unit.

Another one of Gemini's goals was "long duration" flights, which in 1962 meant flights longer than a day. Although Gemini was originally designed for missions of up to a week in duration, matching the minimum time it would take to go to the Moon and back, NASA soon decided to try missions of up to two weeks. Because batteries that could last two weeks would

The paraglider, or "Rogallo Wing," was proposed for use in the Gemini Program. It would have allowed Gemini to make precision landings on land, rather than in the water. But the wing suffered a number of problems. The biggest problem was getting it to deploy properly and reliably. The plan was cancelled. Courtesy NASA.

be too heavy, Gemini would be powered by a new and untried energy source, fuel cells. Fuel cells operate using a chemical reaction joining hydrogen and oxygen to produce energy and water.

•

Like any program that incorporates new concepts, Gemini had many challenges to overcome. One of the biggest and most controversial challenges faced by Chamberlin was landing the spacecraft on the ground using the paraglider. The paraglider, a flexible wing that could be folded into a canister and then unfurled and made rigid using an inflated frame, was developed by an engineer at Langley named Francis M. Rogallo. Many people at NASA were interested in using his concept to recover spacecraft, and early in 1961, three companies were given contracts to study the paraglider for use after Mercury. In November 1961, North American Aviation of Downey, California, was given the contract to continue the development work on the paraglider for the Mark II-Gemini program. Both Chamberlin and his boss Bob Gilruth were among the enthusiasts for the paraglider. Armitage, an early critic of the paraglider, said Gilruth did not mince words in voicing his support of the paraglider when Armitage gave him a critical report on ground landing for Gemini. Despite Gilruth's support, the paraglider program operated under a cloud almost from the beginning. A few days after North American won the paraglider contract, it also won the far bigger contract for the Apollo command and service modules. And North American already had the contract for the second stage of the Saturn V rocket that would send Apollo to the Moon. The result was that North American's resources were spread thin, and the paraglider became a casualty of this problem.

"Within 24 hours, I had discovered a whole lot of key people in the paraglider program had got shifted," said Rod Rose, the ex-Avro engineer who Gilruth had assigned as manager to work with the contractor on the paraglider. "But even if that original crew had stayed on, I don't think we could have made it because the paraglider is a finicky item. It was an R&D program, and fundamentally, we were tying an R&D program to what was essentially a production capsule. They could go on with the capsule a lot faster than we could come up with the R&D. It was a very optimistic program to start off with."

Through 1962 the paraglider and even its parachute backup system went through a series of mishaps. Boilerplate models of Gemini suffered hard landings or crashes due to a variety of failures, many of them preventable. Many people at MSC opposed the use of the paraglider, and by the end of the year, parachute splashdowns in the ocean were planned for early Gemini flights, although there was still hope that the paraglider would be ready later.

Chamberlin also had to deal with other problems. Despite its many departures from Mercury, Gemini was to use several components from the earlier program to save time and money. For example, two Mercury environmental control systems were to be put together to maintain Gemini's small atmosphere. It soon became clear that this scheme wouldn't work, and a new environmental system had to be developed for Gemini. Mercury's heatshield was to be used for Gemini, but fabricating the bigger heatshield proved difficult, and McDonnell turned to a design using honeycomb filled with ablative material.

Gemini's maneuvering thrusters, known as the orbital attitude and maneuvering system, were more advanced than Mercury's thrusters, and Chamberlin was forced to deal with this system's many teething problems at Rocketdyne, the subcontractor that built the thrusters. The development of Gemini's fuel cells was proving difficult, so batteries would be used in early flights. Problems cropped up with the ejection seats. And test flights of the Titan II booster revealed several defects, notably a problem common to most rockets: pogo. This bouncing of the rocket in flight is not an issue when warheads or satellites are the payload, but it is a problem when astronauts are on board. In an early test, the Titan first stage bounced at two-and-a-half times the force of gravity, 11 times a second. The ride would be like sitting in a gigantic paint shaker. To reduce pogo, engineers began working on the fuel feeds and piping that caused the problem.

Despite these issues, Gemini made good progress during the first months of 1962. Lindley recounted that the components of Gemini were built into a mockup that faced an engineering review in August, the first thorough review of Gemini's status since the project officially started. "We presented it then to the balance of NASA plus Air Force and some other interested agencies .... And I think we came out of that mockup review fairly well. It's quite true we had a fairly large number of requests for changes, but they were all in the nitpicking area, with the exception of one, ... and that was the provision of a drogue parachute for re-entry." A month later, Lindley showed the mockup to a visiting President Kennedy, who "seemed to be really taken by this thing."

But as the president inspected the next step in the ambitious space program he was promoting, NASA in general, and Gemini in particular, was facing a budget crisis. The price for Gemini was rising rapidly. McDonnell's September cost estimate for the Gemini spacecraft was nearly a fifth higher than the company's estimate the previous April. As well, cost estimates for Titan and the Atlas-Agenas were also rising. In September, Gemini was 25 per cent more expensive than estimated in May. At the same time, Congress delayed its appropriation of funds for NASA, forcing the agency to put restraints on payments to contractors. When Congress settled NASA's appropriation later that fall, the figure was lower than NASA had hoped, and NASA's leadership decided that the entire shortfall would have to be absorbed by Gemini, rather than being shared with Apollo. For much of October and November, NASA managers were occupied in an exercise to make Gemini fit within its fiscal limits. Test programs were slowed or cancelled, an investigation into using Gemini for lunar flight that had come from NASA headquarters was ended, and flight schedules were pushed further into the future. Chamberlin and others fought back an effort to remove Agena from the Gemini program and replace it with joint flights with "rendezvous evaluation pods" deployed from Gemini's adapter. As a result, Gemini survived its first full year fully intact.

But this respite was brief. New flight tests of the Titan II showed that attempts to fix the booster's pogo problem were having only limited success, and another problem reared its head. The tests showed that the Titan's second-stage engine might be susceptible to unstable burning. Titan was an Air Force program, and the Air Force was showing growing reluctance to fix the pogo problem, since pogo would not stop Titan IIs from delivering nuclear warheads. Chamberlin spent much of March 1963 trying to persuade the Air Force to deal with the problems of Titan.

Meanwhile, the paraglider program was continuing to have problems. A drop test in October had succeeded, but another test in December had ended with a descent using the emergency parachute. Tests of half-scale wings on January 8 and March 11 ended with failures of the wings and of the emergency parachutes.

As Chamberlin's troubles piled up, a report went to NASA headquarters on March 8 showing that Gemini's costs had risen again, without warning, above $1 billion in pre-inflation dollars. On the 14th, NASA associate administrator Robert Seamans and U.S. defense secretary Robert McNamara headed a high-powered group from Washington that went to Houston to look at Gemini. Seamans was surprised and alarmed to learn of the problems with Titan II. As well, Gilruth had another shock just before McNamara's visit when he asked Walter Williams, his associate director, to brief McNamara. "He said, 'I don't want to go. I don't believe in it. I think it's in trouble and I don't think it can go.' That scared the hell out of me, because I didn't think it was that bad." Five days later, on March 19, Gilruth removed Chamberlin from Gemini and assigned him as his senior engineering advisor. Chuck Mathews came in as acting Gemini project manager. He soon became full manager and held that position until the end of Project Gemini.

Said Gilruth: "Chuck was brought in about this time and he brought a great deal to the program because he was not only a fine spacecraft engineer but he understood operations pretty well, having worked in that area."

Both Chamberlin and Mathews held high opinions of each other. "There were a tremendous amount of development problems that had occurred, most of them not Jim's fault," Mathews recalled later. "There were a couple of things where his futuristic look at things got things into the spacecraft that were too R&Dish that produced some of the schedule problems. But in general, you have to say that he left the program in pretty good shape."

"Jim was utilized in the program office as the initializer, the guy who came on and organized all the engineering aspects of the program. He ran the program until it was becoming operational. Then they brought in an operations guy, Chuck Mathews," was the way Armitage, who came from Avro with Chamberlin, saw the transition.

Like many other people who worked with him, James Rose felt that Chamberlin's poor communications skills helped lead to his demise at Gemini, especially when new managers were hired at NASA headquarters who didn't know Chamberlin as well as his colleagues at STG.

As for Chamberlin himself, he later wrote that many of Gemini's problems resulted from having to develop hardware much faster than Apollo and on a budget that was constantly tightened because Gemini had not been provided for in NASA's plans as Mercury or Apollo had been. "Gemini was started after Apollo, and had systems that were in most cases of comparable sophistication, and yet had to have its flight program finished before Apollo. If we waited to get accurate costs before proceeding, we could not possibly beat Apollo. Thus we went ahead as fast as possible with whatever funds could be scrounged. Later, when Apollo had its own problems, the situation on Gemini appeared quite comparable and hence tolerable. Even as it was, the rise in Gemini costs was nothing like those in Mercury or Apollo." Without the tight schedule and budget, Gemini's problems would have appeared routine, he wrote, adding that "the program was sufficiently well established to survive its problems without significant alterations under Chuck Mathews' excellent leadership."

The NASA history of Gemini sums up Chamberlin's departure this way:

When Chamberlin left Gemini, an era ended. In the large and complex undertakings of modern high technology, one person can seldom be credited with so large a share in the shaping of a project as Chamberlin deserved for Gemini. Much of the ultimate success of the project had its roots in Chamberlin's brilliance as a designer and skill as an engineer, but so did some of the current harvest of troubles. The talented engineer can always see new ways to improve his machines, but the successful manager must keep his eyes on costs and schedules, even if that means settling for something good enough instead of better.

Chamberlin's colleagues in and out of NASA deeply respected him as an engineer and designer but also saw his flaws as a manager and recognized the difficulties of the situation. His sudden and unexpected departure was thus not the blow to project morale that it might have been. The shock was also eased by the identity of the man who replaced him. Mathews was well known and widely esteemed. He took over a program that did seem to be in trouble.[6]

The problems that dogged Gemini didn't end with Chamberlin's removal from the program. Mathews had to tangle with problems involving Titan II, Agena, and Gemini's thrusters, fuel cells and ejection seats, as well as the issues surrounding Gemini's schedule and budget. The first unmanned test of the new spacecraft, Gemini-Titan 1, blasted off from Cape Kennedy on April 8, 1964, and flew, as planned, for three orbits. No recovery was attempted. The second unmanned test was dogged by a series of problems, including hurricanes, a lightning strike at the launch complex and an emergency shutdown of Titan's engines during one launch attempt. But finally Gemini-Titan 2 flew on January 19, 1965, in a suborbital test of the spacecraft's heatshield and re-entry equipment. The flight ended with a successful splashdown in the Atlantic, clearing the way for flights with astronauts on board.

•

With one exception, the ambitious aims that Jim Chamberlin set out for his spacecraft were tested during the Gemini program. The exception was the paraglider, which Mathews removed from Gemini in 1964. Although the concept was never used for the recovery of spacecraft, similar wings, using metal stiffeners rather than inflated stiffeners, eventually populated the skies of the world as hang gliders.

Both Chamberlin and Gilruth regretted the end of the paraglider. Chamberlin refused to blame its troubles on North American's winning of the Apollo contract and believed that if paraglider research had been continued, it would have flown on Gemini. "I think we demonstrated that paraglider could have worked, but the restrictions on money and what had gone before and this lack of doing the research with half-scale [models] – that was fatal, absolutely fatal."

"Basically, I would say we were about two years too soon on the paraglider," Gilruth said. "If we had started two years later, we'd have had land landing in that program, because all that was required was to take the booms off the paraglider and it becomes a reliable thing. It was only subsequent to that that Rogallo and some of the people at Langley developed the paraglider without the boom that deployed just like a parachute and was just as simple as a parachute. I always felt that the stimulus we gave that whole gliding parachute technology was probably worth the effort."

Another of Chamberlin's goals was to fly Gemini near the Moon. He said he considered Gemini, at least in its early days, as competition for Apollo. The idea of flying Gemini to the Moon outlasted Chamberlin's tenure as manager and was resurrected as late as 1966, near the end of the program. By then, Chamberlin was deep into his work as a technical troubleshooter on Apollo.

[6] Hacker, Barton C., and James M. Grimwood, *On the Shoulders of Titans: A History of Project Gemini*, National Aeronautics and Space Administration, Washington D.C., 1977, pp. 129-130.

David Baker, in his comprehensive *History of Manned Spaceflight*, called Chamberlin: "an engineer's engineer, probably one of the most brilliant men ever to work for NASA, pruning and honing the design to pack within an incredibly small space a whole generation of new technology, transforming Mercury into a two-man operational spaceship from a tiny one-man capsule. He was the single most important link in the chain that gave NASA the tools it needed to pave Apollo's way, but in March 1963 he was seen as an inappropriate choice for senior management, a technical man with a personality very different to that which was required for project management. Chamberlin had been in charge too long, in fact there were some that said he should never have been afflicted with the post, but only because events had outpaced his ability to keep them under control. In another type of program he may have performed an excellent role in project management, but Gemini was a beast of its own from inception to fruition and needed a tough ringmaster to keep it tame."[7]

James C. Elms, who was Gilruth's deputy as center director, said Chamberlin "was an ideal guy to get the thing going but not the right guy to keep it going. It's almost impossible to find a man who's the right man to get a highly technical program like that organized and started, who can also complete it." Although Elms helped remove Chamberlin, he later told a Canadian audience that Canada's biggest contribution to the U.S. space program was to send Chamberlin to NASA.

"He got it rolling," James Rose said of Chamberlin and Gemini. "He got Gemini rolling and he fought for it. And we had some fights to go through, by golly, and his good perseverance and his good, sharp thinking and having the material ready at the right time and knowing what to have ready at the right time, I think, made this program."

When asked if Chamberlin could be called the chief designer of Gemini, Rose said: "No question in my mind, whatsoever. There wouldn't have been a Gemini if it hadn't been for Jim Chamberlin."

His colleague Fred Matthews summed up Chamberlin's contribution this way: "He was the brains behind Gemini. The joke was, it's not Gemini, it's Jim-and-I."

George Harris receives an award after inspecting the Apollo tracking station at Honeysuckle Creek Australia. Photo courtesy George Harris

[8] Collins, Michael, *Carrying the Fire*, Ballantyne Books, New York, 1975. P. 452.

7 Gemini Flies

George Harris was one of the people at STG who didn't move to Houston in 1962. Instead of leaving NASA, Harris followed his interest in tracking spacecraft to the Goddard Space Flight Center outside of Washington, where NASA's tracking networks, including those for Gemini and Apollo, were headquartered.

Although almost everyone except Jim Chamberlin had originally intended Gemini to be a souped-up Mercury capsule, the new spacecraft represented a major departure from its predecessor, and its communications systems were different, too. The Agena rocket that Gemini would dock with sent its telemetry to Earth via digitally modulated signals rather than the frequency modulated or FM signals that Mercury used. It was, said Harris, a "much better signal." So NASA decided to change Gemini to the new digital modulation, known officially as post code modulation, and George Harris was called in to help out.

Harris headed the team that wrote the specifications for the new equipment, and after the contractor had installed it, his boss called him in and brought up his work testing the Mercury tracking stations with DC-4s carrying Mercury equipment. "He said to me, 'You ran the aircraft programs so well on Mercury, go find some more aeroplanes, set them up to look like Gemini, and let's go fly against the network.' Finally I had three Super Constellations. The third one was General Douglas MacArthur's aircraft, *Bataan*, which I found in a yard in, of all places, Tucson, Arizona. So we had it refurbished and used it. So there we were, tooling around the world in MacArthur's old aeroplane." This was the same aircraft MacArthur had flown back to the U.S. in 1951 after being forcibly retired as commander of U.S. troops in the Pacific and United Nations forces in Korea.

Along with communications, the control setup for Gemini also got a major upgrade. NASA concluded that the Mercury Control Center at Cape Canaveral wouldn't do for Gemini or Apollo, so in July 1962, NASA administrator James Webb announced that the future control center would be built as part of the Manned Spacecraft Center in Houston. This decision, made 10 months after he had announced the move of STG to Houston, followed controversy inside NASA over the issue of whether the control center should stay at the Cape or be moved to Houston where the spacecraft were being designed. A third option was to set up the control center at Goddard, where the tracking network was run. "I guess the way that worked out in the end was that it was better for us to be closer to the design people in order to have a true operational understanding of the vehicle than it was to be close to the launch facility," John Hodge, one of the participants in the decision, recalled.

Chris Kraft and the team that helped design the Mercury Control Center, including ex-Avro engineers Tecwyn Roberts, Dennis Fielder and Hodge, went to work drawing up functional requirements for the new control center, requirements Roberts described as "the concept of the operation, the manner in which we were going to operate, and the tradeoffs that had to be made."

Although Mercury's basic control concepts were sound, Hodge said: "It was obvious that the control center down at the Cape really wasn't going to be enough. We needed some changes to the network because we were dealing with more sophisticated telemetry systems and communications systems, so that had to be changed.

"Dennis did a lot of the work that was associated with how did the control center fit into the rest of the world, which was really the communications part of it, and in addition to that, he really brought forth the simulation system in the control center."

John Shoosmith worked on setting up IBM 7090 computers at NASA's temporary offices in Houston and procuring newer computers for MSC's permanent home at Clear Lake. (Later he was involved in work on the onboard Gemini computer, the first to fly aboard a manned spacecraft, until he decided to return to the Langley Research Center just as Gemini was beginning to fly in 1965.)

Mission Control Center Houston during the flight of Apollo 7. Courtesy NASA.

The consoles in front of each controller in Houston were far more complicated than the consoles used for Mercury, Hodge said, because 10 times more data would be sent back to Earth during a Gemini flight than a Mercury one. "Chris Kraft used to talk about the flight director's job as being one of conducting an orchestra rather than running it. Because everybody had their own thing to do, and that's essentially the way we put it together. In Mercury, we had these remote sites that were independent of each other. And even in Gemini, we realized that communications were not good enough for us to have total central control. And it wasn't until Apollo that we were able to take the chance of remoting all the stations so that there was really only one place to make decisions. That gradually evolved into more central control than we had in the original mission.

"You had to have a VIP room, because there were always visitors," Hodge continued, explaining the layout of the Mission Operations Control Room or MOCR. "Then you had to have something for them to look at. Really, the big screens at the front were more associated with people not in the mission being able to keep up with the mission than it was for the people in the control center, because they had that stuff on their screens anyway."

The new Mission Control Center included more than the famous MOCR, where controllers worked. In fact, there were two MOCRs built on separate floors, so one could be prepared for Apollo as Gemini continued or be used to prepare for upcoming missions during a flight. As well, the control centers were surrounded by a number of "back rooms" where management, experts and other needed staff supported the controllers sitting at the four rows of control consoles in the MOCR. Also nearby were rooms where simulation supervisors could put both astronauts and controllers through simulated missions before the spacecraft left the ground. Those built in the early 1960s would be used, with upgrades, for 30 years until new control rooms were built for shuttle and space station missions in the 1990s.

The fact that the original setup lasted so long had something to do with the concepts that went into it. Roberts, who more than any other individual gets credit for the design of the new control center, remembered that the Mercury Control Center was built specifically for that program, and NASA soon learned that flexibility was needed for future control rooms. "The control center at Houston was designed and built to provide a focal point for flight control of a manned space flight operation without getting tied to major requirements of a specific project." Gone was information displayed in analog format on strip charts. In its place, each controller had a cathode ray tube that could display statistical information on any system in the spacecraft, or a television picture from the launch pad. To make this possible, the data from space and the ground was processed by the new IBM 360 computers. "In the control center at Houston, all activities hinged on the real-time computer complex," Roberts explained. "The single biggest difference between the [Mercury] Control Center at the Cape and the one at Houston was the emphasis on computer usage at Houston." The new dependence on computers was almost mandatory because unlike Mercury, Gemini could change orbits, introducing new layers of complexity to the operation.

"One of the things I had pushed for was the need to distribute data within the flight operations team in the flight control center. The television distribution system emerged from this proposal. Also there was an obvious need to process systems data whereas in the past we tended to use meters and strip charts and things of this kind. The way we had been doing this represented a very inefficient use of people because we could only look at 15 or 16 parameters that way," Hodge said.

The new control center allowed astronauts and controllers to prepare for flights with more realistic simulations. In Mercury, simulations were run using pre-scripted information prepared on tapes. "With the advent of the control center at Houston, the trainers were tied in to the computers to provide a much greater degree of closed loop simulation. Action initiated by the crew and corresponding reaction by the ground led to the next sequence of activity," Hodge said.

Roberts said the construction of the new control center was complicated by the fact that three contractors were involved. The U.S. Army Corps of Engineers put up the building even before NASA and the other two contractors, Philco for the electronic equipment and IBM for the computing complex, had agreed on what should go inside. "There were many arguments, some of them heated, over such matters as carpeting for the operations room. We were pushing for carpeting because of the inherent high noise level. We had learned from Mercury that unless we went out of our way to keep the noise level down in an operations room, it was killing. The facilities people didn't understand – after all, this wasn't the center director's office."

•

The new control room was used for the first time to monitor but not control the Gemini 2 mission, and once again when the first manned Gemini – Gemini 3 – left Pad 19 at Cape Kennedy on March 23, 1965. These flights were the last times the Mercury Control Center at the Cape was used, and Chris Kraft ran the operation at the Cape while Hodge was flight director in the backup center at Houston. The crew of Gemini 3 was Mercury veteran Virgil "Gus" Grissom and John Young, a naval test pilot who was the first to fly from the second group of astronauts, which had been selected in the fall of 1962. As the first Gemini commander, Grissom had worked long and hard to bring the new vehicle to flight status. As a result, Gemini was so identified with Grissom that the spacecraft was informally known as the "Gusmobile." Taller astronauts, who didn't fit as well in the cockpit as the short Grissom, were apt to blame him for their predicament. Despite all their preparation work, Grissom and Young were only allowed to fly Gemini 3 for three orbits. During that time, they became the first astronauts to change the orbit of their spacecraft. Mindful of his post-splashdown problems in Mercury, Grissom nicknamed his spacecraft "Molly Brown" after the unsinkable heroine of a movie about the *Titanic*, and Gemini 3 lived up to its name after a successful flight.

Gemini 3 had flown in the shadow of another Soviet "first" in space. Five days earlier, the second of what was labeled as the second generation of Soviet piloted spacecraft had blasted off from Baikonur. Voskhod 2 carried the first two-member crew, Pavel Belyayev and Alexei Leonov,

Launch of Gemini 3 Courtesy NASA.

and during the flight, Leonov stepped outside Voskhod 2 and became the first person to "walk" in space. Leonov's space walk almost ended with his death when he could not fit back into the inflatable airlock that was attached to Voskhod. He finally let some air out of his space suit so that he could fit back in, but Leonov's ordeal was kept secret for years, while the success of his space walk was trumpeted immediately. Voskhod 2 was also the last of the Voskhod series, which began the previous October with a three-man crew flying aboard Voskhod 1. It later came out that the Voskhod craft were not second-generation vehicles at all, but modified Vostok capsules. The cosmonauts who flew on the two Voskhod missions placed themselves in great danger because neither craft had a launch escape system to carry them away if their booster rockets had exploded. Soviet Premier Nikita Khrushchev, who was ousted within hours of the Voskhod 1 flight, had ordered the hazardous missions to maintain the illusion that the Soviets remained ahead of the U.S. space program. When Voskhod 2 returned to Earth after a day in space, it marked the beginning of a two-year period without a single flight of a Soviet cosmonaut.

Leonov's space walk had been prompted in part by the announcement that the second manned Gemini flight, Gemini 4, would include what NASA called an Extra-Vehicular Activity (EVA). Originally, Gemini 4 pilot Ed White was scheduled to open his hatch and stand up in his seat without leaving it. The first Gemini astronaut wasn't scheduled to exit his spacecraft until Gemini 6, but Leonov's space walk spurred NASA to draw up more ambitious plans for Gemini 4.

One of the first to hear of these plans was the chief of the medical office at Houston, a 39-year-old aviation medicine specialist from Hamilton, Ontario, named Dr. Dwight Owen Coons. Coons had joined NASA in 1963 after 15 years as a physician in the Royal Canadian Air Force.

Dr. D. Owen Coons in 1963. Courtesy NASA.

During his time with the RCAF, he had gone to the Avro plant at Malton to consult on medical aspects of the Arrow, including the cooling system for the pilot and navigator, and the Arrow's ejection capability. Coons got his medical degree at the University of Toronto, and in the 1950s he earned a postgraduate degree in public health at Harvard. While studying at Harvard, Coons became friends with a U.S. Air Force physician named Charles Berry, who later joined NASA and become the leading physician at the Manned Spacecraft Center during the Gemini and Apollo programs. In 1963, Berry talked Coons into joining him at NASA.

Shortly after Voskhod 2 and Gemini 3 flew, Coons, by then the head of the medical office at Houston, was leaving his office to fly to Downey, California, to consult on the Apollo command module when he got a call from George Low, the center's deputy director. "I got my bag and went to his office, and he said, 'Do you reckon we could do a space walk on Gemini 4?' The plan was to do it on 6. Leonov had walked, so we were behind the eight ball again. I said we could do it if we have the right gear. We have the suit and we need an umbilical cord and an emergency supply of oxygen. So he said, 'All right, when you come back, come see me.' So I went down there and came back, and then I got together with Dick Johnston, head of the crew systems division. So Dick just sketched on paper what we might do with the gear that we had and things that we could make. And he put it together in his shop and we tested it surreptitiously at night in the chamber in the acoustics building. Ed White lived next door to me, and we'd slip out at night, about 10 or 10:30, and go over to the center and do the testing. I'd be the monitor and Ed would be in the chamber, and the crew systems guy would be running the chamber and checking the gear. It all tested out and seemed to go fine. When we were satisfied and George Low and Bob Gilruth were satisfied, they went and informed headquarters that they wanted to do a walk on 4." Headquarters approved, and a few days before Gemini 4 was due to lift off on June 3, NASA announced plans for White's space walk on the second orbit.

Coons remained concerned by dangers that the space walk presented to the two astronauts, who would both be exposed to open space when they depressurized Gemini 4's cabin and White opened his hatch. The pressure suits astronauts wear are in fact sophisticated balloons that protect

Dr. D. Owen Coons at the Manned Spacecraft Centre in 1966.
Photo by Bruce Moss courtesy National Archives of Canada. (PA-211348)

the astronaut from the vacuum of space. A reduced air pressure inside the suit gives the space-walking astronaut more freedom to move than is allowed by the air pressure levels used in spacecraft or found at ground level on Earth. With White scheduled to leave the capsule soon after launch, Coons worried that White and commander James McDivitt were in danger of getting "the bends" from nitrogen breathed in with the air on Earth before launch. As divers know, this is a danger because nitrogen dissolved in the blood could form painful and possibly deadly bubbles under the reduced air pressure planned for the space walk. Although Gemini and its suits used pure oxygen, the space walk was due to begin barely an hour after launch – not enough time to clear the nitrogen from their bloodstream. Coons had to face down chief astronaut Al Shepard to make sure that the Gemini 4 crew breathed pure oxygen out of masks while they suited up to ensure that they wouldn't get "the bends" in space once the pressure in their suits was lowered for the space walk.

The space walk was not the only reason NASA doctors such as Berry and Coons were focusing on Gemini 4. McDivitt and White were due to fly four days in space, which was more than three times longer than the longest U.S. flight - 34 hours by Gordon Cooper in Mercury - but still short of the Soviet record of five days. Gemini 4 was the first of three "medical" flights designed to prove that astronauts could endure weightlessness as long as was needed to go to the Moon and back. A basic Apollo Moon-landing flight would take eight days, and the longest planned Apollo mission was two weeks. Flights of these lengths would be tried in Gemini if everything went well on Gemini 4.

•

Gemini 4 also included other "firsts." It was the first flight to be controlled from the new control center in Houston, and the first to be controlled using three shifts of flight controllers, rotating each day. "What we tended to do was split the three shifts into three different kinds of shifts," Hodge recalled. "You ended up with the prime shift, which was the first shift, working on doing things while the spacecraft was going over the concentrated part of the [tracking] network. The second shift kind of solved the problems that came up on the first shift. They were looking at the systems aspects of things. The third shift was really saying, 'Now what did we do yesterday and what are we starting out to do today? And what do we need to do tomorrow?' They were a kind of planning shift." For Gemini 4, Kraft, the mission director, was also flight director on the first shift. The second shift worked under a 31-year-old newcomer to the ranks of flight directors, Eugene Kranz. The third shift was under Hodge, and Glynn Lunney served as backup flight director. Each flight director took a colour that would be used to designate his team. Hodge chose blue, making him blue flight, in charge of the blue team.

Hodge, then 36, stood out in this first group of flight directors with his English accent, his already graying hair, and the pipe and tweed jackets he preferred. Born in Leigh-on-Sea, England, in 1929, he studied engineering at the Northampton Engineering College, University of London, graduating in 1949.

"Most things happened by accident. Because in high school, my big thing was biochemistry, which is what I wanted to get a career in. I graduated at the end of the war, when the way the Brits handled the education problem was to set aside 90 per cent of the positions to ex-servicemen. This meant that the people in high school only had 10 per cent of the availability and so I couldn't get into biochemistry and I ended up being an aeronautical engineer." He went to Vickers-Armstrong for a 12-month apprenticeship, followed by another two years in the aerodynamics department. His interest was in supersonic flight, and when the British government moved away from supersonics, Hodge decided to move himself.

The original four flight directors - Eugene F. Kranz and Christopher C. Kraft Jr. (foreground) and Glynn S. Lunney and John D. Hodge (rear) in 1965. Courtesy NASA.

In 1952 he came to Avro Canada, where almost from the beginning he worked on the Avro Arrow. "I did the [jet engine] intake and the ram, all the inlets. Then the guy I was working for left to work for the government, and they needed a guy to take over the airloads group. I was put in that job." Later on he did flight testing on the Arrow, and it was this experience that led to his being assigned to operations when he came to NASA in 1959. He was in flight control almost from the beginning, and in late 1963 he became chief of the flight control division as NASA prepared for Gemini. At the time, he and his wife Audrey had a son and two daughters, and a second son was born in 1967.

As head of the MSC's flight control division, Hodge was a top manager supervising a large group of engineers. He had the responsibility of building up the group of flight controllers for Gemini and Apollo, and so he was hiring people "left, right and center." Between flights, the flight controllers he directed were busy planning missions; drawing up documentation, flight plans and mission rules; and closely reviewing results from previous flights. As well, they worked on simulations of flights, mainly possible flight emergencies, with the astronauts in spacecraft simulators hooked up to the flight control rooms.

Gene Kranz, who served as Hodge's deputy and as a fellow flight director, described Hodge in his memoirs as someone who sought more consensus and was more philosophical and thoughtful than his peers. Hodge, said Kranz, was effective at long-term planning and setting down a vision.

•

Edward White takes America's first space walk during Gemini 4 mission. Courtesy NASA.

Gemini 4 left Pad 19 at the Cape as planned on the morning of June 3, but soon problems cropped up. A rendezvous with the second stage of Gemini's Titan II rocket was scratched when it became clear that rendezvous in space was far more challenging than had been thought when the exercise was added to the flight plan at the last moment. As a result, White's space walk was shifted from the second to the third orbit. Using the gear developed for the space walk, White spent 19 exhilarating minutes floating at the end of his umbilical cord, until mission control ordered him back into his spacecraft. In an eerie echo of Leonov's problems a few weeks earlier, White had difficulty closing his hatch due to a balky handle. But he succeeded, and Gemini 4 settled into a routine for what was then a marathon of four days.

"One of the things that happened in Gemini 4 that was interesting," said Coons, who was the flight surgeon on duty during Hodge's blue shift, "was that the crew thought they could maximize the use of opportunities for looking at the Earth and other things if they worked for four hours, rested for four hours, worked for four hours, for four days. They did that for one day, and that was it. You can't shift people's circadian rhythm around like that and get away with it. You just can't sleep on cue, especially in a situation like that, which is charged with stuff anyway."

After Gemini 4's success, Gordon Cooper and Charles "Pete" Conrad were to attempt a record-breaking flight of eight days aboard Gemini 5 in August. Because of the record, the flight caused even greater concern to the doctors than Gemini 4. In preparation for this flight, one of the younger doctors in Coons' medical office, a Canadian who had come from the aviation medicine program at Ohio State, volunteered to take a more active role in Gemini 5 that went beyond working in one of the back rooms of mission control as he had done for Gemini 3 and 4. Dr. William R. Carpentier, who was born in Edmonton, Alberta, in 1936 and had grown up there and in the Cowichan Valley on Vancouver Island in British Columbia, would be on the scene when Cooper and Conrad returned to Earth. Having earned his M.D. at the University of B.C. in 1961 and an M.S. degree at Ohio State University three years later, Carpentier completed his residency in aviation medicine at the MSC medical office, and he had been offered a job, but at low pay. "I don't care. If I need to, I'll pay you," was his reply. "It was just impossible for me to visualize a job that could possibly be more exciting or any place that I would rather be." Astronaut Michael Collins described Carpentier, a slim, athletic man who was both reserved and adventurous, as "a delightful person with an offbeat sense of humor." An example of this humour was his description of a flight surgeon as "someone to hold your hand until the doctor gets there."[8] When he entered Victoria College, now the University of Victoria, in the B.C. capital, Carpentier had wanted to be an engineer. But the example of his sister, who was then a nurse, caused him to change his career aspirations to medicine. At about the same time, the sight of a Canadian naval aircraft ascending over Cadboro Bay near Victoria inspired the young Carpentier to learn how to fly floatplanes.

NASA wanted a doctor who was prepared to leap out of a helicopter, and the physician who was assigned to the Gemini 4 recovery team declined to jump. Even if it was not necessary for the doctor to enter the water to treat an injured astronaut in his spacecraft after splashdown, the doctor had to practice splashdown simulations with the frogmen of the U.S. Navy underwater demolition teams (UDTs).

"They called me in and asked me, 'How would you like to have the job in Gemini 5 of riding in the helicopter and jumping in the water and rescuing an injured astronaut?' Carpentier said. "Well, what could be more fun? I couldn't think of anything that would be more fun. It was a world that was so incredibly exciting that I never thought it would be open to me.

"So I had a chance to start working with the crews, the Gemini 5 crew, and get to be a real flight surgeon and not just a trainee. So somebody said to me, 'Have you jumped out of a helicopter before?' I said, 'Well, no.' He said, 'Well, you're going to have to know how to do that because you're right under the jet engine, and besides, those guys jump out of that helicopter at

[8] Collins, Michael, *Carrying the Fire*, Ballantyne Books, New York, 1975. P. 452.

40 feet going 20 knots.' So I said, 'Well, I guess I'll have to learn how to do that.' So I went and talked to somebody and they arranged for one of the Coast Guard helicopters to go down to Galveston, and they would take me out in the bay and I would jump out a few times.

"So I went down there and met the helicopter pilot, the co-pilot, the crew chief and the mate, and they said, 'What do you want to do?' and I said, 'I want to jump out of your helicopter, and I want to jump out at 40 feet going 20 knots.' And they said 'You want to do WHAT?' and I said, 'That's what they said I had to do.' And they said, 'Nobody's ever done that before that we know of,' and so they took me out in the bay and said, 'Well, we'll start out slow here,' and they started at 10 feet going 5 knots, and I'd jump in the water, and they'd haul me out in the horse collar. And then they'd try 20 feet at 10 knots, and I jumped out and got hauled up. And they'd say, 'You sure you want to keep going?' and I'd say, 'Well, if those UDT guys can do it, I sure as hell can.' I didn't have a mask or snorkel. I just had on a wetsuit jacket. So we finally got it. I jumped out at 40 feet going 20, 25 knots. Then I got out to the ship and then I met the officer from the UDT, and he said, 'Have you ever jumped out of a helicopter before?' And I said 'Yeah, I went out and got some training with the Coast Guard and now I can jump out from 40 feet at 20 knots.' And he said, 'You did WHAT? We jump out at 10 feet going zero.'"

Thus prepared for recovery duty, Carpentier flew out to the recovery ship *Lake Champlain,* where he jumped in the water in practice runs as Cooper and Conrad got ready to spend eight days in space. Gemini 5 was the first spacecraft to fly with fuel cells, though shortly after launch on August 21 the fuel cells began to suffer from low pressure when the oxygen heater failed. A rendezvous exercise with a pod ejected from the spacecraft was cancelled, and Kraft faced his

Gemini 5 recovery on USS Lake Champlain, *August 29, 1965. Dr. William R. Carpentier's head is visible over astronaut Gordon Cooper's right shoulder.* Courtesy NASA.

The fuel cell problem, which developed early in the Gemini 5 space flight, is discussed during a group conference held around the flight director's console in mission control. Left to right are astronaut James A. McDivitt, spacecraft communicator; Richard D. Glover, electrical, environmental, and communications officer; John W. Aaron, environmental systems officer; Astronaut Elliot M. See Jr., Gemini 5 backup crew pilot; Christopher C. Kraft Jr. (hand at chin), flight director; Eugene F. Kranz (foreground), flight director; and John D. Hodge (far right) flight director. Courtesy NASA.

first real crisis as a flight director in Gemini. He decided to keep Gemini 5 flying in a powered down state. Hodge and his planning shift were kept busy redesigning the flight plan, including a rendezvous exercise on the third day of the flight in which they flew to a point in space where a "phantom Agena" theoretically awaited. "That really gave us a lot of experience, because it showed that most of the [flight control] ideas we had were going to be able to work," Hodge said. The astronauts spent days of drifting flight in a powered down, cold spacecraft with as much interior space as the front seats of a Volkswagen Beetle.

When Gemini 5 returned to Earth after eight days, giving U.S. astronauts the longest space flight for the first time, Carpentier didn't have to jump in the water. Instead, he met the astronauts when they were hoisted into the helicopter and returned to the carrier for their postflight medical examinations.

As he would do in future Gemini and Apollo recoveries, Carpentier changed in the helicopter from a wet suit to an orange flight suit in which he followed the astronauts to the ship's sick bay, where the examinations took place. "On all those recoveries, you can always recognize me in my orange flight suit. I was the only one who ever wore an orange flight suit. Back when I started doing this, everyone had an orange flight suit. It was an international colour. When you went

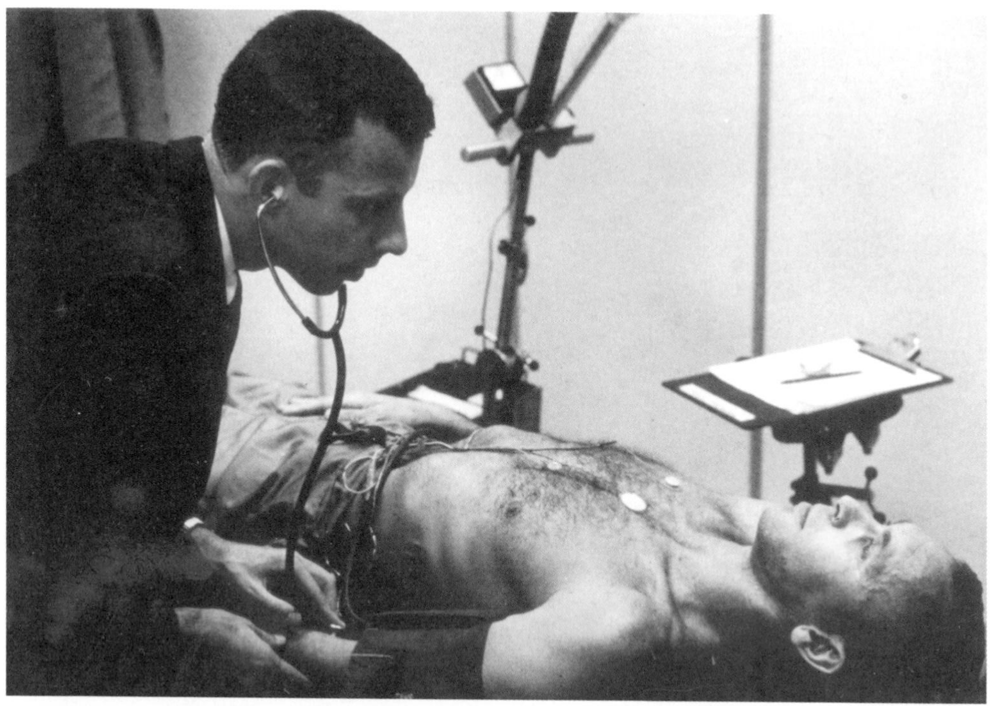

Dr. William Carpentier tests equipment on Dr. John Droescher, Manned Spacecraft Center, Houston, Texas, 1966.
Photo by Bruce Moss courtesy National Archives of Canada. (PA-211349)

Dr. William Carpentier, Houston, Texas, 1966.
Photo by Bruce Moss courtesy National Archives of Canada. (PA-211347)

down, everyone would see it. But the orange flight suits were flammable, and they found that most of the problems were from fire and not from not being found. So they went to an inflammable flight suit that was kind of olive drab. When the navy changed, the astronauts all got NASA blue flight suits. They gave me a NASA blue flight suit. But I couldn't wear it in public. I felt it made me look like I was pretending to be something I was not. I couldn't bring myself to wear an astronaut flight suit in public. So I kept wearing my orange flight suit."

The next long-duration mission was Gemini 7, which returned to Earth on December 18 after two weeks in space. Commander Frank Borman and pilot Jim Lovell got a little relief from the cramped confines of Gemini thanks to lightweight space suits that they were allowed to remove for much of the flight. But Carpentier, who was waiting aboard the USS *Wasp* for them, was still very worried because of the results from a test that was then being given to every returning astronaut. In this tilt table test, the astronaut lies down on a table, which was then rapidly tilted up 70 degrees, showing how the cardiovascular system adapted to a sudden change of position. The heart rate data followed a disturbing trend. "Gemini 7 was a very big concern for me," Carpentier said. "I had looked at the Gemini 3 data, the Gemini 4 data, the Gemini 5 data, and it went from a few hours to four days to eight days [in space]. When they were put on the tilt table on the ship after recovery, their heart rate was going higher and higher and higher, and blood pressure was dropping lower and lower and lower. The results from Gordon Cooper's flight, MA-9, 34 hours, was on the same line. And it was just going up. If you took that line out to 14 days, there was absolutely no way anybody could sustain that heart rate or be able to stand up at that time in that shape."

Overhead Gemini 7 recovery, USS Wasp, *December 18, 1965. Dr. William R. Carpentier, wearing NASA overalls, is behind and to the right of astronaut Frank Borman, and is in front of the two naval officers.* Courtesy NASA.

This worrisome set of data represented deconditioning of the heart. The effects of weightlessness on the heart are similar to what people experience when they lie in bed for long periods. When the Gemini 7 astronauts returned to Earth, Carpentier worried that their hearts wouldn't be up to the task of pumping blood under the pull of gravity. Instead, blood would just pool in the astronauts' legs and the astronauts would collapse. "During that flight, I was practicing with the swimmers, trying everything we possibly could: How fast could we get somebody out of the spacecraft, how fast could we get their heads down, how are we going to resuscitate somebody in a life raft when we've got 10-foot seas? But they landed, they came up to the helicopter, they were certainly no worse off than the Gemini 4 crew was, and they walked across the carrier deck. Everything was fine. Then we set Jim Lovell on the tilt table and he passed out. We put Frank Borman on and he did fine." The heart rate data from the two astronauts showed that Carpentier's fears were unfounded, thanks in part to an inflight exercise device installed in Gemini 7 and cuffs that put pressure on the legs, forcing blood up into the body and making the heart work. Gemini 7's duration record stood for more than four years, and subsequent long-duration flights, some of which have been longer than a year, proved that exercise helps prepare space travellers for the shock of returning to Earth's gravity.

•

As 1965 wore on, more attention was being focused on Gemini's goal of rendezvous and docking in space. On October 25, Wally Schirra and Tom Stafford entered their Gemini 6 spacecraft on Pad 19, while at Pad 14, where three years before Schirra had roared into orbit atop an Atlas, the countdown for the first Agena launch of Gemini came to its conclusion. The plan was for Gemini 6 to blast off as the Agena completed its first orbit with a pass over Cape Kennedy. Once in orbit, Gemini would begin its pursuit of the Agena. What happened was that the Agena exploded shortly after its main engine ignited. The Gemini 6 launch was scrubbed while NASA and McDonnell managers tried to figure out how to salvage its mission. Within hours they advanced several ideas, including one to have Gemini 6 rendezvous with Gemini 7. To do this, the Gemini 6 spacecraft and its Titan II launch vehicle would have to be taken down from Pad 19, then returned, reassembled and tested after the launch of Gemini 7.

The first true space rendezvous between Gemini 6 and 7. Courtesy NASA.

Chris Kraft and the members of Hodge's flight control division discussed how to track and control two Gemini spacecraft at the same time, which couldn't be done on existing systems. They decided that during the day that Gemini 6 was in space, Gemini 7 would be treated as a Mercury spacecraft, with data teletyped to Houston rather than using the advanced telemetry systems, which for that one day would be used for Gemini 6. Kraft approved, and three days after the Agena failure, the daring plan for the rendezvous of two Gemini spacecraft was announced.

After the launch of Gemini 7 on December 4, it took only eight days to get Gemini 6 ready. The engines of its Titan II booster ignited at the appointed time and then immediately shut down. Schirra correctly sensed that liftoff had not occurred, and he opted not to pull the abort handle. The cause of the malfunction was found, and three days later, on December 15, Schirra and Stafford headed off in pursuit of Gemini 7. After six hours, the two spacecraft were flying in formation while mission controllers briefly celebrated the first rendezvous in space. Gemini 6 and 7 flew in formation for three orbits, and then separated. The next day, Gemini 6 made the first successful controlled re-entry of the Gemini program and splashed down near the *Wasp*, which two days later recovered Gemini 7.

March 16, 1966, dawned with the same air of expectation as the previous October 25. Gemini-Titan 8 stood ready to chase an Agena poised nearby atop an Atlas rocket. The pursuit would start off a planned three-day flight that was to include a space walk. This time, the Agena entered orbit without a hitch, and Gemini 8 blasted off right on time, commanded by the first U.S. civilian astronaut, former NASA test pilot Neil Armstrong, and piloted by the first member of the third group of astronauts to fly, David Scott. For the first time, someone other than Chris Kraft was lead flight director. John Hodge took the spot when Kraft decided to devote more time to the Apollo program, which was beginning its operational phase.

The seriousness of the decision to terminate the Gemini 8 flight is reflected in the faces of these four NASA officials working in mission control. From left to right are: William C. Schneider, mission director; John D. Hodge, flight director; Robert Thompson, recovery chief; and Christopher C. Kraft., head of flight operations at MSC. Courtesy NASA.

For four orbits, Armstrong and Scott played catch-up with the Agena, and 6 hours and 34 minutes after leaving Earth, Armstrong slid Gemini's nose into Agena's docking collar, achieving the first docking in space. The maneuver took place over a tracking ship in the South Atlantic, and the Gemini-Agena combination moved into an area with little tracking coverage as Armstrong and Scott tested Agena's maneuvering system. Nearly a half hour after the docking, the docked craft banked without explanation. This was initially brought under control, but the craft soon began to spin. The crew of Gemini 8 first shut off the Agena and finally cut it loose, thinking the Agena was the source of their problem. But Gemini began to tumble and spin faster, the spin rate rising to nearly 360 degrees a second, a rate that would soon cause the crew to lose consciousness.

While this drama unfolded, Gemini 8 was out of contact. But when a CAPCOM aboard a tracking ship in the Pacific re-established communications, he heard Armstrong say, "We've got serious problems here. We're tumbling end over end up here. We're disengaged from the Agena."

Hodge called the ship, trying to make out the garbled transmission from Gemini. "Did he say he could not turn the Agena off?"

"No, he says he has separated from the Agena and he's in a roll and he can't stop it," CAPCOM Jim Fucci replied.

"Did I hear him say he had a stuck hand controller?" Hodge asked, responding to a statement by Scott. The hand controller wasn't at fault, but a short-circuit on one of Gemini's thrusters was.

The two astronauts turned off their main maneuvering thrusters and activated the re-entry thrusters to bring Gemini back under control. Hodge was faced with the first life-and-death emergency situation in the U.S. space program. The mission rules said Gemini had to return to Earth as soon as possible once the re-entry thrusters were activated, and Hodge asked quickly about fuel use on the re-entry thrusters. The astronauts soon isolated the balky thruster that had caused the problem, but any hope that the rules could be bent was dashed because of low levels of thruster fuel in the main system caused by the stuck thruster. If the activated re-entry thrusters leaked fuel, there would be no way to control Gemini on the way home. Hodge decided that Gemini 8 should return to Earth. "Neil Armstrong did a wonderful job of getting it under control, but in the process had used up half of the re-entry reaction control fuel. And the decision we had to make - we had about 20 minutes between the [tracking ship] and Hawaii, the last we would see of them for a long, long time, we decided to bring them back in," Hodge explained.

The next question was when and where Gemini 8 would return. It was in its fifth orbit, and the only remaining landing opportunities that day were on the sixth and seventh orbits. Failing that, Gemini 8 would have to wait until the next day to return. Hodge decided to bring Gemini 8 back on the seventh orbit, near a destroyer in the western Pacific south of Japan. "By the time they came over Hawaii, we had all the data ready to give them, the retrofire time and the angle and all that kind of stuff. They disappeared off the network and we waited." Hodge's blue team had been on duty for nearly nine hours of flight at that point, so he handed over control for the re-entry to Gene Kranz's white team, which was originally due to handle the scheduled re-entry. Naturally, Hodge and his team remained in the control room until the flight was over. Gemini 8 fired its retrorockets over Africa, and it landed near the target point. After three hours of bobbing in the Pacific, Armstrong and Scott were picked up by the destroyer *Mason*. Gemini had achieved its goal of docking, but just barely, and Dave Scott's space walk was lost with the curtailment of Gemini 8.

•

While men like Hodge were working long hours on both Gemini and Apollo, the pressure extended, inevitably, to their wives and families.

"I very often took on the night shift in the control center, because my mind works on planning as much as anything else, and that's when you did the planning, on the night shift," Hodge said. "So I'd be on a night shift, which meant that I was home to sleep during the day, so my kids got to believe that what I did for a living was sleep. So my two daughters would play

'Mommy and Daddy,' you know, and one of them would be lying down on the couch fast asleep, and the other one would be wandering around saying, 'Shh. Shh.'"

George Watts was set straight once when he was reminiscing with his wife Norma about the go-go days at Avro and NASA. "I remember, I guess about 15 years after we were married, we were talking at a New Year's party or something, I said, 'You know, those first 15 years were the happiest of my life.' And out of the dead silence in there, my wife said, 'They weren't for me.' Sitting at home, settling arguments between two little children in the dead of winter, with no car sometimes. It was awful. And here I thought I was a very sensitive, caring person, and I had never realized that."

When husbands moved to Virginia or Houston in pursuit of work, wives would often have to deal with selling houses and arranging the moves, on top of running homes and raising children with their husbands hundreds of miles away.

Owen Maynard's son Ross remembers that the long hours his father worked, even when he wasn't travelling, meant that sometimes he wasn't home much except to sleep. Paradoxically, during flights, when Maynard was working a strict eight-hour shift in the control center, he was at home more during the day than he was on regular work days. (When Maynard chose to leave NASA in 1970, he acknowledged that one of the reasons was his need to spend more time with his family.)

The marriages of many NASA engineers collapsed under the strain of meeting the lunar goal, though most of the Avro engineers' marriages overcame these stresses. Regardless of what happened, the women who married the leaders of NASA built as well as their husbands did.

"I think Mum was actually quite a frontierswoman," Annette Maynard, the Maynards' youngest child, said of her mother Helen Maynard. "This was the first generation of families to move from home to pursue high-level careers. Mum left all family, friends and support to raise her children in a foreign country for the sake of her husband's career. She really forged new territory, and they left a legacy to the children that you can do anything, anywhere if you just set your mind to it."

With their husbands working long hours at NASA, the women at home turned the housing developments that sprang up around the MSC campus at Clear Lake into communities, as Dona Erb explained. "You have to step out and make friends. Oddly enough, I, a Canadian, worked with the League of Women Voters. It was one of the many things I worked on in the community. We had some 60-odd taxing authorities in our area. Many unincorporated areas, now filled with subdivisions, had to be watched. Girl Scouts and Boy Scouts had to be established for the children. A whole new community was being established. So there was a lot of work going on at the time. The astronaut wives who were there at the time, like Jan Armstrong, worked for the community, too. Jan was an excellent swimmer and taught water gymnastics and water ballet to the young girls in our neighbourhood pool. People like that who worked in the community were really appreciated. It was a fun place for the kids to grow up. I think it worked out."

•

Gemini's problems didn't end with Gemini 8, and the challenges of docking and EVAs remained for the last four missions. In May, Gemini 9 was prepared for a mission involving an Agena docking and the first use of a self-contained astronaut maneuvering unit (AMU) backpack. This time the Atlas failed, dropping Agena in the Atlantic Ocean. A target vehicle called the augmented target docking adapter or ATDA, which had been cobbled together after the Gemini 6 Agena failure, was pressed into service and launched on June 1. A computer glitch held Gemini 9 on the pad, and when Tom Stafford and Eugene Cernan finally lifted off two days later, they discovered that the shrouds protecting the ATDA's docking device had failed to separate. With no chance of a docking, Gemini 9 carried out rendezvous exercises, including one simulating the rescue of a crippled lunar module. Cernan also became the second American to walk in space, but he ran into trouble when he left Gemini 9's cabin and tried to connect to the AMU in its storage spot in the Gemini adapter section. Cernan's exertions overtaxed his suit's environmental system, causing his helmet visor to fog over and leaving him nearly blind. Cernan was nearly exhausted. After two hours, Cernan abandoned the AMU and ended his space walk.

Gemini 10, which flew in July, finally exorcised the docking demon that had haunted Gemini. John Young and Michael Collins not only docked with their own Agena and used it to thrust them into a higher orbit, but they also performed a rendezvous with the Agena left behind by Gemini 8. Collins also opened his hatch three times. The first time he stood up in his seat and took photos. The second time he left the cabin for a full space walk, during which he retrieved experiments from Gemini 8's Agena, but experienced difficulties working on the Agena. Without means of anchoring himself, he floated away every time he made a move. Shortly after this space walk, Collins opened the hatch again to throw out his space-walking equipment.

Pete Conrad and Richard Gordon confirmed this success in September when they rendezvoused and docked Gemini 11 with its Agena during their first circuit of the Earth. The first-orbit rendezvous had been suggested by Dick Carley before Gemini began flying, but the idea was set aside in previous flights in favour of rendezvous after four orbits. When the Apollo program office asked for a rendezvous that more closely approximated the plans for a rapid Apollo rendezvous and docking with the lunar module in lunar orbit, Carley's proposal was put into action for Gemini 11. Once docked, Conrad and Gordon used the Agena's large rocket engine to thrust the Gemini-Agena combination into an orbit that took it 1,370 km above the Earth, a record. Gemini 11 and its Agena later undocked and created artificial gravity as they rotated around each other, attached by a tether. But when he attached the tether during a space walk, Gordon had become so tired, hot and sweaty that Conrad ordered him back inside early. Owen Coons, who was especially concerned about EVAs, said that Gordon's lack of conditioning helped lead to the curtailment of his space walk.

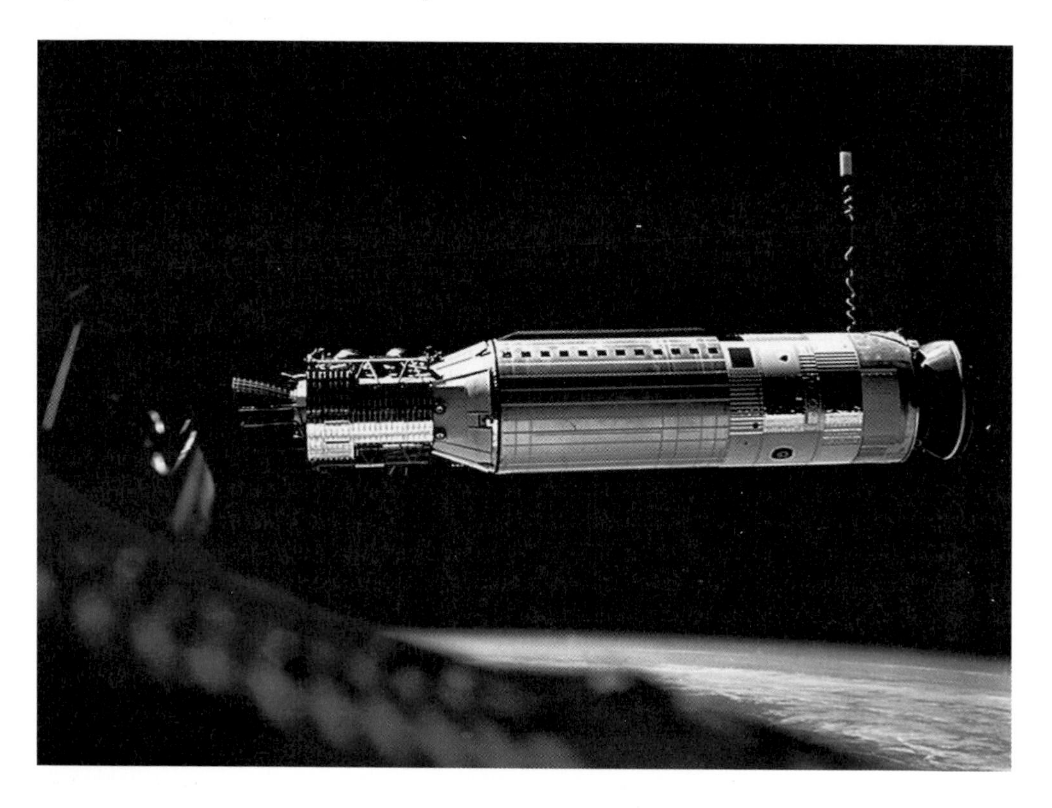

An Agena target vehicle in orbit. Courtesy NASA.

When the last Gemini and its Agena lifted off on November 11, 1966, only one issue remained to be resolved: working outside the spacecraft. Before he began his flight with Jim Lovell, pilot Edwin "Buzz" Aldrin worked hard to prepare for his space walk. He became the first astronaut to prepare for his space walk wearing his space suit as he floated inside a huge water tank where he rehearsed every move he planned to make outside the spacecraft. In addition, many restraints and attach points were installed outside Gemini 12 to allow Aldrin to do his work.

Coons whistled when recalling Aldrin's work on Gemini 12. "Talk about a focused individual. He was determined he was going to pull it off." In two standup EVAs and one full space walk, Aldrin proved that astronauts could work outside their spacecraft. As well, Gemini 12 docked with an Agena and repeated the tether experiment from Gemini 11.

The Gemini program ended with Gemini 12's splashdown on November 15. NASA proclaimed Gemini a complete success, a sense that was heightened by the fact that it had not come easily. Even before it was over, NASA's deputy administrator, Robert Seamans, summed up Gemini this way: "We've done a great deal more with Gemini than originally intended. Gemini has done much more to open the way to the Moon than we could have hoped for five years ago. With Gemini, we've developed our ability to maneuver in space, to change orbits, to inspect other objects in space, to rendezvous and dock, and to use the power of an orbiting rocket as a switch engine in space. Gemini has supported men in space, in good health, for twice as long as it takes to get to the Moon and back. Gemini has enabled the astronauts themselves to demonstrate their ability to function effectively in and out of their spacecraft, and to serve as test pilots in a variety of new space operations, and to attend to significant scientific and technical experiments. We're amassing a great wealth of operating experience and operating skills."

The official report on Gemini's contributions to Apollo, co-written by Owen Maynard, noted, among other things, that the findings from the failure of Gemini 6's Agena were applied to the lunar module ascent engine. "Probably the most significant contributions of Gemini have been the training of personnel and organizations in the disciplines of management, operations, manufacturing, and engineering. This nucleus of experience has been disseminated through the many facets of Apollo and will benefit all future manned space flight programs."

While the astronauts and flight controllers strongly agreed that Gemini was crucial to the success of Apollo, not every design engineer agreed. They wondered whether the resources put into Gemini might have been better used in Apollo. And they questioned whether Gemini taught Apollo many design lessons. Because Apollo and Gemini developed almost simultaneously, they believed that few of the lessons learned in building Gemini were applied to Apollo.

John Hodge disagreed. "An awful lot was learned," he said of Gemini in 1968. "I think an awful lot was learned about spacecraft design, about the sort of things you need hammered down and the sort of things you don't need hammered down. How much you can use the crew and how much you should have automated. How much the ground can participate relative to how much is done on board. A lot of that had not been fed into the Apollo design, because Apollo was pretty well fixed in design while we were flying the Gemini vehicles. Some people say that it would have been better to put Gemini crew, Gemini effort, into Apollo, but I think we couldn't have learned as much as we learned in that time, and I think Apollo is gaining from the results of that."

When Gemini wound up, Hodge was given a NASA Exceptional Service Medal for "planning and directing the flight control aspects of manned space flight missions and in developing highly proficient flight control teams necessary for the conduct of the missions." Shortly before that, Hodge, who had moved to Apollo after Gemini 8, flew to his ancestral home to receive an honorary doctor of sciences degree from the City University of London.

Jim Chamberlin, who had started it all, was given the NASA Outstanding Scientific Achievement Medal for "his outstanding scientific contributions and conceptual design of the Gemini spacecraft and program; for his leadership and technical guidance in the engineering of the basic and underlying design principles of the Gemini spacecraft and for his development of many operational concepts for the Gemini Program." A year earlier, Gilruth had given

Chamberlin a Certificate of Commendation that lauded him "For his outstanding contributions to this nation's manned space flight programs, for the technical direction and leadership of Project Mercury, for his creation and promotion of the Gemini concept, and for his guidance in the design of all manned spacecraft used in the United States exploration of space to date."

•

At the time of Gemini 12, it appeared that Gemini still had more flights ahead. The U.S. Air Force had been strongly interested in Gemini from its inception, and its attempts to take over Gemini had added to Chamberlin's troubles during his tenure as program manager, but NASA had fended off the military's efforts. In late 1963, when another military man-in-space program called Dyna-Soar was cancelled, a program called Manned Orbiting Laboratory (MOL) was established. MOL involved a modified Gemini spacecraft called Gemini B that would fly attached to a cylindrical laboratory operated by military astronauts, who could use the laboratory to conduct visual and photographic reconnaissance of Soviet territory and the territory of other potential enemies. The Gemini 2 spacecraft was refurbished and reconfigured into a Gemini B spacecraft with a hatch in its heatshield to allow astronauts to pass from Gemini to the laboratory. The spacecraft flew on a suborbital flight atop a Titan III rocket from Cape Kennedy on November 3, 1966, just a few days before Gemini 12. Gemini 2 survived its second flight with flying colours and became the first spacecraft to be reused. But MOL was slowed down by heavy spending on the Vietnam War, and finally, in June 1969, the Nixon administration ended the program.

When Chamberlin designed Gemini, he proposed that it be used for flights around the Moon and even as a mother ship for a lunar landing. The success of the program caused others in McDonnell and NASA, including astronaut Pete Conrad, to consider Gemini for lunar missions using various boosters. These proposals were floated as late as 1967, but even after this last proposal was dropped and MOL was cancelled in 1969, Gemini still lived on the drawing boards. As NASA looked beyond Apollo, one proposal under consideration involved building a modified Gemini spacecraft called Big G to ferry astronauts to a space station and back. The two-seat cabin would be retained as a flight deck, and behind it there would be room for up to 10 passengers. In various proposals advanced between 1967 and 1971, Big G served as a stopgap before a reusable shuttle craft was built, or was promoted as a shuttle craft itself. But as NASA won support for a winged shuttle in the fall of 1971, Big G fell by the wayside. The fact that Gemini lived on the drawing boards 10 years after its inception and 5 years after its last flight was another tribute to the vision of Jim Chamberlin.

8 Owen Maynard and the Rise of Apollo

Bright and early on the last Saturday of June 1966, the auditorium at the Manned Spacecraft Center filled up with the people involved in running the Apollo program, including astronauts, engineers, scientists, and managers from throughout NASA and American industry. It was five years and one month since John F. Kennedy had issued his challenge to land on the Moon, and this day, for the first time, the details of how his goal would be met were being presented to the top people charged with carrying it out. By 1966, spacecraft and rockets were being built all over America. Astronauts were already in training for the first Apollo missions, and the launch site for the Saturn V rocket was built. An estimated 400,000 people were working to put astronauts on the Moon.

"This is not the usual kind of symposium," MSC director Bob Gilruth told the crowd as he called it to order. "It's designed more for the people who are putting it on rather than the audience. In other words, it's more like a working session in which we hope to get a hard-core review of the lunar mission and the documentation of it here during the next three days."

Air Force Major General Sam Phillips, the man at NASA headquarters in charge of Apollo, followed Gilruth. "The lunar mission is going to be a real tough mission. There's a lot of hard work that lies ahead in the way of detailed planning to operate within the flexibilities that have been designed into the system, and there are a lot of hard decisions that lie ahead. I think it's particularly appropriate and very timely to review in detail the lunar landing mission itself, to subject the mission to a critical design review. The purpose of this symposium is to ensure that the right things are being done in preparing and planning for the execution of the actual lunar mission."

Joseph Shea, the Apollo program manager at Houston, concluded the introductory speeches by noting that the symposium would "show, in almost nauseating detail, how it's going to get done." The symposium would not cover the missions leading up to the first lunar landing attempt, but only the lunar landing mission itself.

Finally, he introduced Owen Maynard to give an overview of the first lunar landing mission. "Maynard, as you probably know, is chief of the mission operations division within the spacecraft program office here, and he has been the main architect of this overall symposium."

Having joined Apollo at its inception, Maynard had risen quickly to become head of the systems integration division of the Apollo program, and recently he had been transferred to head the mission operations division in the Apollo program office. Now the stocky and authoritative 41-year-old native of Sarnia, Ontario, walked to the lectern.

"It's useful to think of the lunar landing mission as being planned in a series of steps or decision points separated by mission plateaus," he said, introducing one of the major planning concepts that he became known for in Apollo. "The decision to continue to the next plateau is made only after an assessment of the spacecraft's present status and its ability to function properly on the next plateau. If after such an assessment it is determined that the spacecraft will not be able to function properly, then the decision may be made to proceed with an alternate mission. Alternate missions, therefore, will be planned essentially for each plateau. Similarly, on certain of the plateaus, including lunar stay, the decision may be made to delay proceeding in the mission for a period of time. In this respect, the mission is open ended, and considerable flexibility exists. The end points of these plateaus represent major commit points in the lunar mission and are characterized by propulsive maneuvers representing major changes in spacecraft velocity."

The nine mission plateaus that appeared on the projection screen behind Maynard formed an outline of the lunar landing mission. The first plateau was the prelaunch period prior to liftoff at Launch Complex 39 at Cape Kennedy. Three astronauts would be inside the command module atop a Saturn V rocket. The second plateau was the Earth parking orbit that the Apollo spacecraft, still linked to the third stage of the Saturn V, would fly before the stage fired a second time to put Apollo on its course for the Moon. The translunar coast between the Earth and the Moon, which would last nearly three days, was the third plateau. The fourth plateau was the period when the spacecraft was in lunar orbit prior to the lunar module's descent to the Moon, when the LM, with two astronauts aboard, would separate from the command and service module, with a single astronaut remaining aboard. The LM's descent was the fifth plateau, and lunar surface operations, including walks on the Moon by both LM astronauts, was the sixth. The seventh plateau was the LM's ascent from the Moon, and the eighth was the lunar orbit operations after the LM rejoined the CSM, reuniting the astronauts. The ninth plateau was the CSM's coast back to Earth prior to re-entry in the Pacific Ocean.

Maynard took the rest of the morning to explain each phase of the mission in detail. As he returned to speak at the end of a coffee break, Shea quipped: "I see our speaker surrounded and making his way to the podium, breaking away from the adoring crowds. You'd think he was an astronaut." Most of the humour during the three-day symposium was provided during question periods and introductions of speakers in the form of puns and jokes by Shea, the New Yorker who had managed Apollo in Houston for the previous three years. In the afternoon, three talks covered the early phases of the lunar landing mission: navigation and guidance in Apollo, mission planning questions, and lunar landing strategies.

•

The next morning, Sunday, Shea started the day in characteristic fashion as he introduced another Avro veteran: "The episode ended yesterday, you'll recall, with our hardy little band of adventurers just arriving on the lunar surface. This morning we pick up the scenario on the surface, and we're about to answer one of the fundamental questions. This is a question that has been raised by the scientific community: Will the science that is to be done on the lunar mission be good, or will the whole mission be of no avail? We have the answer to that: No. A Vale named Bob is going to tell us about the science."

Robert Vale, who had worked on structures in Mercury, was now the deputy manager of the experiments program office at Houston. The 44-year-old Toronto native was one of the higher ranking people at NASA from the Avro group.

"This morning I'd like to cover what has transpired to date in the planning for the activities on the lunar surface," Vale began. "The planning for these activities is still very much in the embryo stage. There are many, many people that are involved in this, and there are many, many iterations of this planning to go yet. First of all, the very, very broad scientific objectives have been set. Just recently the basic priorities were set, and they are these: First of all, of course, will be the astronauts' observations. The prime objective from the science viewpoint is to come back with as many samples of the lunar material as possible. Secondly, it is desirable of course to have these samples as fully documented as we possibly can. Also, the astronaut will be asked to emplace the lunar surface experiments package on the surface. This package will be operating for one year after his departure and will gather long-term scientific data. The setting of these priorities does set up a somewhat enigmatic situation for the mission planners, insofar as sample collection is also inherent in the fourth topic here, field geology. But as we go along here, I think you'll find out, you'll realize that there is a pretty comfortable period of time allocated to do all the scientific chores that are desired."

Drawing up the science work being described by Vale had not been an easy job. Most of the people working on Apollo were focused on the goal of simply landing astronauts on the Moon and returning them home. What the astronauts did once they got to the Moon didn't seem important. The scientific community, of course, saw matters differently, and scientists working for NASA, the U.S. Geological Survey, and the National Academy of Sciences, dealt with the management of Apollo to make sure that the astronauts who walked on the Moon extracted full scientific value from their perilous and expensive journeys. As Vale explained to the symposium, the scientific objectives had been set at a meeting the year before at Woods Hole, Massachusetts.

Robert E. Vale in 1964. Courtesy NASA.

Robert E. Vale receives service award in 1970. Courtesy NASA.

Vale told his audience that it was planned to have both astronauts make two excursions of three hours each on the lunar surface during the first lunar landing, out of a total stay on the surface of 18 hours and 22 minutes. During the first excursion, the two astronauts would take photos and collect samples, and then unpack and deploy the Apollo Lunar Surface Experiments Package (ALSEP). The ALSEP, powered by a radioisotope thermal generator, contained experiments including a seismometer, magnetometer, solar wind detector, ion detector, and others that would give scientists valuable information about the Moon and its environment. The experiments would require setup time, and the task of moving the plutonium fuel for the generator was potentially dangerous. The second excursion would be devoted to more photography, sample collection and geological work using special tools. At the end of the work, the astronauts would have about 35 kg of lunar samples to take home with them.

"In conclusion, we feel that the planning of the lunar surface activities must not be overplanned," Vale said. "The astronaut must be fully used, his judgments and his flexibility and selectability must be used. We still have quite a long ways to go yet in getting detailed mission profiles. We have been making, I think, fairly large steps in achieving this recently, but we are looking forward to firming these up before the end of the year."

After questions on Vale's presentation, Shea moved on to the topic of the next talk, which concerned the return of the LM from the Moon to the CSM orbiting above. "If you think back some four years, probably the most difficult part of the decision to do lunar orbit rendezvous was the question of the ease of rendezvous, particularly at the Moon. Through the Gemini program and through a lot of additional simulations, I think we are in pretty good shape as far as our confidence in doing rendezvous. This next talk discusses in some detail how the rendezvous actually goes."

With that, Shea introduced another former Avro engineer, Morris Jenkins of the mission planning and analysis division at Houston, who in a very technical presentation described the path that the LM would follow in normal and emergency situations from the lunar surface to the CSM orbiting ahead. During his talk, he discussed how the computers on the spacecraft would be updated after each maneuver, and how the spacecraft and ground established the paths of the two spacecraft. Jenkins also described emergency situations, including those in which the LM was launched at a time that was not convenient for rendezvous with the CSM, or where it was barely able to make orbit and required the CSM to change its orbit to rendezvous with the LM prior to the docking that would reunite the crew members.

For nearly five years, Jenkins and the group he headed inside the mission analysis branch had been working on trajectories for lunar missions, especially lunar landing missions. The job of drawing up trajectories for lunar flight is wonderfully complicated: It involves putting a spacecraft into Earth orbit, an orbit that's tilted in relation to both the Earth's equator and the Moon's orbit. At the right moment, the spacecraft has to be injected into a path that just misses, by slightly more than 100 km, where the Moon will be when the spacecraft arrives in three days. The spacecraft's return path to Earth must put it into a very narrow corridor that ensures that the spacecraft hits the Earth's atmosphere at the right place and the right angle. A tiny error means the crew is lost. If the crew is to land on the Moon, new complexities are added. And on top of everything, it must always be taken into account that the Earth and Moon orbit the sun, are not perfect spheres, and wobble slightly in their orbits.

Those diagrams of the lunar flight that place the Earth and Moon across a page from each other make things look far simpler than they really are. In a realistic diagram, if the Earth were 5 cm or 2 inches across, the Moon would be a ball 1.3 cm or a half inch across, and the two balls would be about 150 cm or 5 feet apart. To even get in the vicinity of the Moon, which is about 384,400 km away from Earth, one must be a good shot.

"The pressing need was to be able to generate lunar trajectories quickly and with certainty," Jenkins wrote later, describing his work. "This was essential even in the early days to breed confidence within and with whomever you liaised concerning your analysis." The base concept he worked with was a figure-8 path that would miss the Moon by just 100 km and return to a safe splashdown on Earth. "Not only could a lunar orbit ensue, but with later mission development, a landing, a subsequent rendezvous, and with transearth injection, a return to the Earth." This trajectory would have to be economical with fuel, he added, and take into account lighting conditions at launch, lunar landing and splashdown.

Over the years preceding this meeting, Jenkins and his group had used a highly involved mathematical estimation method, some outside expert help and brute computing power to prepare trajectories for lunar missions and give the people planning Apollo confidence that the spacecraft could be safely guided to its targets and back home. Another group built on the foundation laid by Jenkins' group to draw up detailed trajectory calculations for each mission and also make the changes that would inevitably be necessary as each mission progressed.

The day continued with papers about Apollo's return from the Moon to the Earth, and possible aborts during the mission, when Shea called Jenkins back to the podium. "The last item on the agenda for today is software compatibility with lunar mission objectives," Shea said. "I think this is a particularly interesting paper, also. It gave me a feel for the first time about how the equations actually are going to be implemented to do all these good things that we keep saying we're doing."

Part of the responsibility given to Jenkins' group was preparing the program requirements to be put on the small onboard computers that flew aboard the command and lunar modules.

"The first consideration in regards to preparing software is making sure that we've got adequacy for every mission phase," Jenkins began. He noted that adding program capability to handle different mission scenarios "would lead you into a lot of complexity which would not be necessary."

Morris V. Jenkins (right) receives Superior Achievement Award in 1969. Courtesy NASA.

In another highly technical presentation, he discussed how the software for the onboard computers on the command module and the lunar module would work in conjunction with the guidance and navigation systems on both spacecraft, which used special sextants devised at the Massachusetts Institute of Technology. The onboard computers for Apollo had about 36,000 words of non-erasable memory and 2,000 words of erasable memory, tiny by today's standards. Jenkins walked through the use of computers at each phase of flight and pointed out that the ground's large computer capability existed to back up the equipment on board the spacecraft. He also outlined where computing capability had been removed from the onboard computers to make sure they would be available for more important work. As well, because programming the onboard computers involved literally handcrafting and then testing the memory, programs for individual missions had to be ready months in advance of the liftoff.

(Tom Chambers, a former Avro engineer who worked on control systems in Apollo, explained that the memory in Apollo's computers was made out of wire. "It was interesting to me that it was made on a looming machine. A loom. The stuff was expensive, but it was very reliable. What was criticized about it was that it was inflexible. It turned out to be a very fine thing to do. One of the problems we [later] had in the shuttle was that the software got absolutely out of hand. It was so easy to change that everybody wants to change it.")

In the end, Jenkins explained that the computers would help increase crew safety and the chance of mission success. "Even with some reductions in the onboard programs, by using these simple targeting schemes which I went through, it looks to us as though we would have no real problems, except the details might kill us," Jenkins joked as he concluded.

At the time of the symposium, Jenkins was 43. Born in Southampton, England, he served in the Royal Air Force as a navigator during the Second World War. During the war he made his first visit to Canada when he was trained in navigation at Rivers, Manitoba. After the war he worked at the Supermarine technical office of Vickers Armstrong for nine years, during which he learned stress, aerodynamics, and stability and control. He joined Avro Canada in 1956 and worked on stability and control aspects of models of the Arrow, and later of the aircraft itself. Once at NASA in 1959, Jenkins said he worked on "odds and ends" in Mercury before moving into his work on lunar trajectories in 1961. Jenkins and his wife Joan, a teacher who later became an engineer, have two children. Jenkins is modest about his work. "It wasn't easy to get this trajectory scheme going. The group that I led did it. It wasn't I who did it."

The third and final day of the symposium, which was run alternately by Shea and Maynard, covered communications, training, space suits and landing site selection.

"I would commend you all for the tenacity you had to sit through so much detail over the last three days," Shea said in his wrap-up. "The way I look at this is what we've been presented today, the last three days, is really a conceptual lunar mission with a large amount of data in the presentations and behind the presentations." He said the symposium showed that software for the mission "is at least as critical as the hardware." He then briefly went through the flight schedule, saying that the first manned CSM might be ready to fly early the next year. Shea added that "as long as there are no reporters in the room," the mission could fly as early as late 1966. Not much more than a year after the symposium, he said, final dress rehearsals in Earth orbit for lunar mission could take place, and the lunar mission could be ready to go in 18 months, at the end of 1967.

•

The ambitious projections that marked the close of Maynard's symposium reflected the mood not only of Maynard and Shea, but of most people inside the room, for June 1966 marked something of a high-water mark for Apollo and the U.S. space program. The recent flight of Gemini 9 had given NASA more confidence in the use of rendezvous. The success of Gemini, even in the face of adversity, gave confidence to those facing problems getting Apollo off the ground. Early that month, as well, the Surveyor 1 automatic probe had soft landed on the Moon, sending back photos and backing up the findings of an earlier Soviet lander that a LM touching down on the lunar surface would not sink below a layer of dust, as some scientists feared. The Surveyor landing vindicated a decision made years before by Maynard and Caldwell Johnson that the LM's landing gear design should be based on the assumption that the Moon's surface was as hard as the surface of a desert on Earth. Surveyor also gave valuable information to Apollo about thermal conditions on the Moon. Bryan Erb, whose expertise on heatshields also extended to other thermal questions, was in the Surveyor science support room the night Surveyor landed, and he found that the temperatures on Surveyor on the Moon were about what was expected, good news for those working on the LM.

A month before the symposium began, the doors of the gargantuan new Vehicle Assembly Building at Cape Kennedy had opened for the first time, and a test article that looked exactly like a Saturn V rocket rolled out to Pad 39 atop a crawler transporter. Funding for NASA was at its peak.

Among those who were most involved in the rise of Apollo was Owen Eugene Maynard. There were few parts of the Apollo spacecraft that did not bear the marks of his handiwork. Working with Max Faget and Caldwell Johnson, Maynard was part of the team that did the initial design work on the Apollo command and service module back in 1960 and 1961. Having helped turn NASA to the use of lunar orbit rendezvous, along with Chamberlin, Jenkins and others, Maynard provided much of the inspiration behind the design of the lunar module. Through the development of the Apollo program, he was the man charged with bringing the many parts of Apollo together into a whole system that worked together harmoniously to deliver astronauts to the surface of the Moon and bring them home safely.

Owen Maynard shows Apollo command module to Prince Philip on March 11, 1966. George Low stands behind Maynard. Courtesy NASA.

Owen Maynard shows Apollo command module to Prince Philip on March 11, 1966. Bob Gilruth stands beside the Prince. Courtesy NASA.

The special skills that Maynard brought to Apollo were honed in a career that saw him help build some of Canada's finest wooden boats and included time flying and building equally remarkable aircraft. Born in Sarnia, Ontario, on October 27, 1924, Maynard grew up on farms near Brigden, Point Edward and Corunna, small towns that surrounded Sarnia, which is just below Lake Huron and across the St. Clair River from Michigan. He grew up dreaming not of aircraft but of large ocean liners such as the *Queen Mary*. After attending schools in and around Sarnia, Maynard was excused from high school in 1940 to work in the war effort. He learned the craftsmanship that would serve him so well when he worked at the Mac Craft boat works in Sarnia, making wooden sub chasers under the tutelage of Joseph Napoleon "Nap" Lisee, known as one of the finest wood turners of his time. Maynard then got work as a machinist at the Electric Autolite plant in Point Edward, where he helped build distributors, starting motors, generators and antennas for the armed forces. In his youth, Maynard remembers reading Buck Rogers comic strips, but felt that outer space adventures were purely in the realm of science fiction and wouldn't apply to him.

A few weeks before his 18th birthday, Maynard followed the example of his brother Glenn and enlisted in the Royal Canadian Air Force. In September 1942, Canada was entering its fourth year in the Second World War, and Maynard went in for flight training at Uplands just outside Ottawa, starting with the famed Tiger Moth biplane. He graduated at the top of his class in 1944, just above another pilot officer who was the son of a high-ranking RCAF officer. The two young pilots were assigned to a course in flying the de Havilland Mosquito, one of the legendary aircraft of the Second World War. The bomber with a plywood body was so challenging that usually only veteran pilots were assigned to fly it. The officer's son wanted to fly the Mosquito, and since Maynard finished ahead of him, both were given the opportunity. The assignment to train on the Mosquito proved fateful for both men. Late in the year, the officer's son perished in an accident while still in training, but Maynard continued and after completing his training early in 1945, Flying Officer Maynard was shipped overseas to Britain and prepared to fly bombing missions on industrial plants in Norwegian fjords. Before he could fly in combat, the war in Europe ended. Shortly after his return to Canada that summer, the war in the Pacific was also over. "The biggest hero maker at the end of World War II was the Mosquito," Maynard said, alluding to the Mosquito's great danger and remembering his friend's sacrifice and those of others. "I was a terribly young and terribly well-trained Mosquito pilot."

Back home, Maynard spent a few months completing high school and then was accepted for studies in aeronautical engineering at the University of Toronto. In the meantime, he needed a job, so he went to the Avro plant at Malton. When Maynard applied for work, the personnel officer was not impressed with his academic credentials, but when he said he'd been an RCAF pilot, the personnel man said, "You're hired." When Maynard added that he'd flown the Mosquito, the reply was, "You're really hired." It wasn't the last time his experience with the Mosquito helped him.

Maynard first worked at Avro for eight months as a loftsman, doing detailed working drawings of aircraft including a propeller-driven aircraft, the CF-100 and the Jetliner. For this work he used his experience as a boat builder. At the end of the summer he started the first of two years of engineering studies at the U of T Ajax campus, a former ammunition plant. He worked in the Avro mechanical shop during the summers between his studies at the U of T, making parts for the CF-100 and the Jetliner, and even working on the Jetliner's interior. "That gave me an entrée into the world of craftsmanship that most engineers don't get," he said. In 1949, Maynard returned to the RCAF, spending a summer in the City of Toronto Squadron at Downsview, north of Toronto and east of Malton, where he flew radar calibration tests, delivery flights and acceptance flights on new aircraft being delivered to the squadron. At the end of the summer he left the RCAF for good.

His final two years of study took place at the main U of T campus, and in 1951, Maynard graduated as an aeronautical engineer and returned to Avro, where he was put to work in the stress department, analyzing designs, starting with a weapons pack being studied for the CF-100. He also took postgraduate courses in advanced aircraft design at U of T's Institute of Aerophysics and in dynamic loads at the Massachusetts Institute of Technology. Maynard moved to the Arrow when that project began, and among his many tasks on the CF-105, he did extensive work on its landing gear, which used liquid spring shock absorbers. "When I went to NASA, I was viewed as a landing systems expert."

A solidly built man with slicked-back light brown hair and intense, narrow blue eyes, Maynard often spoke at great length about the engineering concepts behind his work. He was used to giving orders and having them followed, but he was equally accustomed to taking and carrying out orders. Although he sometimes appeared to be brash, Maynard never hesitated to speak with the greatest respect of those he worked for, such as Bob Gilruth, Sam Phillips and Joe Shea.

He also unabashedly declared his admiration for his parents and his wife. His father, Thomas G. Maynard, was a veteran of the First World War who worked for many years as a postal clerk. His mother, born Margaret Arnold, was a talented amateur artist. "My mother allowed me to have an imagination," he said, and his imagination would help him in his work on the space program. He credits both his parents with teaching him how to focus on reaching his goals, something that would serve him in good stead in Apollo. Maynard was working at Langley when his father retired from the Canadian post office, and his parents were so taken with the area that they moved to Newport News, where the elder Maynard got a new job as a park ranger. When Maynard was working on a model of an early design of the Apollo command module, his mother contributed by making small space suits out of space suit fabric for the astronaut figures in the model.

In 1947, Maynard married Helen Richardson of New Toronto, Ontario. Maynard credited her as being a far better judge of character than he is, and he proudly remembered a NASA reunion in the 1980s where Ruth Cohen, the wife of Aaron Cohen, who was then director of the Lyndon Johnson Space Center after having worked under Maynard in Apollo, introduced Helen Maynard. "This is Helen Maynard – she's the one who taught me how to be the boss's wife." She recounted how Helen Maynard, herself a boss's daughter, had insisted that her husband do his chores at home, such as taking out the garbage, no matter how high and mighty he was at the office. Together, the Maynards raised a son and three daughters.

Through the time of Apollo, the Maynards resided in Friendswood, a semi-rural suburb south of Houston near the Manned Space Center, where there was plenty of room for the family, a horse, dogs, cats and other assorted pets on a one-acre tract that included a pecan orchard and a creek running through it. While other NASA people lived nearby, the home was chosen because it was not in a neighbourhood where everyone worked for NASA, and Helen Maynard tried to raise the family away from the politics of NASA.

During the Apollo years, Maynard's work caused him to leave much of the burden of raising the family on Helen, but he always made a point to be home for important occasions for the children. He also spent a lot of time teaching his children to overcome their fears and to learn how things worked. "If you wanted to know what time it was, he'd tell you how to make a watch," said daughter Merrill Marshall. "He has such a vast knowledge of everything that we have our own kids call him," said another daughter, Beth Devlin.

•

When President Kennedy set the Moon-landing goal for Apollo in May 1961, the design of the Apollo command module was already well along under the supervision of Max Faget, Caldwell Johnson and their group, which included Maynard and Bryan Erb. Although many radical changes in the CM's design for Earth return were suggested by contractors, the CM wound up being a conical vehicle topping a rounded heatshield. Since Gemini hadn't even been designed yet, and because the same people who designed Mercury also designed Apollo, the Apollo CM was more a direct descendant of Mercury than of Gemini. As the summer of 1961 gave way to fall, it was time to find a contractor to build the new spacecraft. During the summer, 14 aerospace firms had been asked to bid on the Apollo command and service modules, and potential bidders had received information on Apollo at a special conference. An 11-member source evaluation board, headed by Faget and including Jim Chamberlin, was appointed to evaluate the proposals.

Five aerospace giants decided to bid on the Apollo contract, and they delivered their proposals, each one containing a mass of documentation, to the board at the Chamberlin Hotel near the Langley Research Center on October 9, 1961. After each proposal was given initial study by the board and the more than 100 NASA engineers assisting it, each bidder made an oral presentation on the third day. The board then worked for several weeks, evaluating the proposals, and in late November the Martin Company received the highest rating from the source evaluation board, with North American Aviation Inc. in second place. But in one of the most important and controversial decisions in the history of U.S. aerospace, NASA administrator James Webb, deputy administrator Hugh Dryden, and associate administrator Robert Seamans awarded the contract to North American on November 28. Allegations soon surfaced that politics had overruled quality in the decision, but Webb, Dryden and Seamans said North American's long relationship with NASA and NACA helped tilt the balance in its favour.

The CSM contract was just one of several major Apollo contracts awarded at the time. North American, which was based in the Los Angeles suburb of Downey, California, also had the contract to build the second stage of the Saturn V. The Boeing Co. had the Saturn V first stage, and Douglas Aircraft won the contract for the Saturn's third stage, which was also the second stage for the Saturn IB, whose first stage was made by Chrysler. The instrument unit that controlled the Saturns was built by IBM.

When the decision to go with lunar orbit rendezvous was finalized the following summer, NASA had to choose who would build the spacecraft then known as the lunar excursion module or LEM. North American, which believed before the LOR decision that its spacecraft would land on the Moon, lobbied NASA to allow it to control the building of the LEM as well, but was rejected. Nine companies, not including North American, later bid on the LEM contract, and on November 7, 1962, Grumman Aircraft Engineering Corporation of Bethpage, New York, was chosen to build it.

In 1961, while he and others were promoting the cause of lunar orbit rendezvous, Maynard set to work with his associates on drawing up the first specifications for what would become the lunar module. During this time, he was working under Chuck Mathews. "I would say they did an outstanding job of it," said Mathews, "because we were able to write the work statement and specifications for this lunar module, put the request for proposal out on the street, I think, about in early September, and we had a handshake with the contractor before Christmas, something that would be unheard of today. But they had done such a good job of defining this thing that we were able to do the technical evaluation of all the contractual work very rapidly in selecting a contractor, which happened to be Grumman. And I give Owen the credit for not only coming up with many of the ideas, though I'm sure others in the group came up with good ideas, but he was the one who organized the effort and brought it to that fruition. That involved tricky things like the first definition of what the lunar surface might be."

In January 1962, Maynard was appointed head of the spacecraft integration branch of the Manned Spacecraft Center, supervising 40 engineers. This branch worked on preliminary studies of the LEM, space stations, and missions to Mars. Much of his work during this time involved designing the LEM and selling LOR through NASA. Early in 1963, with Grumman on board and work on the LEM beginning in earnest, Maynard moved to the Apollo spacecraft program as the manager of the LEM systems office.

Maynard said that the name lunar excursion module originated with him and Laurence G. Williams, an engineer who worked alongside Maynard through his time in the Apollo program. "I couldn't even stomach the name lunar module. I had to put the word excursion into it .... We don't want to give anyone the connotation that this thing's going to go land on the Moon and do great wonderful things on the Moon. It's only going to get down there and it's going to land if it can land, and if it can't, it's going to come home," he explained. "Until we know we can land, we are looking at just an excursion."

While the difference between LM and LEM may appear subtle to an outsider, to Maynard it was important and reflected the way he managed in Apollo. While managing according to a set goal became commonplace in the 1990s, it was still novel for many people in the 1960s. Maynard and his colleagues had a clear goal set by President Kennedy: "landing a man on the Moon and returning him safely to the Earth" before the end of the decade. Maynard often referred to this goal when he talked about Apollo. "It was actually a long time before we threw in the requirement that you had to get out of the thing and walk on the lunar surface. Kennedy didn't say anything about getting out and gathering Moon rocks," he said. "That's what my job was. Don't make it too complex, don't make it too much of a compromise." And so it was that Maynard helped coin the LEM name. Eventually, NASA headquarters decreed that the spacecraft would be known as the lunar module, possibly to dismiss the idea that the vehicle could make "excursions" around the Moon after it landed. But the "LEM" pronunciation of the acronym stuck.

Maynard had already made major contributions to the LM's design by his early work in 1962 when he was helping to sell LOR around NASA. NASA's Apollo history sums it up this way: "Thomas J. Kelly had directed Grumman's Apollo-related studies since 1960, earning for himself the title, 'father of the LEM,' but the vehicle that finally emerged was a 'design by committee,' that included significant contributions from the Houston [technical] panels, notably Owen E.

Maynard's group."[9]

When asked who his closest equivalent at NASA was, Kelly said Maynard. "Well, you can see from the time we proposed on the job until we actually built it, every detail of the lunar module design changed. And in the first year, these changes were coming thick and fast. Owen was always part of that. He was the approving authority. I accused him of having this complete finished LM drawing in his desk drawer that he wouldn't show us. When we came up with a new set of changes and we had a revised drawing of the LM that showed those changes, I'd say, 'Well how are we doing, Owen? Are we getting close to what you've got in your drawer there?' He'd look at it and he'd say, 'Yeah, you're getting pretty close.'"

Maynard denied it. 'They used to accuse me of coming back and looking in my desk drawer and saying, is it right yet? They'd always figure that I had a picture in my desk drawer. But I didn't. It started out with five landing gear legs, I think it was six tanks in the descent stage, four in the ascent stage, four reaction control system quadrants, two docking tunnels, two crew stations, and an ascent and descent stage, of course, and it evolved … to four tanks in the descent stage, [and] two tanks in the ascent stage." As well, he pointed out, the number of legs became four, the RCS thruster quads were moved, and one of the two docking tunnels was eliminated in the first months after Grumman got the LM contract.

Said Kelly of Maynard: "He was really the lead [NASA] guy on the program for the lunar module, at least for the first two years or so, and then he kind of branched out to do other things. Owen was still involved. He never really left the lunar module side of the program. He had a lot to do with it."

Maynard recalled many discussions on LM design. The number of tanks was reduced to simplify the LM and save weight. Because the oxidizer is heavier than the fuel, the fuel tank in the ascent stage had to be placed farther out from the cabin than the oxidizer tank, giving the ascent stage a slightly lopsided look. The four thruster quadrants were moved so that the exhaust from the thrusters wouldn't affect the landing legs. The second docking tunnel, at the front of the

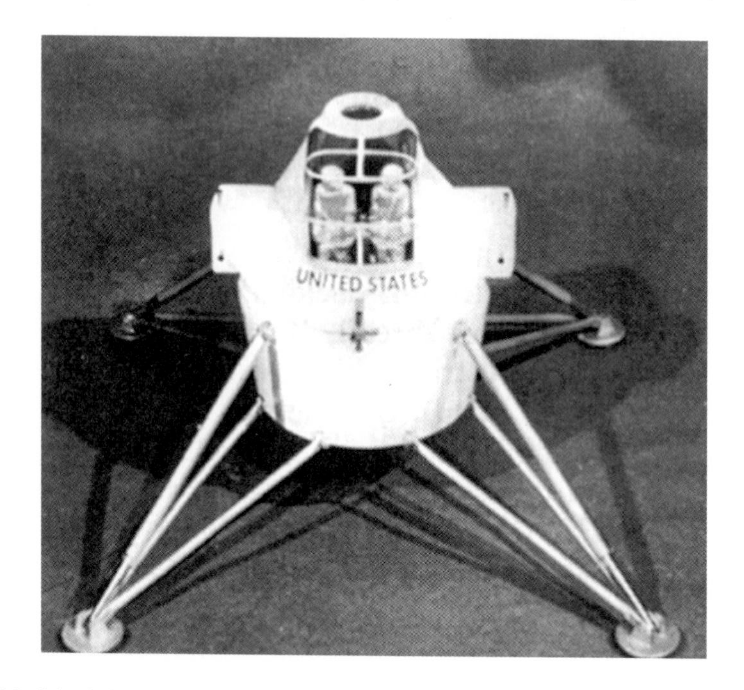

Model of the lunar module as it was first presented in 1962. *Courtesy NASA.*

[9] Brooks, Courtney G., James M. Grimwood and Loyd S. Swenson, Jr., *Chariots For Apollo: A History of Manned Lunar Spacecraft*, National Aeronautics and Space Administration, Washington D.C., 1979.

LM ascent stage, would give the LM astronauts a direct view of the CSM through their front windows during docking as they returned from the Moon. But there was already a docking tunnel at the top of the LM where the CSM would dock with the LM on its way to the Moon, and the weight of a second docking structure could be cut by putting a small window in the ceiling of the LM and having the commander lean back, look up through the small docking window, and use the usual LM controls to dock.

But the biggest change concerned the front windows and the crew stations in the LM. In NASA's and Grumman's original proposals, the two astronauts would sit in the LM and look out large domed helicopter-like windows as they landed the vehicle. The large windows added weight to the LM, would be difficult to fabricate, and would cause lighting and temperature control problems inside the LM. Then someone got the idea to have the crew stand up inside the LM. Seats were not a necessity because the LM would only operate in weightlessness or in the Moon's one-sixth gravity. "Going to the standup configuration simplified so many things," Maynard said. "We got rid of seats in the small cabin." The astronauts would have their eyes right up against the windows for good visibility during landing, he said, and the LM's front structure was stronger and easier to build. "I think that one of our great good fortunes was that that standup crew station worked so well. It tended to simplify a lot of things."

In March, 1964, Grumman unveiled test mockup 1 (TM-1), which allowed testing of crew mobility and movement. The round front hatch on the mockup proved awkward to crawl through with a life-support backpack, and this problem contributed to the decision to eliminate the front hatch's function as a docking port and make it a square hatch. Tests on the mockup also helped lead to the decision that astronauts would move from the LM cabin to the lunar surface by way of a ladder on the front leg of the LM. The TM-1 review board was chaired by Maynard, and included Faget, Kelly, and astronaut Deke Slayton.

Through all these debates and many more on questions about such things as landing radar, instrument displays, and the placement of equipment to balance the spacecraft, Maynard was always in the heart of the discussion. Characteristically, he is quick to share credit with those he worked under or supervised, and above all, with the contractor. "Grumman went through the design studies with our participating in the review capacity and concurring and suggesting changes here and there, but largely it would be a Grumman configuration." Although he called the LM "my baby" in passing, when asked what he thought of when he looked at the Moon and considered his role in the LM, Maynard said: "I look at the Moon and think, aren't I glad that Grumman was the contractor." Maynard explained that Grumman and North American worked in different ways. When NASA wanted a change to the CSM, North American would simply make the change. "Grumman would never do what we wanted to do unless and until they were convinced that it was right," an attitude that Maynard said in retrospect was "wonderful." While others in NASA preferred the obedience of North American, Maynard didn't insist on everything being done his way without question "because I wasn't that smart." When informed of Kelly's comments about his role in designing the LM, Maynard is prepared to take some of the credit. "It turns out that my studies turned out to be closer to the final product than most other people's drawings. But maybe that was because I was the customer."

While changes were being made to the LM in 1964, Maynard was promoted to the position he would hold for most of his time in the Apollo program: chief of the systems engineering division in the Apollo spacecraft program office. Maynard was assistant chief of the division for a few weeks after he was transferred as part of a series of management changes made when Joe Shea became program manager. He soon became division chief when the original chief left NASA. He was one of six (and later eight) division chiefs reporting to Shea. Some people describe Maynard's position as the Apollo program's chief engineer, although the people he worked for, including Shea, Gilruth and Phillips, were also engineers. Maynard's nine-page job description began by saying that he was "the primary point for the establishment, coordination and control for design, specification and development for Apollo flight and ground systems."

There would be many demands placed on limited resources of weight, power, space, computing capacity and electrical power in the spacecraft, and Maynard had day-to-day responsibility for dealing with these competing demands.

Maynard said one of the reasons he was promoted was his relationship with Shea. "He was an unmanned kind of guy. He wasn't a pilot, he wasn't an aeronautical engineer. All of his previous work had been in unmanned programs. The astronaut community and the Manned Spacecraft Center were all pilots or aeronautical engineering oriented. He figured that I had the entrée into that whole system. That's why I kept getting those jobs at those high levels that I thought were above my capabilities. He was trying to work himself into the community that I led. That's why he wanted me as the chief. He had a fair amount of confidence in me, and he also believed that I respected him. I didn't hesitate to ask his opinion on certain things. So I was sort of the perfect interface between him and this hostile group of manned people."

In his position, Maynard continued to oversee, from a higher level, the lunar module. But now he also was involved in developing the command and service modules beyond their initial design phase. He was already familiar with them from his work for Faget and Johnson, including the awkward attempts to design a CSM that could land on the Moon, and he had been involved in evaluating the proposals in the bidding process for the CSM contract. "I think the interesting thing, historically, is that there were, at that point in time, as many contractors in the country able to conduct studies of this nature," Maynard said. "Even more amazing is that there were contractors – even more contractors – that were able to propose to build the system."

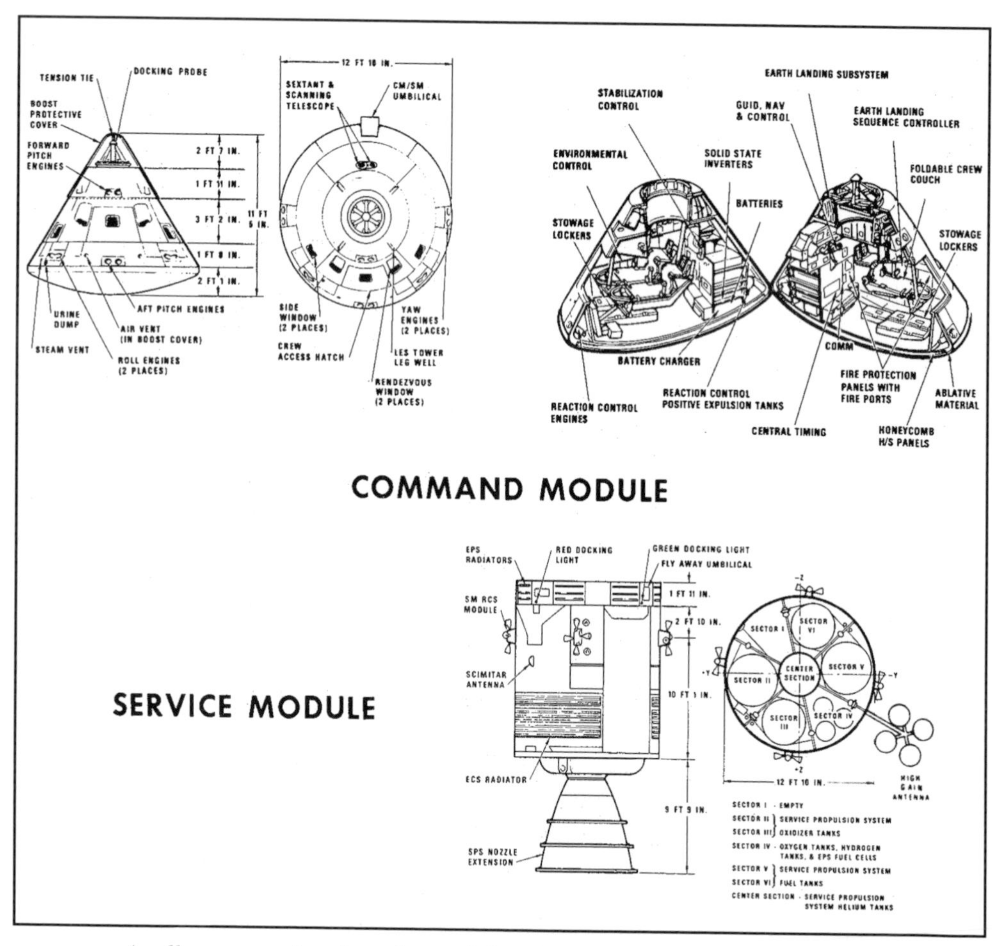

Apollo command and service module schematic circa 1969. Courtesy NASA.

By 1964, North American was getting ready to build the Block I spacecraft that was designed for early Apollo flights without the LM. At the same time, detailed design work was going on for the Block II CSM that would fly to the Moon with the LM. In April 1964, Maynard was assigned to co-chair, along with his opposite number at North American, Norman J. Ryker, the effort to work out the design and ground rules for the Block II CSM. The same month, Maynard served on a review board examining the mockup for the Block I CSM. That fall, he co-chaired a comprehensive design review for Block II, during which issues from the design of the docking tunnel to the question of where the lunar sample boxes would be placed were discussed. This related to another part of Maynard's job: negotiating final specifications for the CSM and LM. This work involved teams of engineers from Maynard's office and the contractors who would review written specifications before coming to an agreement. "Those specification negotiations were the most traumatic thing I ever went through in my life," he said in 1970. "They were really something." Sometimes talks would be held up while experts in an area would study what the specifications should be. And often contractors and NASA would disagree, necessitating more talks. Once the specifications were set, any changes that were made to the spacecraft had to pass through a change control board to ensure all changes were logical and consistent with the goals of the program.

Maynard was also responsible for making sure that the CSM and LM could work together, not only with each other but with the launch vehicle, communications network and recovery forces, among other things. He assigned one of his top people, Larry Williams, to deal with the interfaces between the two spacecraft — the docking system, the tunnel through which astronauts would move between the two, and the various electrical and system connections between them. And then there were issues like control panels. "We recognized that we had two different spacecraft, two different crew stations that were going to have to be flown by the same crew. So we put one man in charge of seeing to it that these two crew stations were sufficiently identical that the crew would have a good capability of getting out of one vehicle into the other one and flying it without completely changing their thought processes." This work he gave to Joseph P. Loftus. These efforts involved bringing the two contractors together, one an Air Force contractor from the west coast and the other a Navy contractor from the east coast, and setting up common rules for the interfaces and control panels.

•

At the time NASA started work on Apollo in 1960, another Canadian engineer from Avro was moved to the new program along with Maynard. R. Bryan Erb, a tall, bespectacled and handsome engineer who in the 1960s sported a moustache and a Vandyke, was one of the key members of the team that developed the heatshield for the command module. One of the more thoughtful members of the Avro group, Erb's technical orientation stood out even in this group of hard-driving engineers. Throughout his career, Erb has always been eager to promote the future applications of space exploration.

Erb, who was born in Calgary in 1931, remembers that as a child he was a Buck Rogers fan and liked to play with a chemistry set. "When I was in the fifth grade, when I was 11, we had one of those people come to our school who was billed as an explorer. What stuck in my mind was his prediction that man would someday go to the Moon. I remember telling my parents. That was at Earl Grey School in Calgary. My father had really wanted to be a doctor, and he was pushing me towards medicine. That was interesting enough, and I was willing to go along with that until the moment I pulled out my application to go to university. I was sitting there with the application form in my typewriter, and it said, what faculty do you desire to enter, and it dawned on me like a bolt out of the blue that I had no interest whatsoever in medicine. And I started thumbing through the catalogue, and I found a little poem at the front of the engineering section about a man who builds a bridge is as much a poet as a man who writes a sonnet. And I thought, 'That's sort of neat, I think I'll go for engineering.' Not a particularly well-planned choice, but I've found engineering very satisfying and I'm very grateful for that career choice."

R. Bryan Erb, acting manager of the Lunar Receiving Laboratory, 1969. Courtesy NASA.

Erb took his engineering training at the University of Alberta in Edmonton. "I have a background in civil engineering, but I was always more interested in the fluid mechanics side of things. I was a private pilot during my last year at university, and that got me interested in aeronautics. So when I had the opportunity to do a scholarship in England at the College of Aeronautics at Cranfield, and took a degree in aerodynamics, I had the good fortune to work under a fellow who was a keen member of the British Interplanetary Society, and he was concerned about what was then called the re-entry problem, even though nothing had entered, so I chose to do a thesis topic on the subject of re-entry heating. That got me into the heat transfer area."

Erb returned to the U of A, where he obtained a master's degree in engineering. At this time he married his wife Dona, a mathematics teacher also graduating from the U of A, and after a short job as a civil engineer, Erb went to Avro in 1955. "During my time with Avro, I was basically doing aerothermodynamics and thermal analysis, addressing how hot the back end of the Arrow gets from the combined effects of heating from the engine and heating from aerodynamic heat. That was in the days before computers. It involved a lot of mathematics," he recalled. The cancellation of the Arrow he called a "lucky disaster" for him.

At NASA during Project Mercury, Erb developed a computer program to analyze the performance of the Mercury heatshield, but he was soon moved to Apollo work. At the time MSC moved to Houston, he was head of the thermal analysis section within the structures and mechanics division. Later he became chief of the thermo-structures branch and subsystem manager for the Apollo heatshield. In this role he was responsible for managing a large staff and various contractors charged with developing a heatshield that would do things no heatshield had

Robert Gilruth awarding Certificate of Commendation to Bryan Erb for the first lunar landing, July, 1969. Courtesy NASA.

ever done before. His direct work on the heatshield carried through the critical years until around 1965, when he became assistant chief of the structures and mechanics division with continuing responsibility for Apollo thermal work, but assuming higher level responsibilities for other Apollo systems as well, including structures, parachutes and materials.

Speaking of his work on the Apollo heatshield, Erb said: "What I was working on at the time was, knowing we were going to have to come back from the Moon, and the re-entry problem was very different because now we had speeds of 36,000 feet per second [11 km/sec] instead of 25,000 feet per second [7.6 km/sec], and the difference came at the higher speed, the shock wave gets a great deal hotter. I guess my early analysis led, among other things, to the conclusion of how the corners of the command module had to be rounded so you minimize the shock layer heating. You have it too sharp, you have too much convection; you have it too blunt, you have too much radiation, so you had to have an optimum [shape] in there." Others have claimed that the CM had a rounded edge because the diameter of the Saturn rocket had been reduced, but Erb points out that the CM heatshield simply could not have the sharp edge that the Mercury and Gemini heatshields had. And the Apollo CSM was attached to the Saturn rockets by a tapered adapter that contained the LM, so the diameter of the Saturn didn't directly affect the diameter of the CM.

The shape was only the start of the technical challenges for the Apollo heatshield. Like Mercury and Gemini, the Apollo heatshield was an ablator, made of material that burned away, carrying away the heat with it. "The real problem was, how do you manufacture an ablator and

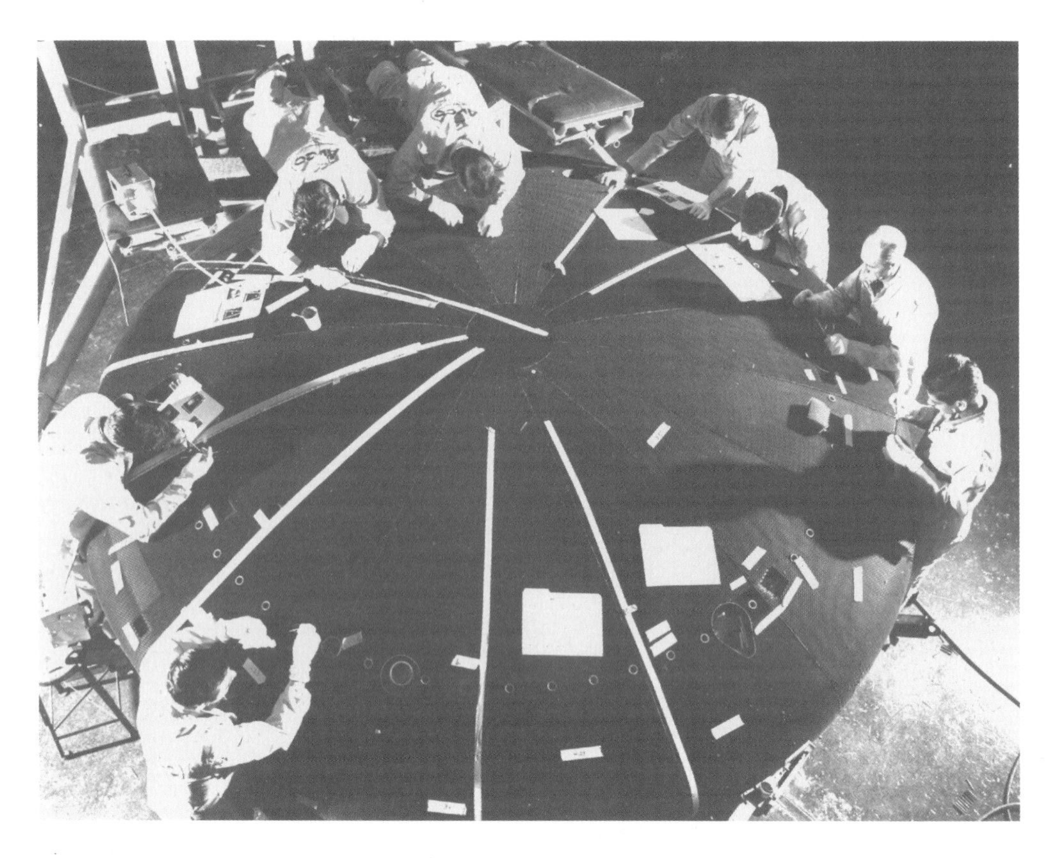

Technicians at Avco (in Everett, Mass.) working on the blunt end of the Apollo heatshield, checking the bond of the honeycomb and filling the 350,000 cells with ablator. Courtesy NASA.

Command Module from Apollo 4 flight, first Block II CM, after successful reentry test. Courtesy NASA.

make sure it was glued on to the [spacecraft structure]. We started off with what was almost the shuttle approach, with tiles. They were fabricated and bonded on. But we could never come up with a way where we could be sure that the bond would hold. In the end analysis, we went to bond an open honeycomb onto the [spacecraft] and then we looked at the cells and checked that the bond was solid. Then we would stuff ablator into the honeycomb cells. That was the manufacturing approach that we ended up taking. This required injecting the ablator into 350,000 honeycomb cells, which is a monumental manufacturing task."

More than one ablative material was analyzed and tested. Erb and his team looked for another ablator material to test in case the first material selected didn't meet the test of Apollo. And because of Apollo's higher heat loads, all the materials were harder to test on the ground.

"Just as one example of the problems we had gotten into, Langley did some testing on material that they were promoting, and they were testing it in an arc jet, as it turned out, at too high a pressure. The material was eroding like crazy, and Langley was waving the flag that the Apollo ablators would vanish in a few seconds or a few minutes at most. We decided we had better do a flight test, so we did a special test with a Scout rocket, a five-stage solid rocket. Three stages boosted it up, then it flipped over and two stages drove back down to get a higher entry velocity. We built a special cap of the Apollo ablator material about 11 inches in diameter, and it was shaped like a heatshield. And I was perfectly comfortable that it would be a satisfactory test, based on the analysis that I'd done. And Langley was equally convinced that it was going to fail miserably. It survived without any problem at all."

Another member of the Avro group left Houston to help supervise construction of the Apollo CSM. David Ewart, who had taken his engineering at Cambridge University, had been engaged in assessing the bids for the CSM. When the contract was awarded to North American, MSC set up a resident office at Downey, California, staffed with NASA engineers who would work with their contractor counterparts on building the CSM. Ewart, who had just moved to Houston, accepted an offer to be one of the NASA engineers resident at Downey. Another reason Ewart accepted was that he had enjoyed working in aircraft plants at Fairey Aviation in England and at Avro Canada.

"This was like going back home to get into the environment of a hands-on hardware-building place," he said. "That's one factor of many I had to consider in a center such as the Manned Spacecraft Center at the time. You're working in a semi-research environment and it's a little different from the industrial environment. Anyway, the task of what was then North American Aviation was the design and building of the command and service modules for the Apollo program. Again, we went through an elaborate system of development tests. Lots of full-size boilerplate models of the spacecraft." Some were built for drop testing at El Centro, California, and others for launch at White Sands, New Mexico, atop Little Joe II, which was a follow-on vehicle to the Little Joe of Mercury. There were also boilerplates that flew on the Saturn I rocket before production spacecraft flew atop Saturn IB and Saturn V rockets. "We had a lot of boilerplates in the Apollo program. Some became more sophisticated as we went along, with more and more of the operational systems. We were involved in the general testing and subsequent certification and qualification testing."

During this time, Ewart worked as a subsystems engineer, overseeing the testing, installation and integration of the guidance and navigation, and stabilization and control subsystems into the CM. Much of the guidance system was developed at the Massachusetts Institute of Technology and built by AC Electronics. By 1965, Ewart was chief of the NASA quality engineering office, where he oversaw acceptance tests of command and service modules.

•

During this time, Maynard was faced with a welter of decisions, some of them involving whether a piece of equipment should be included in the spacecraft or not. His instructions are recorded in numerous memos sent around the program office and to the contractor. For example, a memo ordered a redesign of the awkward and complicated command module hatch used in

Block I to allow space walks from the Block II CM. Other Maynard memos recorded disputes with contractors over issues such as cooling in the CM environmental control system.

And Maynard informed his fellow managers of problems plaguing Apollo. For example, in September 1965, Maynard wrote Shea that the major problems facing Apollo spacecraft designers and builders at that time included:

- Weight growth, an eternal problem in aircraft and spacecraft programs. In this case, weight growth exceeded predictions "by a serious margin."
- Criteria for lunar landing.
- Integration of scientific experiments in Apollo.
- Water landing criteria when the CM returned to Earth.
- Land landing. At the time, plans still existed to land the CM on the ground, although these were later dropped.
- Conflicts over the thermal design of the spacecraft involving temperature control and spacecraft attitudes relative to the sun during flight. During the translunar and transearth coasts, the spacecraft would be in constant sunlight, and if it were in a steady attitude, the side facing the sun would get very hot and the side away from the sun very cold. One solution advanced by Erb and others, and eventually chosen, was to slowly spin Apollo in "barbecue mode."
- Propulsion performance in every engine was at that point still below expectations.
- Design of the space suit and the portable life-support system was subject to unexpected changes, and these changes affected the design of the Apollo spacecraft.

Maynard also made basic decisions about how to categorize missions or parts of missions, as in the "plateau" concept outlined at the 1966 symposium. In April 1965 he announced that objectives, which until then had been classified into first, second, and third orders, would be replaced by primary and secondary objectives. A mission could be held back or cancelled if a malfunction interfered with a primary objective, while secondary objectives would be considered desirable but not mandatory.

In early 1966, Maynard was shifted to be chief of the mission operations division in the Apollo program office. This division had been absorbed into the systems engineering division shortly after Maynard took over systems engineering, but was now again separated as Maynard found himself spending more time defining the upcoming mission requirements. His deputy at systems engineering, Robert W. Williams, took over that division. Among Maynard's tasks at the mission operations division was organizing the June 1966 lunar landing symposium. Maynard said the idea for the symposium came from Sam Phillips, the program director in Washington. "This was above and beyond my capability to put that symposium together, but I guess General Phillips and Joe Shea decided that I was the guy who could put it together. Periodically, I would hear from Joe Shea that Sam Phillips was asking how the symposium was going. I was finally able to dedicate a fair amount of time to the symposium when Bob Williams became available as a deputy and then took over the systems engineering division." Maynard was spending much of his time in that division defining requirements for the lunar landing mission and the flights that would lead up to it. He was assisted by project engineers who worked on specific parts of the missions. The actual mission design work was done by the mission planning and analysis division in MSC's flight operations directorate, where Morris Jenkins worked.

By the beginning of 1967, Gemini was over and the spotlight was exclusively on Apollo. Two CSMs had flown atop Saturn IB rockets in test launches in 1966. The first flight with astronauts on board, known unofficially as Apollo 1, failed to meet Joe Shea's hopes for a 1966 launch due to a series of problems with the Block I CSM being used for the flight. In spite of this, confidence was still high in the new year, and Apollo 1 was set to fly on February 21 with astronauts Gus Grissom, Ed White and Roger Chaffee on board.

At this time, Maynard was getting ready for a role working with Chris Kraft's flight controllers in supporting missions, since mission operations fell within Maynard's responsibilities in the program office. The last week of January, Maynard felt sufficiently good about things to take time out for an interview. "I was being interviewed by a Canadian journalist with a movie camera. I had him out in my sailboat and told him how wonderful everything was. We were having a pretty good time and being pretty cavalier. Our arrogance was showing all over the place. We thought we had the world by the tail."

9 From the Fire to Lunar Orbit

Every Apollo countdown was preceded by at least one dress rehearsal, where astronauts put on their space suits, went to the launch pad and entered the spacecraft. At the same time, controllers took their places both in the blockhouse at the Cape and at the control center at Houston, just as they would on launch day. And so it was with Apollo 1 on Friday, January 27, 1967. At 1:20 p.m. astronauts Gus Grissom, Ed White and Roger Chaffee entered their spacecraft atop a Saturn IB rocket on Pad 34, and the dress rehearsal countdown for their February 21 launch was on. The "plugs out" test was due to end with a practice escape from the spacecraft by the astronauts.

Sitting at the flight director's console in Houston that day were Chris Kraft and John Hodge, who, along with Gene Kranz, were slated to be the flight directors for the mission. As in the early Gemini flights, Kraft was lead flight director, Kranz ran the systems shift and Hodge the planning shift. Since Gemini 8, 10 months earlier, Hodge had sat at the flight director's console for two Saturn IB launches in 1966 that tested Apollo equipment in preparation for this flight. Dr. Owen Coons was seated nearby at the flight surgeon's console. Owen Maynard was a little farther away, in a back room called the SPAN or spacecraft analysis room just outside the mission control room. Because mission control doesn't actually take charge of a mission until after liftoff, the countdown rehearsal was a relatively undemanding exercise for those in Houston. As in real countdowns, the controllers in Houston were monitoring spacecraft systems, familiarizing themselves with the quirks of each vehicle that show up in the data.

To enter or leave the Block I CM, the astronauts had to use an awkward three-part hatch that was secure but took at least 90 seconds to open. The hatch design was in part a legacy of Grissom's Mercury suborbital flight, when the hatch of his Liberty Bell 7 capsule blew prematurely after splashdown. One of the battles that Maynard had fought and lost in developing Apollo was for a hatch that opened more easily than the hatch on Apollo 1. The Block II CMs

The crew of Apollo 1: (l to r) Virgil Grissom, Edward White, Roger Chaffee.
Courtesy NASA.

were due to have a different hatch that would allow space walks, but that would mean nothing to the crew of Apollo 1. Now the hatch was closed, and the spacecraft's pure oxygen atmosphere was pressurized.

The test dragged on all afternoon, plagued by delays due first to a smell inside the cabin and later to problems with the communications system. "It was getting to be about 5:30 in the afternoon in Houston, and we were having a lot of communications problems," Coons recalled. "A lot of static on the line. Gus [Grissom] is getting frustrated, to say the least. We're in a hold to try to sort out the communications. They haven't had a thing to eat all day. Gus comes up and says, 'God damn it, you guys, how the hell do you expect to get to the Moon if you can't talk to each other down here?' And seconds later, there is this white noise in our headsets. Just terrible static. Then everything goes silent."

At that moment, Hodge heard a jumble of voices and then silence. When he called to alert the Cape that Houston's telemetry was lost, he was told to stand by. He soon realized that something serious had happened.

"I was in the SPAN room that day as the operations manager," Maynard said. "You could hear most of what was going on. You could hear, 'Fire in the spacecraft.' I don't remember all the details, but I do remember hearing that there was a fire. That triggered the catastrophe plan. It was fairly clear on the network that enough bad things had happened, that it was a catastrophe. It wasn't clear at the time whether they had been killed, but we knew enough that we had to use the catastrophe plan. Everything went blank after that. I wasn't in the middle of anything after that."

Back at the Cape, technicians struggled to open the hatch that the three astronauts had been unable to open. Fuelled by pure oxygen, the fire burned until the CM's pressure vessel ruptured, causing toxic smoke to pour into the white room that enclosed the spacecraft. In spite of the smoke and the danger from the explosives in the escape rocket above Apollo 1, the technicians finally forced the hatch open and found astronauts Grissom, White and Chaffee dead.

"And then some time later, the pad leader comes up on the emergency com system, his voice about two octaves above normal, and says, 'It's terrible, it's terrible. There's been a fire.'" said Coons. When his deputy, Dr. Fred Kelly, verified from the Cape that the astronauts were dead, Coons headed home to deliver the terrible news to Ed White's wife Pat, who lived next door.

"By the time I got home, there were TV people in front of the house. Fortunately, the driveway in our house was on the side away from Ed's house. So I just drove in, and went out the family room door and through the back yard. Neil and Jan Armstrong were already there. That's what happens at Edwards when something happens. I'm sure Pat knew by the look on my face what had happened."

Once they had preserved their data, the traumatized flight controllers in Houston went to a local watering hole to deal with their shock at losing a crew. The following Monday, Hodge and Kranz called a meeting of flight controllers where Hodge started the meeting by bringing everyone up to date on what had happened. Kranz, a more emotional man who would later gain fame for rallying his controllers during the Apollo 13 crisis, angrily told the controllers that a lack of toughness had contributed to the disaster, and from then on in, flight controllers would demand a higher standard of readiness in equipment and crews.

Within hours of the fire, NASA launched an investigation that included astronaut Frank Borman and experts from within and outside NASA that would go on until it reported in April. Although the precise cause of the fire was never explained, the investigation found ample evidence of why the fire had such tragic consequences.

"We've never explained it satisfactorily," Coons said. "My personal view is that there is a cable run on the lower deck [of the spacecraft]. If the leg rest is opened, all the instrumentation is there. The cable run was covered by a three-side aluminum duct as protection. This thing had been stepped on and stepped on and stepped on. I think the ducting was cracked, and it caused a short in that cable. The cable would burn, and I think that's where the fire started. It went right into the environmental control system. All the equipment at couch level was covered with polyurethane foam, which was flammable. It has never been explained to my satisfaction. It had to be electrical in association with a material that could burn in the 100-per-cent oxygen atmosphere."

Hodge spent a month at Cape Kennedy helping in the investigation, and when he returned, Bob Gilruth assigned him to work on preparations for the mission that would replace Apollo 1 to make sure that such a catastrophe wouldn't happen again. "So we did a nuts and bolts look at the safety aspects of the thing," he explained.

The fact that NASA was left to conduct its own investigation was a tribute to the political skills of James Webb, the NASA administrator, but the U.S. Congress still held a series of hearings on the fire where the top brass of NASA had to submit to a grilling. Maynard went to Washington to back them up, but since he was then in mission operations, most of the work involved in making the changes to Apollo to prevent another fire fell on his successor at systems engineering, Bob Williams. The Block II CM underwent changes including a redesigned hatch that would be even easier to open than the hatch that Maynard had designated for that spacecraft. All the flammable material in the spacecraft, including space suit material, was replaced, and systems, wiring, launch pad procedures and quality control were changed and upgraded to prevent a repeat of the fire.

The congressional hearings and probes by the media reopened the controversy that had attended the awarding of the CSM contract to North American back in 1961. The fire led to personnel changes at the contractor. One of the Avro engineers who watched the controversy with particular interest was George Watts, who had left NASA in 1964 and gone to work for Lockheed in the Los Angeles area.

"I was disgusted with the amount of money they were charging to build the command and service module on Apollo," said Watts, whose feelings contributed to his decision to leave NASA. "McDonnell was okay, but North American was trying to grab whatever bucks they could. I would say that the public was charged 10 to 100 times as much as they should be." Watts, who did loads analysis on the Mercury and Apollo spacecraft, charged that a boilerplate Apollo capsule cost NASA $14 million, when it should have cost around $100,000. "That was just an example. There were many, many examples. If you're working for the federal government, you are a steward of the public funds. Basically, that's your job. You've got morals and principles, and if you don't live by them, you've got problems. Sooner or later."

Despite the problems that were unearthed after the fire, Watts's viewpoint remained among the minority.

"It came to the point of, anything that was questionable, we were going to fix, and it now was going to be perfection," Maynard remembered. "The whole attitude kind of changed. No more shipping things and fixing them at the Cape."

Dave Ewart, who worked for NASA at the North American plant at Downey, noted that the Cape's reputation for manufacturing as well as launching spacecraft contributed to the habit of shipping uncompleted spacecraft there.

While in retrospect a hatch that took 90 seconds to open and the extensive testing of spacecraft with a pure oxygen atmosphere should have been seen as a lethal combination, Ewart said there were few signs at the contractor's plant that Apollo was headed for tragedy. "There was considerable schedule slippage. There's no getting around that kind of schedule slippage on a high visibility program. There were so many people involved, assuring that we'd thought of everything." Ewart was assigned to an audit team that toured contractors' sites in search of design changes that needed to be made beyond the elements that had contributed to the fire. This allowed improvements to be made in the CM guidance system and the spacecraft computer.

•

During the months after the fire and the investigation, Maynard found plenty to do: "The hiatus after the fire gave us a chance to hone up what we were doing, an awful lot better than we had. We took all of those things that had to be certified and flight tested, which was everything, every functional and performance requirement that we had ever defined. We put those things together in a large series of test objectives and said, 'Okay, those things that are associated with certifying the command and service module, let's see how many missions it will take to verify that. Like in the Avro Arrow, there were several aircraft assigned to do the test program, and it would take many months to pull that off. We had the hiatus to let us sit and think of these things. I would sit with not only my people, but I would also sit with Gilruth and talk these things over. It was great. We weren't nearly as overstressed as we were before the fire. Our schedule wasn't nearly as tight."

Of all the test plans Maynard and his group drew up during that time in 1967, the one he is best known for was the A to G sequence for Apollo missions, leading from the next major test through to the lunar landing. The A Mission, an unmanned flight of the Apollo command and service module atop a Saturn V rocket, was scheduled to fly later in 1967, and the B Mission, an unmanned lunar module flight, was due to follow with LM-1 launched aboard a Saturn IB. The first manned CSM flight, which would achieve the goals originally set for Apollo 1, was designated as the C Mission. The D Mission was a joint manned flight of the CSM and LM in low Earth orbit, launched by a Saturn V. Following a successful manned LM flight on the first or second try, the sequence could move to a manned high Earth orbit mission with the CSM and LM launched atop a Saturn V, the E mission. The final link before the long awaited lunar landing G mission was the F mission, manned operations with the CSM and LM in lunar orbit. The A to G sequence, which was promulgated in September 1967, was drawn up with the assistance of many people in the flight operations directorate, notably Rod Rose and Carl Huss. Earlier versions of the mission sequence from Maynard's group did not include a lunar orbit mission, which was included in the sequence at the insistance of flight operations. The debates over the lunar orbit F mission were far from over, however, and the actual missions flown departed from this sequence.

While Maynard worked on mission sequence and requirements, the effects of the fire were still working their way through the halls of the Manned Spacecraft Center. "Because people were killed and the equipment was in shambles, the egos were straightened out very quickly. People had been uptight with each other and lost confidence in one another. I can't cite a case where anyone got fired. For many reasons I don't know about, people were resigning and leaving, and going back where they came from, in droves." Among those who resigned were Bob Williams from the systems engineering division, and the program manager, Joe Shea, who was succeeded by George Low.

"One of the things that George Low decided was that we shouldn't accept pay raises that year. This is our own penance. George Low is left without a chief of the systems engineering division. There may have been a period when I may have been a potential guy to get blamed. Then they went back and remembered that they had gone against my guidance on the flight hatch.

"One morning I was building a corral for my horse. George Low was coming back from his jogging in the mist and he sees me there digging post holes. He stops and we start talking. And he said, 'I want to ask you if you'll come back and take over the systems engineering function as well as the mission operations function [in the program office]. I would handle the administrative problems that we've got and you handle the engineering problems. Would you do that?' So we talked about that, and I said, right there - I guess I had anticipated this coming - that I would assume that responsibility. I said I would do this and you can count on me to do that. But I will do it through the second lunar landing attempt, and whether or not we are successful, I will then resign and go back to industry. I figured that if we weren't successful after two attempts, I would be a nervous wreck. I wouldn't trust myself to continue." With that, Maynard was at the helm of the effort to fix the Apollo spacecraft.

The CM underwent the changes it needed to be made safe for astronauts, and the LM was even at the time of the fire running behind schedule. The LM, which many consider the first true spacecraft since it operated only in the vacuum of space, presented many challenges not faced by builders of previous spacecraft. For example, when the ascent stage lifted off from the Moon, would the blast of the ascent-stage engine on the descent stage being left behind on the Moon damage the ascent stage? Trials in elaborate testing rigs at White Sands showed this wouldn't be a problem.

Another problem was weight, and Grumman, with Maynard's blessing, had resorted to some unorthodox means to save it. The LM's fuel tanks were pressurized by supercritical helium, an exotic, difficult and untested technology that represented a major weight reduction from nitrogen, the agent usually used to pressurize fuel tanks and force fuel into the engine. "We had to pull a few risky technologies out of the hat to save weight, and one of them was supercritical helium. We had to do thermodynamic tests at Grumman to convince ourselves that we had a hope. We were able to do that with pretty high confidence in ground tests." The technology involved storing helium under high pressure and at a low temperature. "I had to convince Owen on that one," Tom Kelly of Grumman said about the helium. "When I had him convinced, he took on the job of selling it within NASA." Maynard's toughest selling job was with Max Faget, the top engineer at MSC, who "just about exploded" when he heard at a mockup review that Maynard had agreed to try it. "I talked him into it finally," Maynard said. Supercritical helium wound up working as hoped.

As the first flight of the LM neared, the craft was still overweight. Under prodding from Maynard, Caldwell Johnson and others at NASA, Grumman instituted programs called "scrape" and SWIP (super weight improvement program) to bring the LM down to the prescribed weight. For each pound removed from the LM, Grumman got a bonus. One result was the LM's flimsy exterior covering. Erb remembers the charge from Shea to work hard on weight reduction. After exhortations to do sharper analysis and better testing, he said, "if you can save weight at $20,000 per pound, do it. If it's going to be $40,000 per pound, come and talk to me."

Weight is a problem in most aircraft and spacecraft programs, and the CSM had its own weight loss program. Even the doctors were enlisted in cutting weight. "We were looking for every possible way to take weight off the spacecraft," Coons remembered. "For every ounce you take on the round trip in the command module, you have to have 500 ounces at launch. We were counting aspirins."

•

In the fall of 1967, Apollo was returning to flight status. First up was a launch that is little remembered today but was one of the most daring and critical flights of the Apollo program. Four years earlier, when George Mueller became NASA's associate administrator for manned space flight, he decided that Saturn V would not be tested in the conservative, step-by-step fashion favoured by Wernher von Braun and his group at Huntsville. Instead, the first flight of the Saturn V would be an "all-up" test with all three of its stages live. Only the third stage had flown before, as the second stage of the Saturn IB. And so, on November 9, 1967, at 7:00 a.m., the Saturn V, with an Apollo CSM and a dummy LM as a payload, lifted off Pad 39A at Cape Kennedy. In the wake of the fire, a failure during the flight of Apollo 4 may have spelled the end of the Apollo program. But the Saturn V and the CSM worked perfectly during what was the A flight in Maynard's A to G sequence. The Saturn V shook everything within miles of the pad, and after a workout of rocket and spacecraft, including the all-important service module engine, the Apollo 4 CM splashed down in the Pacific after 8 hours and 37 minutes of flight. Mueller's daring gamble had paid off, and Apollo was back on its feet, but more flights remained before Saturn V and the Apollo spacecraft would be fully proven for use by astronauts.

The Saturn V was one of the most remarkable feats of the Apollo program. Standing 110 m high with the Apollo spacecraft, the Saturn V was a three-stage vehicle that even today remains the most powerful rocket ever built by the United States. The five F-1 engines in the first stage together gave the Saturn V 7.5 million pounds or 35 meganewtons of thrust. The second stage had five F-1 engines, and the third stage used one F-1 that could be restarted to allow Apollo to fly one or two orbits of Earth before its second burn put the spacecraft on course for the Moon. What was then the world's largest building by volume, the Vehicle Assembly Building, was

The first successful launch of a Saturn V carrying the Apollo 4 unmanned capsule.
Courtesy NASA.

required to assemble the Saturn Vs at the Cape, and they were moved to the launch pad atop gigantic crawler vehicles. The Saturn V was the third and last member of the Saturn family of rockets, which was the only major rocket family used exclusively for civilian applications up to that time. The Saturns – two-stage Saturn I and Saturn IB rockets, and the Saturn V - were developed by von Braun's team at the Marshall Space Flight Center, and in contrast to their influence on the Apollo spacecraft, engineers from Avro had only a modest role in developing these behemoths.

Mario Pesando, who had headed up flight test operations at Avro, had gone to RCA near Boston after the Arrow was cancelled. Pesando was systems manager for a year-long program that began in 1961, working with Marshall to develop an operational flight control program (OFC) for the Saturns. The OFC concept was that RCA would establish what functions of the rocket needed to be followed in real time during flight and set out the sequence of commands to give in real time to the rocket if a malfunction occurred and an alternate mission was required. Although many of the technical concepts outlined in RCA's OFC study were picked up by NASA and the contractors that built the Saturns, the follow-on OFC work that RCA had hoped for didn't materialize. Pesando said one reason was that OFC was a solution in search of a problem. Fred Matthews, who left NASA and joined RCA and the OFC study in 1962 after it had started, said Huntsville had hoped to control the Saturns through their flights. But when NASA decided that full control of the flights would be handed to mission control in Houston as soon as the rocket cleared the launch tower, Huntsville's control role was diminished, and the OFC work dried up.

Matthews also worked on another RCA contract involving Saturn, analyzing data coming from the rocket in flight. The Saturns were the first rockets to send back data in digital form, and it came back in such a large quantity that it was difficult to analyze in a short period of time. Matthews said he and his group developed a computerized approach by which NASA could quickly identify where the Saturns had departed from expected performance. One area where this helped out was in determining fuel levels, because sloshing fuel sometimes led to confusing readings. Matthews' approach was used by NASA and the Saturn contractors to follow the performance of their rockets during flights through the life of the Saturn program.

A senior engineer from Avro who went to Boeing Co. also contributed to the Saturn. Carl Lindow, a British Columbia native who was engineering project manager at Avro in the last year of the Arrow program, went to Boeing in Seattle, Washington, after leaving Avro in 1961. The next year he became development program manager for launch systems at the aerospace giant, and he held senior posts involving launch and propulsion systems for five years, until he moved back into aircraft. His work included proposals and contracts involving many different types of launch vehicles, including rockets fuelled by solid, liquid and nuclear fuels. Lindow was heavily involved in Boeing's unsuccessful bids for the lunar module and for the first stages of the Saturn I and IB rockets. Boeing did succeed in landing the contract for the gigantic first stage of the Saturn V. The first stage required many advances in fabrication technology to build, but it was developed with relatively few problems, especially when compared to the other two stages, which both used liquid hydrogen as a fuel, something more powerful but trickier to use than the kerosene fuel in the first stage.

•

On April 4, 1968, the second Saturn V blasted off in what was supposed to be a repeat of the previous November's success with Apollo 4. But Apollo 6 departed from NASA's script almost from the moment of liftoff. During the first-stage burn, the rocket suffered from severe pogo. As well, the exterior of one of the four panels in the adapter that covered the LM test article and connected the CSM to the rocket fell out early in the flight. Shortly after the ignition of the second stage, two of its five engines stopped firing. As a result, the second stage did not give Apollo 6 the planned velocity, and the third stage had to fire longer than planned. Two orbits later, when mission control ordered the third stage to restart to boost the spacecraft into a higher orbit in a simulation of translunar injections for later missions, it did not fire, and the Apollo CSM separated from the stage. The service module's engine was used to boost the CSM into a higher orbit, and Apollo 6 splashed down in the Pacific after 10 hours of flight. Although Americans heard little about Apollo 6's troubles because civil rights leader Martin Luther King was gunned down that day in Memphis, Tennessee, the Apollo team at the Cape, Houston and Huntsville had a serious set of problems to fix before Apollo could go on. There were less than 21 months left before the deadline Kennedy had set.

Studies of telemetry and tests with Saturn hardware led to fixes for all the problems. The shaking of the rocket known as pogo was a familiar problem to Gemini veterans, but Saturn V was supposed to be immune to this problem because it used different fuels. A cure was found by charging the first-stage engines with helium gas to allow a smooth flow of fuel to the engines. A redesign of the liquid hydrogen fuel lines in the second- and third-stage engines overcame a problem that didn't show up in ground tests but appeared on Apollo 6 in the vacuum of space.

Jim Chamberlin tackled the problem of why the adapter panel had blown out. Five years before, when he had been removed as Gemini program manager, Chamberlin was assigned to be senior engineering advisor to MSC director Bob Gilruth, and later was manager for design and analysis. In effect, Chamberlin was Gilruth's troubleshooter for Apollo. He was appointed to review boards for various mockups and spacecraft, looking for defects or changes that needed to be made. Among other things, he was appointed in 1964 as chair of a space suit design review board that reviewed and approved space suits for Gemini and Apollo. The board's other members included Maynard and astronauts Jim Lovell and Michael Collins. Chamberlin was also involved in looking at particular problems, such as how well the CM would fare when it splashed down or if it struck land. As the LM became a matter of concern, he became a regular visitor to Grumman's plant in Bethpage, New York, providing advice to Grumman and NASA on problems such as whether the LM's docking equipment could stand the strain of docking, the thermal blankets' level of protection against thruster firings, the possibility that the tailoff of the descent engine after lunar landing might damage the LM, and whether the thermal stresses of flight could cause LM wiring to fail.

"As I recall, Jim Chamberlin got involved in looking at what we were coming up with [to reduce the LM's weight] and kind of passing second opinion judgments to NASA on whether these changes were acceptable or not," Kelly said. "I also think that he got involved in the combustion instability problems on the LM ascent engine. And that was a long, tortuous decision where we ultimately replaced the Bell rocket engine with a Rocketdyne engine, and it took a lot of testing and analysis to finally reach that conclusion, and the final decision on that was made by George Low in NASA. But I think Jim Chamberlin was pretty heavily involved in that."

Following Apollo 6, the first reaction of Apollo's engineers to the problem with the adapter panel was relief that the entire panel had not failed, which would have meant the total loss of Apollo 6. Then they assumed that the panel's covering had shaken loose as a result of the first-stage pogo. Using what Apollo manager George Low described as "back of the envelope" calculations, Chamberlin came up with the "first consistent story" to explain the panel's problem. The honeycomb structure of the panel had absorbed moisture in the wet and humid Florida environment, and when the panel heated up during launch, the moisture caused the panel covering to blow out. After some discussion, technicians fixed the problem by applying a layer of cork to the adapter panels to absorb the moisture. As well, they drilled holes in the panels to allow pressure inside the adapter to escape. Although these fixes took months to perfect, by the summer of 1968 Apollo management began to think about flying the next Saturn V with a crew on board.

Between the two Saturn V tests, the first lunar module, LM-1, flew January 22, 1968, on what was known as the B mission or Apollo 5. LM-1 rode into Earth orbit atop the Saturn IB that had been slated to launch Apollo 1. Eugene Kranz acted as primary flight director for Apollo 5, and John Hodge ran the second shift. Due to a programming error, the LM's descent engine shut down early on its first firing, but the LM's descent and ascent engines worked properly on subsequent burns, including a firing of the two parts of the separated LM, and the modified mission ended after nine hours of flight. Mindful of the highly rigorous checkout all spacecraft went through at Cape Kennedy before launch, Maynard himself had what at first appeared to be a modest goal for Apollo 5. "I said to myself and all the guys in my division that my objectives for that LM is for Grumman to get the thing built at Grumman, take it to the Cape, get it put on the launch vehicle, get it launched, and if it falls in the ocean, it's 100 per cent successful. I don't care what happens after it leaves the ground. As long as we get it through that maze of stuff at the Cape. It was the first time Grumman had done anything like that."

Apollo management decided that the next LM would fly with astronauts in it. But there were still development problems with the LM ascent engine, and a series of other problems that were slowing production of LM-2. The first manned Saturn V flight was supposed to fly before the end of 1968 with both a CSM and a LM on board, but the LM's problems were becoming the biggest threat to meeting that schedule and the lunar landing goal.

Meanwhile, North American, now North American Rockwell, had delivered a trouble-free Block II CSM to the Cape, and on October 11 it blasted off atop another Saturn IB with the crew of Apollo 7 on board. Mercury and Gemini veteran Wally Schirra led rookies Donn Eisele and Walter Cunningham on an 11-day flight that proved that the CSM was ready to fly to the Moon. Fighting off colds, the astronauts tested all CSM systems, rendezvoused with the second stage of their Saturn IB, and broadcast the first television shows from an American spacecraft. The C mission in Maynard's sequence had now been flown.

Even before Apollo 7 flew, it was becoming clear that the wait for a second LM was threatening Apollo's tight schedule for the D mission with the CSM and LM. And during that time, intelligence information came in showing that Apollo was racing with more than the end-of-the-decade deadline set by President Kennedy.

•

Reconnaissance satellite photos of the Soviet launching center at Baikonur showed construction of two huge launch pads for a rocket that could only be used to compete with the Saturn V. In the fall of 1967, a test model of the gigantic rocket, which the Soviets called N-1, showed up in the reconnaissance photos. Furthermore, in 1968 the Soviets had launched a robot spacecraft named Zond atop a Proton rocket that appeared to be capable of carrying cosmonauts around the Moon. This knowledge, which was not made public at the time, led the leadership of NASA to take one of its most dramatic decisions ever.

To the general public in 1968, it appeared as if the race to the Moon had never really been engaged. In human space flight, the U.S. had taken a clear lead over the Soviets during the Gemini program. During those two years, no cosmonauts flew. Only Soviet robotic probes, such as Luna 9 in 1966, which sent the first photos from the Moon's surface, and Luna 10, which was the first spacecraft to orbit the Moon, kept up the appearance of a Moon race. Then in April 1967, Cosmonaut Vladimir Komarov flew into space aboard a new spacecraft called Soyuz, and rumours abounded that he would dock with another Soyuz to be launched the next day in a Soviet space spectacular. But from the moment he entered orbit and one of the Soyuz solar panels refused to unfurl, Komarov was in trouble, and the second launch was cancelled. When Komarov returned to Earth a day later, his parachutes tangled, failing to slow his descent, and he died on impact. Soviet cosmonauts were grounded until the fall of 1968.

It was only after the Cold War ended two decades later and Soviet secrecy was lifted that the full story of the Soviet side of the race to the Moon came out. Sergei Korolev, the chief designer of the Soviet space program, was planning flights to the Moon by cosmonauts as far back as the time of Sputnik. In the early 1960s he began work on the N-1 rocket. But the N-1 was hampered by a lack of interest from the Soviet leadership, who believed that Kennedy's lunar landing goal was little more than a bluff. The politburo placed a higher priority on closing the U.S. lead in the nuclear arms race. Even when Korolev managed to convince the Soviet leadership in 1964 that the Americans were serious about the Moon, they never agreed to spend enough money to properly test the N-1 before it was launched.

As well, the lunar landing effort suffered from rivalries among the leaders of competing design bureaus. After Korolev battled with the top engine designer, Valentin Glushko, over propellants for the N-1, Glushko refused to have anything more to do with the rocket. The result was that the N-1's first stage was powered by 30 engines that were much smaller than the mighty F-1s of the Saturn V. Korolev also feuded for control of the lunar program with Vladimir Chelomei, whose bureau designed the Proton rocket. Until Nikita Khrushchev was deposed as Soviet leader in 1964, Chelomei enjoyed an advantage in the infighting because his bureau employed Khrushchev's son Sergei. Chelomei's design bureau started its own circumlunar program, which would use Protons to boost cosmonauts around the Moon, but Korolev wrested control of the program from Chelomei in 1965.

In January 1966, Korolev died during surgery, and the leadership of his bureau fell to Vasily Mishin, who did not have Korolev's influence or management abilities. Under Mishin, the Soviets continued with the landing program and with the circumlunar program, which involved a stripped-down and modified Soyuz spacecraft launched by a Proton rocket. The Zond launches of 1968, some of which carried biological specimens, were test flights in the circumlunar program.

Although in 1968 the Soviet chances of landing on the Moon before the Americans were dimming, they continued their landing program. Korolev had decided to use lunar orbit rendezvous to reach the lunar surface. And although on paper the N-1 was the match of the Saturn V, the American rockets and spacecraft were vastly superior in many ways. The Soviets planned for two cosmonauts to blast off aboard a mother ship atop the N-1. After coasting to lunar orbit, one of the cosmonauts would space walk to the lunar cabin attached to the mother ship. The cabin would separate with a rocket stage, and the stage would slow the cabin until it separated 2 km above the lunar surface. The cosmonaut would only have seconds to select his landing site. If the landing were successful, the cosmonaut would be able to walk alone on the surface using a space suit equipped with a device resembling a hula hoop that allowed the cosmonaut to right himself if he fell on his back. After the lunar walk, the cosmonaut would re-enter the lunar cabin, blast off and dock with the mother ship, space walk to rejoin his colleague, and then return to Earth in the mother ship, which was also based on the Soyuz. While similar to Apollo, this mission would have been far more hazardous.

There was one more element to the Soviet competition for the Moon. They were also developing a robotic spacecraft that would land on the Moon, drill for a sample of lunar soil, and then seal the sample in a return capsule that would return to Earth.

The full dimensions of these programs didn't become known for decades due to secrecy in both the Soviet Union and the United States. James Webb, the NASA administrator, told congressional leaders in secret about the N-1, which acquired the nickname "Webb's Giant." The secrecy surrounding the Soviet lunar program contributed to a lingering sense in the years that followed that there had been no Moon race and that Apollo had been unnecessary in terms of fighting the Cold War. As early as 1968, NASA's future was in question not only due to the Apollo fire, but also because of growing civil strife in America and escalating disquiet over the war in Vietnam. NASA's funding was falling and no program had yet been set up to follow Apollo.

•

In August 1968, Apollo program manager George Low was faced with the intelligence reports on the Soviet lunar plans, the return to good form of the Saturn V and the CSM, and the prospect of further delays in the LM. He proposed that instead of flying a repeat of Apollo 7 or waiting months for a LM, Apollo 8 should orbit the Moon before the end of the year. He soon got agreement from many of his colleagues, although Webb initially withheld his approval until Apollo 7 was successful. Nevertheless, Webb gave Low the green light to begin planning a lunar orbit flight for Apollo 8. The crew training for the first CSM-LM flight opted to remain in training for that mission, so the next crew in line, Frank Borman, Jim Lovell and William Anders, began training for a lunar orbit flight on Apollo 8.

Maynard and others were concerned that without the lunar module, the astronauts' lives would be dependent on only one spacecraft a quarter of a million miles from Earth. "We always wanted that [lunar] module there. Since it was not a requirement, we never thought about flying such a mission [to the Moon without the LM]. We went back and said, well, can it do it? And we said, yeah, we can do it." The new plan for Apollo 8 didn't have a spot in Maynard's A-to-G sequence, so it was called the C-prime mission.

"I was not privy to that kind of classified information as to what the Soviets were or were not up to," Maynard said. "From Gilruth's point of view, I was not permitted to even dream about it: 'Do not make any decisions based on being in a race with anybody. We hired you to engineer this thing. We want the near-optimum engineering solutions to whatever the mission is. Don't go busting your tail to save money. Don't go busting your tail to risk the program in any way - engineering or otherwise - to do something earlier. That's not your business, Owen Maynard.' We redid all the documents that established all of the test objectives, the flight plan, mission rules, ran the simulations, supported the flight controllers in the SPAN room. We put on our overalls, went to work and made it happen. We didn't badmouth it. George Low would go by jogging, and he'd say, 'You see that Moon? In three months, we're going to be there. Isn't that something, that we're going there?'"

For flights to and near the Moon, Apollo's communications and tracking network had to be vastly changed from the Gemini network. Both George Harris and Tecwyn Roberts, who were now at the Goddard Space Flight Center, had been preparing for the changes that Apollo would bring. Not all the Mercury and Gemini stations were needed, but tracking ships and aircraft were required to provide communications coverage for Apollo's launch into Earth orbit, its injection into a path to the Moon, and its return to Earth. Far out in space and near the Moon, Apollo was tracked by three 26 m, unified S-band dishes at Goldstone California; Canberra, Australia; and Madrid, Spain. Harris said he had to do a great deal of work designing smooth "handovers" of communications between the three deep-space dishes. "When we developed the unified S-band system and built NASA's fleet of tracking ships which were deployed on two oceans, we had the tracking capabilities needed to send the astronauts to the Moon and return them safely to Earth," Roberts said.

By the time Apollo flew, communications satellites were in place over the Atlantic and Pacific to permit the transmission of high-speed data to the control center from any tracking station, Roberts said. "Between Gemini and Apollo the greatest switch in emphasis was not so far as the trajectory information was concerned, but in the systems information, in the amount of telemetry data and the method by which it was handled." Indeed, so much data was coming in that the pertinent information needed to be extracted from the stream for the use of controllers, he explained. Because the Houston climate was hard on his health, Roberts had moved to Goddard during Gemini. Had Roberts remained in Houston, Kraft planned to make him a flight director. At Goddard Roberts eventually became director of networks, and because Goddard was the nerve center for virtually all incoming data from NASA spacecraft, he remained at the heart of communications for Apollo and the programs that followed it.

Len Packham worked on the spacecraft end of radar and communications systems. His work involved design and testing of radar equipment, communications antennas, television cameras, and even the headsets that the astronauts wore. Packham, a native of Saskatoon who took mechanical and electrical engineering at the University of Saskatchewan, served as an aircraft mechanic in the RCAF during the war before he joined Avro. An avid golfer and father of one son, Packham and his group faced the challenges of developing equipment that could communicate at lunar distances. With help from Packham's group, Apollo 7 was the first U.S. spacecraft to broadcast live television from space. Black-and-white cameras were used for early Apollo missions, before giving way to colour TV on Apollo 10. Packham was also heavily involved in developing the radar systems on board the lunar module.

After the success of Apollo 7, the C mission, NASA's new acting administrator, Thomas Paine, agreed in November to fly Apollo 8 as the C-prime lunar orbit mission and made the announcement. In the meantime, the Soviets were busy.

In September, Zond 5 returned to Earth intact after a loop around the Moon, although a gentler re-entry path would have been required if cosmonauts had been on board. In October, the Soviets proved their Soyuz spacecraft in a new manned mission, and the day before Paine signed off on Apollo 8, Zond 6 was launched. Zond 6 flew around the Moon but crashed in the Soviet Union when a parachute deployed prematurely. The Russians were able to salvage photographic film, including a photo of the Earth above the lunar horizon, allowing them to cover up the failure. The photos fuelled rumours that the Soviets would attempt to send cosmonauts around

Leonard E. Packham (left) receives award from James C. McLane Jr. in 1969.
Courtesy NASA.

the Moon before Apollo 8's scheduled launch on December 21. But the Soviet mission was off due to the crash of Zond 6.

•

Most of the people who worked to make Apollo 8 a success didn't concern themselves with the Russians. The question on their mind was, could Apollo 8 be flown safely? Rod Rose, a member of the Avro group who made one of the most important contributions to Apollo, remembers how his work on Apollo 8 began. "One day Chris Kraft stuck his head in the door and said, 'What do you think of a C-prime in lunar orbit?' I said, 'Well, let us take a look at a few things.' We looked at a few things and said, 'Yeah, it sounds okay.' So he said, 'Well, I'd like to have a mission plan.' So I got all the team together and we started thrashing it out. Not from scratch, but assembling the pieces. We'd been doing Apollo 11 flights in mission planning by that time. An awful lot of the pieces were basically there. It was a matter of making sure they all played together. In those days, we weren't involved in any of the politics. As far as we're concerned, it was a technical challenge."

Rose was in the flight operations directorate, working under Chris Kraft, planning missions. At the time of Apollo 8, Rose was 41, a native of Huntingdon, England. A bespectacled man whose friendly face and soft voice belie a set of firm opinions and a strong ability to argue them, Rose once described his work on Apollo as going to thousands of meetings. But speaking about those meetings, he said, "I was never a head nodder. If it needed to be pointed out, I'd point it out."

Kraft said Rose was one of the more contentious people at MSC. "I saw Rod Rose was capable. I put him on my staff, and whenever I had a problem, whenever I wanted an answer, whenever I wanted anything done right, I gave it to Rod Rose. That impact made people dislike him even more, because he was speaking for me, and I wanted him speaking for me. But he and

Rod Rose in 1960 at Langley Research Center. At the time, Rose worked as project engineer on the Little Joe program for Mercury. Courtesy NASA.

I got along just like that. He knew where to draw the line with me. He saw where I was going and he would back off. He wouldn't take me on. Sometimes that was good, sometimes that was bad. I knew how to use him."

As a student in wartime and postwar England, Rose had attended the Manchester College of Technology while serving a five-year apprenticeship at A.V. Roe in Manchester. "You went to every department in the shops, research, flight sheds, and drawing office, the materials research, structures," Rose said. "It was hard work, but I think I went to a total of some 27 departments. Anywhere from machine shop detail fitting to assembling, pipe bending, toolroom, heat treatment and press shop, forging, and then on to the A.V. Roe airfield at Woodford, where I did flight controls, landing gear, engines, fuel systems, things like that, flight testing. Came back, did some time at the drawing office, stress office, materials research, and then on to aerodynamics."

Upon graduating from the college and completing his apprenticeship, Rose won a scholarship to the newly established College of Aeronautics at Cranfield, which he attended from 1949 to 1951. He then went to Supermarine, where he worked on aircraft performance, loads and engine performance. By 1957, Rose concluded that he had gone as far as he could at Supermarine, and he was frustrated by a conservatism in the British aircraft industry that restricted who could supply vital parts and who could do what. After reading in the paper about Avro Canada, Rose crossed the Atlantic with his wife Leila and their two sons, the youngest then 11 months old, aboard the SS *Homeric*. During his 23 months at Avro, he worked as an aerodynamicist on the CF-100 and the Arrow, and also in flight test and in advanced design. Five years into his career at NASA, he moved to flight operations for Gemini and quickly became a key member of Kraft's team.

Much of Rose's work in Mercury and the early Gemini program was as a project engineer, and he continued this work in Apollo, representing the Flight Operations Directorate during many trips to North American's plant at Downey, California. "Walt Williams pressed this idea for a long, long time, that you need to get the operators of the equipment, not just the client, but the operators, involved in the design process right from the word go. If you can get the subsystem designers, the engineers, the flight operators, and get that all integrated right at the beginning, you can save yourself a lot of pain."

In Gemini, Rose and his flight operations colleagues began to put together what became known as flight operations plans or FOPs. These were detailed plans of everything that was to go on during a mission and when it would take place. By the time Apollo began to fly, the FOPs were key documents used to set out how the missions would be carried off.

"Basically, I conceived the idea that in order to get a successful mission from the operations plan, all the players needed to be aware of what everyone else was doing, and making sure that everyone else was playing from the same sheet of music. So to that end, I said I'll have a set of meetings where a representative from every organization that's affected is going to be there with the authority to commit their organization. And we're going to review the whole mission and all its support and everything else to make sure what we're doing is right, and if it isn't, let's make it right or is there a better way of doing it? As an example, at first the medical doctors would complain like heck about coming to these meetings. They couldn't see any good coming out of this. It was a waste of time. When the doctors came and they heard what was going on in the meeting, they said, 'Hey, this affects us. You shouldn't be doing this and this and this.' I said, 'Hey guys, this is why you're here. That's exactly what these meetings are all about. It's a means of exposing all our plans to everybody involved to make sure they all square up.' And from then on, folks were regular attenders."

Rose chaired these meetings, which in the time of Apollo would usually draw 60 people, including the flight crew or their backups, to put together the FOPs. FOP meetings for individual flights took place monthly during the period leading up to each flight, so when flight schedules were heavy, there would be several FOP meetings each month to deal with all the upcoming flights.

"We stood back and took a broader picture. Here are the mission objectives, here's the mission we've come up with. Now let's start filling in the squares, looking at how all this was going to tie together. What is the network going to do, aircraft and ships and ground stations. What's the communications, what's the radiation, what's the training for the ground and the crew, what data is required when? The earlier operations plans covered everything. We covered where the ships were going to be, what the aircraft schedule was going to be, who communicated when, the timeline on board. When they were going to sleep and when they weren't and what they were going to do when they're awake. It really took the whole mission."

Sometimes big engineering issues would raise their heads at FOP meetings. For example, there was the question of what happened to the adapter panels on the Saturn V third stage that protected the LM when the CSM separated from the stage. Originally the panels were to remain connected but would open up to allow the CSM to pluck the LM from the Saturn. Rose said many people in FOP meetings questioned whether it would be that easy to extract the LM with the panels still attached. Finally, after a panel didn't open properly on Apollo 7, Apollo managers decided that the panels would be cast off as soon as the CSM separated from the Saturn stage, ensuring that the LM could be pulled out without difficulty.

After the FOP was drawn up, the mission controllers took the framework from the FOPs and worked on minute-by-minute operations, especially as they affected navigation and flight control. Kraft gives Rose and Howard (Bill) Tindall, another top engineer at Houston, credit for pulling together flight plans for the Apollo missions.

•

For Apollo 8, Rose said the original idea from the program office was to fly three-quarters of the way out to the Moon, or perhaps do a loop around the Moon, but not enter lunar orbit. The decision to enter lunar orbit on Apollo 8 hinged on the service propulsion system (SPS), the big engine on the service module that would put the spacecraft in lunar orbit and bring it out again. "The argument with that was that you were three days away from home anyway. And you better have confidence in the SPS, because without it, the whole program is built on paper. If you're going to go that far, why not go whole hog and go around the Moon and into lunar orbit?"

When NASA decided that Apollo 8 would go to the Moon, Rose and his panel were able to assemble a plan in short order. "A lot of people said, 'Hell, how did you get a mission plan together so quickly?' They had to realize that we had been doing all the flight planning, navigation, etc., etc., for going to the Moon for some time. In fact, Morris Jenkins and his people had all the equations and software perfected early in '68. In fact in '67. In order to get a mission plan for Apollo 8, basically what you had to do was get the liftoff time and targeting into a set of software that already existed. And that's how we were able to put together a C-prime flight operations plan so quickly. Because all the ground work had already been done."

Owen Maynard was also hard at work making sure that the spacecraft itself was ready to go. Maynard was involved in a confrontation during Apollo 8's countdown demonstration test on December 11, 10 days before the scheduled liftoff. The atmosphere was tense because Apollo 8 was the first Saturn V flight after the failure of Apollo 6, and the first with men on board. A communications problem caused a hold at T-minus six minutes, and mission commander Frank Borman expressed his exasperation with the hold as the problem was being solved. Then one of the batteries powering the command module began to fail, and Maynard, who was in the SPAN room off the mission control room at Houston, came on the line and called for a switch in batteries. The countdown resumed without the battery being changed, and Maynard again went on the line. Under brusque questioning from the test conductor, Maynard reported that some of his engineers were concerned that the low battery could damage the spacecraft and be a "crew hazard" due to possible outgassing of hydrogen and oxygen that could lead to a fire. The test conductor ordered another hold at T-minus 3 minutes 14 seconds, shortly before the automated final phase of the countdown was due to start. For the first and only time, Maynard got a countdown halted. "I wonder why we ended up with such a bad battery?" an annoyed Borman asked. The astronauts were having trouble reaching the control panel that would allow them to change the batteries. After some argument about the batteries, the countdown was picked up, the batteries were powered down, and the countdown test completed. Even Apollo program manager George Low got involved in the argument, which was highly unusual. When the real countdown started a few days later, the spacecraft was ready with a new battery. "If I ever would have resigned over anything, I would have resigned about that situation if I hadn't made a commitment," said Maynard, who later had a frank discussion with Low about the dangers of the situation. The emotions he felt in this incident remained vivid: Thirty years later, Maynard had a heart attack while listening to a tape of the countdown demonstration test.

Apollo 8, with astronauts Borman, Anders and Lovell aboard, blasted off on time at 7:51

The crew of Apollo 8: (l to r) William Anders, James Lovell, Frank Borman
Courtesy NASA.

a.m. from pad 39A at Cape Kennedy on Saturday, December 21, 1968.Within minutes it was clear that the first big gamble of Apollo 8 had been won: the first two stages of the Saturn V worked perfectly, and the third stage put Apollo 8 in orbit. On the second orbit, the third stage lit up a second time for the burn that put Apollo 8 in the history books. This translunar injection burn sent Apollo 8 on its way to the Moon, the first time humans had moved into space beyond low Earth orbit. Apollo 8 separated from the Saturn stage, and for the next three days it flew out toward the Moon. During that flight, the astronauts looked through their windows to see the Earth recede to the point where it could be covered by the tip of a thumb. The crew shared this view with Earthbound viewers during black-and-white TV broadcasts from on board the spacecraft.

On the morning of Christmas Eve, Apollo 8 was given a "go" for lunar orbit and passed into radio silence behind the Moon, where the SPS had to fire nearly perfectly to ensure that the crew of Apollo 8 did not crash into the Moon or fly into a path that would take them away from Earth. If the engine failed to fire at all, Apollo 8 would simply loop around the Moon and head back to Earth. When Apollo 8 came around the side of the Moon on its first orbit, a relieved mission control learned that the engine worked as planned.

After another SPS burn to change Apollo 8's orbit at the start of the third orbit, Frank Borman called to mission control and asked for Rod Rose.

Rose was expecting the call, because he and Borman had been working on a special task. The men were friends and neighbours in El Lago, the subdivision near the Manned Spacecraft Center. At the local Episcopal Church, St. Christopher's, both were members of the vestry, the group of lay advisors who helped run the church. Two Sundays before Apollo 8 blasted off, Frank Borman found out that he was on the duty list as a lay reader for the Christmas Eve communion at St. Christopher's. Borman, knowing that he would be circling the Moon at that time, arranged with the minister, Jim Buckner, to deliver his reading from lunar orbit for recording and later playback at the service. Rose selected for Borman's reading the "Prayer For Vision, Faith and Work" by G.F. Weld from *Prayers for the Church Service League*, published by the Diocese of Massachusetts. Rose gave a copy of the prayer to Borman. "We decided to call it experiment P 1, and Frank agreed to give me one lunar [orbit] notice before he read the prayer so all the recording could be finalized."

A few minutes after Borman's call to Rose, CAPCOM Mike Collins said, "Rod Rose is sitting up in the viewing room. He can hear what you say."

"I wonder if he is ready for experiment P1?" Borman asked.

"He says thumbs up on P1," Collins replied.

"Rod and I got together and I was going to record – say a little prayer for our church service tonight," Borman said.

When Collins gave the go-ahead, Borman said, "Okay. This is to Rod Rose and people at St. Christopher's, actually to people everywhere.

> Give us, O God, the vision which can see Thy love in the world in spite of human failure. Give us the faith, the trust, the goodness in spite of all of our ignorance and weakness. Give us the knowledge that we may continue to pray with understanding hearts, and show us what each of us can do to set forth the day of universal peace. Amen.

"Amen," Collins replied.

"I was supposed to lay read tonight and I couldn't quite make it," Borman concluded.

"Roger, I think they understand," Collins said.

Earthrise as seen from Apollo 8, December 1968.
Courtesy NASA.

Although Gordon Cooper recorded a prayer on his onboard equipment during his Mercury flight, this was the first prayer broadcast from space. That evening, the crew of Apollo 8 sent their second television broadcast from lunar orbit, showing views of the Earth and the lunar surface. As the broadcast neared its end, first Anders, then Lovell, and finally Borman read the first 10 verses of Genesis as the gray, battered surface at lunar sunset appeared on the television picture. "And from the crew of Apollo 8, we close with good night, good luck, a Merry Christmas, and God bless all of you – all of you on the good Earth," Borman said, concluding both the reading and the broadcast.

Rose went to St. Christopher's with tapes of both the prayer that Borman had recorded that morning and the Genesis reading, and both were played at the service. While the service was going on, Apollo 8 went behind the Moon again at the end of its 10th orbit, and the SPS fired its critical burn to put Apollo 8 on its way back to what Borman had memorably called the good Earth. Rose had arranged to get a phone call at the church when Jim Lovell confirmed that Apollo 8 was homeward bound. "The timing was beautiful, because we could give that to the minister just as he was giving his final dismissal to the congregation," Rose said. "That was a great way to start Christmas Day."

The prayer became the subject of an unsuccessful lawsuit brought by Madalyn Murray O'Hair, the well-known atheist, who disputed the use of government facilities for a religious observance. "Subsequently, we had many letters about the prayer. We had a letter from a Reverend Roth in New York City, who was a Catholic priest who also was a musician. He wanted permission to make an anthem out of the prayer. We told him as far as we were concerned, that was great. Lo and behold, he did it, and it was called the prayer from outer space, and he dedicated it to St. Christopher's Episcopal Church. For the next several years at Christmas service, the St. Christopher choir sang the anthem from outer space."

The crew of Apollo 8 spent most of Christmas Day and the next day resting after the work they had done during their 10 circuits of the Moon. They were bringing home many photos of the lunar surface, including shots of potential landing sites for Apollo and of the Earth rising over the lunar surface.

On December 27, Apollo 8 returned to the vicinity of the Earth and the last of its big tests. Bryan Erb, who was winding up his work on the Apollo heatshield, waited anxiously for the re-entry, the fastest ever. "When we brought in Apollo 8, it was the first time we had done 36,000 feet per second [40,000 km/hr]. With a crew on board. That was the real test as far as I was concerned." Erb remembered a conversation months before with his division chief, Joe Kotanchik. "He said to me, 'Bryan, when the first Apollo mission is on the way back from the moon, and when they come out of lunar orbit and you know they are committed to entering the earth's atmosphere at 36,000 feet per second in three days' time, are you going to sleep well at night?' I thought briefly and said, yeah, I will, and I did." The heatshield, the parachutes and the rest of the recovery systems worked perfectly, and Apollo 8 and its crew were picked up in the Pacific by the crew of the aircraft carrier USS *Yorktown*.

For many people both inside and outside the space program, Apollo 8 represented the climax of the Apollo program, not the lunar landing that came later. For the first time, human beings had moved out from the Earth and into the influence of another celestial body. Although the Earthrise photos and other pictures of the Earth weren't part of the flight plan, they became one of the greatest legacies of Apollo's flights to the Moon. The photos gave all humans a new perspective on the true dimensions of the fragile planet we live on. Many believe that Apollo's Earth photos contributed to the emergence of the environmental movement in 1970.

Apollo 8 also came at the end of a year of great strife and bloodshed in the world. Some of the events included growing anger over the war in Vietnam, the Soviet invasion of Czechoslovakia, tension in the Middle East, increasing separatism in Canada, student revolts in many countries, racial problems in the U.S., the assassinations of Martin Luther King and Robert F. Kennedy, the end of the Johnson administration and the election of a Republican administration under Richard Nixon. In a vivid counterpoint, the views of Earth floating in the darkness and the reading of biblical passages that hold deep meaning for many of the world's religions constituted some of Apollo's greatest moments.

For NASA, the gamble it took with Apollo 8 had paid off.

"Somebody asked me if we could fly the equivalent of Apollo 8 today, and I said 'I doubt it, because we'd have 10 committees that would have to approve it before we could do it. We'd never get it done.'" Rod Rose said. "When you look back on it, it was an extraordinarily gutsy decision. And it came off, fortunately."

Flight Director John D. Hodge in 1968.
Courtesy NASA.

10 Reaching the Goal

As 1969 began, the Saturn V and the command and service modules were proven vehicles, and the lunar module was a few weeks away from launch on its first shakedown flight with astronauts. But another craft was causing headaches at the Manned Spacecraft Center, and it wasn't even going into space. It was the lunar landing training vehicle (LLTV), one of the strangest flying machines ever built. Designed to allow astronauts to simulate landing on the Moon, the LLTV was a mass of pipes, struts, engines and tanks perched atop four legs. The pilot sat in a cabin at the front with a simulated LM window. Similar to "flying bedstead" aircraft built in the U.K. in the 1950s to test vertical takeoff and landing, the LLTV had a jet engine pointing downward that lifted five-sixths of the vehicle, simulating lunar gravity, which is one-sixth of Earth's gravity. The remaining sixth of the vehicle's mass was lifted by two hydrogen peroxide rockets, and 16 smaller hydrogen peroxide rockets were mounted in quads to simulate the LM thrusters. The rockets chuffed and puffed during the flight of the LLTV, adding to the spectacle. Bell Aerosystems built five of the vehicles. The first two were called lunar landing research vehicles (LLRV) because they tested the whole concept. Although a similar vehicle was tethered in a gigantic test stand at Langley Research Center, many people at NASA believed that the Langley facility was inadequate and that the LLTV was necessary to give the astronauts the experience they needed to properly fly the LM.

One of the people who didn't like the LLTV was Owen Maynard. "It was a very difficult machine. I don't think it killed anybody, but it scared the hell out of a lot of people." Maynard was up against the feelings of those who knew how different flying in one-sixth G was from flying on Earth. "When you pitch [the LM] over at one-sixth G, you get a lot less horizontal acceleration [than on Earth]. You have to get used to that. You need a lot of experience in one-sixth G, but the few seconds you get in the lunar landing training vehicle didn't seem to be worth it."

The lunar landing test vehicle in flight.

Courtesy NASA.

One of the two LLRVs was brought to Houston in early 1968, where it was flown by astronauts Neil Armstrong and Pete Conrad, who knew by then that they would likely be commanding lunar landing missions. On May 6, 1968, Armstrong was flying his 21st flight in the ungainly craft when it pitched over and resisted his attempts to right it. Armstrong ejected and parachuted safely as the vehicle crashed. The cause – a problem with the helium used to pressurize the rocket tanks – was found, but the next such craft, the first LLTV, crashed in December 1968. The test pilot on board successfully ejected. This crash was linked to wind conditions, but the LLTV program was in crisis with Kennedy's deadline a year away.

Another of the ex-Avro engineers was brought in to work on the LLTV - Peter Armitage, who was then deputy chief of the landing and recovery division at MSC. Armitage had already chaired a committee to investigate the ground flight control procedures for the LLTV. "This was an 'out of the blue' assignment to make recommendations to prevent any more flight accidents. My committee submitted its report, we were commended for our work and our recommendations were incorporated in the training program.

"After this job I was back in the landing and recovery work when yet another 'out of the blue' requirement came up. I was asked to report to Bob Gilruth's office, and for what reason I had no idea. Bob asked me to sit down and said, 'Peter, you were a flight test engineer at Avro.' 'Yes, but that was some years ago,' I said. It was not a time for excuses, I soon realized, as Gilruth told me that I was to take over the management of the LLTV program and immediately report to Deke Slayton." Slayton, the grounded Mercury astronaut, was then director of flight crew operations. "His LLTV group, however, were not too thrilled to have the guy from landing and recovery come in to direct their program. There was not much time, the LLTV had to be put through a whole new envelope of flight certification tests, a step-by-step flight test program to show that the aerodynamic capabilities of the vehicle were sound. We quickly pulled together a test plan, presented it to the management committee chaired by George Low. All of the top MSC

Peter J. Armitage in 1969.
Courtesy NASA.

engineering and operations management were at the meeting. After going through my briefing, showing the complete LLTV flight test envelope and the decision structure I had put together, George asked the key question. 'If you get a test result that is not within the planned flight envelope, what will you do?' Everyone looked at me. I said, 'I will inform Deke of the problem and ask that he reconvene this management committee and let you decide what to do next.' George said, 'That's the answer I wanted to hear.'

"I went to NASA headquarters in Washington to brief the administrator for manned spaceflight, and our flight test program to qualify the LLTV program was approved. The tests went fine over the next several weeks, and the only remaining LLTV was cleared to resume Neil Armstrong's lunar landing training. Again, I went back to my job in landing and recovery. Being used in a troubleshooting capacity was demanding but very satisfying."

Armitage was born in 1929 in the industrial city of Leeds in Yorkshire, the son of a tool-and-die maker and a seamstress. As a youth, Armitage learned from an uncle how to machine metal parts. The depression put his father out of work, so his mother supported the family until they moved to the south of England, where his father found work in the aircraft industry. As World War II broke out, the Armitage family lived in Hamble, a small village just outside of Southampton. There the young Armitage merely had to look up to see the Battle of Britain in 1940, and during the war he was selected to be an "aircraft spotter" for his school. Spotters looked for aircraft when the air raid sirens sounded while other students went on with their lessons. When they saw aircraft, Armitage and his fellow spotters rang bells, and the students entered the air raid shelter. He spent most of the war in Hamble, except for a year when the family moved back north to Leeds to avoid the dangers of air battles and bombing.

By war's end, Armitage was working as a trainee draftsman in the aircraft industry. He attended Southampton University as a part-time student, and in 1948 he got work at the Cierva Autogiro Company designing helicopters. He worked on the Skeeter light observation helicopter that was used by the British Army. In 1950 he was drafted into the Royal Air Force, and after flight training, Armitage was posted to the RAF 617 squadron, which had won fame in the war as the "Dam Busters." "I left the RAF when my two years' service time was up, after much consideration to sign on for another four years to stay with my crew. We had been assigned to do photographic mapping missions in Kenya. I really was torn between going with my crew or leaving the RAF and getting back on my design career. I finally decided to leave the RAF, which turned out to be the right decision as my crew were all killed about one month later, having flown into a Kenyan mountainside."

He got work at Folland Aircraft, where he worked on the wing structure of a lightweight fighter called the Midge, which later was known in RAF service as the Gnat. "On the drawing board next to mine was a fellow who one day persuaded me to come with him and be interviewed in Southampton for a possible job at Avro Canada. I didn't want to go to the interview, as I had just left the RAF and was quite happy to be restarting my design career. To humour him, I said, 'O.K. I'll come to the interview.' He wanted to go to Canada very much. In late November 1952, I found myself on the RMV *Ascania* sailing from Liverpool to Halifax, Nova Scotia. From there I went by train to Toronto to join Avro Canada as a flight engineer testing CF-100 aircraft. My friend on the next drawing board who wanted to go to Canada so bad didn't get offered a job!"

Armitage wasn't sure at first how long he would stay at Avro, but he remained there until the Arrow was cancelled, working mainly in flight test. His stay in Canada was interrupted in 1955 and 1956 when Avro gave him a scholarship to study at the College of Aeronautics at Cranfield back in England, where he earned his master's degree.

While serving at RAF Binbrook, in Lincolnshire, Armitage met his future wife June, and they were married in Toronto in 1954. The first of their four boys was born while the Armitages lived at Cranfield.

In the first 11 years of his time at NASA, Armitage worked on recovery systems, which involved preparing spacecraft, astronauts and the U.S. Navy for the challenges of splashdowns and recoveries. Once NASA decided that Gemini and Apollo would splash down like Mercury, Armitage and his colleagues had to have the means at close hand to prepare for more splashdowns.

"We thought that a military LCM [landing craft, medium] would meet our requirements, and our contacts with the U.S. Army showed that an LCM could be made available to us on a permanent loan basis. Convincing our own upper management that we should have our own 'navy' was not so easy. We put together a briefing, which I gave to our center director, Bob Gilruth. He listened to our pitch and finally said, with a twinkle in his eye, 'OK, go ahead and get your LCM.' So we started to talk more closely with the U.S. Army for an LCM. It came to light that the Army people in Norfolk, Virginia, had a very good LCU [landing craft, utility] which could be made available to us. Now an LCM is essentially a 60 foot [18 m] long floating 'bathtub' used to transport soldiers to the beach. An LCU is quite another thing. LCUs transport tanks, trucks and other heavy equipment and are 115 feet [35 m] long with sleeping and galley facilities for a 12-man crew. We decided not to reopen the issue, but go ahead and get the LCU on the basis that our management had moved on to other more important things and that they would not notice a change from LCM to LCU. So we got our LCU, the 'NASA Navy' as it was often dubbed, and we hired our own captain who was an ex Army LCU 'driver.' When our LCU arrived in Clear Lake by the NASA Manned Spacecraft Center it was all painted in NASA blue and had a complete new flying bridge, heavy lift equipment, new galley and modern communications equipment. It turned out to be a wise choice, since many of our water recovery tests in Galveston Bay and the Gulf of Mexico required overnight trips. The LCU served us well until finally becoming surplus to our needs at the start of the shuttle program."

•

As the time of lunar landing approached, a major new challenge was thrown into the recovery plans for the Apollo lunar missions: the need to protect Earth from any germs or microorganisms that might exist on the Moon. Although many in NASA believed that the chances of such contamination were too small to be taken seriously, pressure from a national-level Interagency Committee on Back Contamination (ICBC), with representatives from the National Academy of Sciences and the Departments of Public Health, Agriculture and Interior, caused NASA and Congress to authorize the construction of a special facility at Houston to receive returning astronauts and lunar samples. The committee required both an effective quarantine and demonstration that the lunar material contained nothing that would be harmful to Earth life forms. As well, special procedures and equipment were required to get the astronauts and lunar samples from their splashed-down spacecraft in the Pacific Ocean to Houston. These measures were unpopular in Houston, but in 1965, firm instructions were passed to MSC to deal with the contamination concerns, and a memo from Maynard that year referred to these measures as something that his engineers were "morally obligated" to fulfill.

During 1967 and 1968, the lunar receiving laboratory (LRL) was built, outfitted and tested. Toward the end of 1968 there was a very unsatisfactory simulation, and then an operational readiness review (ORR) that involved three of the former Avro engineers: Hodge as chair of the review, Armitage as executive secretary and Erb as the representative of the engineering and development directorate. Many of the problems the ORR board identified soon became Erb's to solve when he became the LRL's assistant manager.

Armitage recalls: "As the lunar program approached, John Hodge asked me to serve as executive secretary to his committee to certify the operational readiness of the newly constructed lunar receiving laboratory. This job was in addition to my job in landing and recovery division but was fitting, since in landing and recovery we were developing the quarantine equipment and facilities to house the astronauts after returning from the Moon. We developed the special air transportable trailer systems built from Airstream Trailers, and biological suits to keep the returning flight crews from contaminating the world with 'Moon bugs.' As it turned out, this LRL experience would prove invaluable to me later on."

Once the astronauts landed on the Moon, the crew was under quarantine. There were many arguments about how to recover the astronauts. Many wanted the spacecraft hoisted onto the recovery ship with the crew still inside so they could move directly into the specially modified Airstream Trailer, known as the mobile quarantine facility (MQF), in the recovery ship's hangar bay, where they would be isolated from the outside. But in a controversial decision, NASA managers decided to remove the crew from the spacecraft in the water immediately after splashdown, which meant that the astronauts would have to don biological isolation garments (BIGs) after splashdown and keep them on until they were inside the trailer.

Only a technician and a doctor would be inside the trailer with the crew. After the trailer was removed from the ship and flown in a cargo plane to Houston, its occupants would spend the rest of the three-week quarantine period inside the LRL, which like the MQF was sealed. The LRL also included special isolation facilities to handle the lunar samples the astronauts brought back with them.

The quarantine preparations involved the Canadian doctors, Owen Coons and Bill Carpentier. Coons worked on Hodge's review committee just before he resigned from NASA in early 1969 due to disagreements with his boss, Dr. Charles Berry, about the future organization of the medical office at MSC. Coons and his family moved to Dallas, even then a major airline center, and established a successful aviation medicine practice that he continued until his death in 1997.

Carpentier was chosen to be the doctor inside the MQF with the astronauts, so when recovery forces were deployed in March to recover Apollo 9, Carpentier was part of a group on the recovery ship that rehearsed the recovery of the first lunar landing mission. "We went out on the ship with the mobile quarantine facility and went through a simulation to work out all the bugs for lunar quarantine," said Carpentier. "We had three guys who were simulated astronauts, the recovery engineer and me. The five of us spent five days in the mobile quarantine facility on board the ship going through all the procedures that we would [carry out] - getting the rock boxes, making sure the sterilization procedures were actually working, making sure the waste tanks would support five people for five days, the food system, and the environmental control system, which had a negative pressure like the lunar quarantine facility."

•

Apollo 9 was the D mission, the test of the CSM and the LM in Earth orbit. The mission got underway March 3 when a Saturn V carried the first full Apollo spacecraft into space. The flight of Apollo 9, with astronauts Jim McDivitt, Dave Scott and Russell Schweickart, was one of the most important steps on the way to the lunar landing, yet it is nearly forgotten today. During the 10-day flight, the astronauts put LM through its paces in Earth orbit. McDivitt and Schweickart flew their LM, call-signed Spider, through a demanding set of tests in Earth orbit, firing both its descent and ascent engines. For the first time, astronauts flew a spacecraft incapable of returning to Earth. The test of the LM ended with a successful rendezvous and docking with Scott aboard the CSM, call-signed Gumdrop. Before they flew the LM, the crew of Apollo 9 tested the CSM and LM in joint flight, and Schweickart took a space walk to test the space suit and portable life-support system for astronauts walking on the Moon.

With the D mission successfully completed, Maynard's A to G sequence called for two more missions before the first lunar landing attempt. But the successes of Apollo 8 and Apollo 9, along with concerns about flying orbits that would linger in the Van Allen radiation belts, led to the abandonment of the idea of flying the E mission, a test of the CSM and LM in high Earth orbit. The Apollo 8 C-prime mission called into question the need for the E mission and, for some, the F mission as well. "Our concern now was the operational concern of working with [the CSM and LM] in lunar orbit," Maynard said. "We weren't going to get any smarter about that until we got to lunar orbit."

This left the F mission, a test of the CSM and LM in lunar orbit without landing, for the next mission, Apollo 10, which had a veteran crew commanded by Tom Stafford. But some people, notably Maynard, believed that Apollo 10 should attempt to land on the Moon. As the Apollo managers savoured the success of Apollo 9, the arguments about Apollo 10 were in full swing.

"I got convinced that once we had put the guys through the risk of the launch, and checking it out in Earth orbit, and going through the risk of solar events, and do everything but land, I've already taken several years off the guy's life just sort of scaring him to death, and now I've got to, so to speak, take the same time off the lives of another set of guys to actually accomplish the objective," Maynard said.

During its lunar orbits, Apollo 8 had in Maynard's view achieved one of the goals of the F mission – gaining new information about the Moon. Apollo 8 brought back photos and other information about the Moon. It showed that the mascons - mass concentrations of heavy material that lay under the Moon's surface like raisins inside a pudding - have stronger gravitational effects on the paths of orbiting and descending spacecraft than had been believed when they were discovered by Lunar Orbiter spacecraft in 1966 and 1967. "So we thought that, well, now that we know that, the rationale of going to Apollo 10 without the landing was not much of a rationale now that we know about the mascons. Now we have pretty good confidence that we could go down to land. We knew where we were going, even if the mascons were pushing us off. There's lots of time and there's always the capability of lighting up that ascent stage and going back to orbit."

Strongly opposing this point of view was the flight operations directorate, especially Chris Kraft and Rod Rose. "I have a memo from Chris to George Low dated June 1, 1967 - it was one I wrote - laying out the whys and wherefores of why Apollo 10 should be a rehearsal flight with no landing," Rose said. "I spelled out all the things the crew had to do coming and going, plus all the things we had turned up in the lunar surface operations plan, all the things they had to do and learn. It's a whole new thing there. Operating the LM, communicating through the LM, descent communications. It was a tricky, tricky situation. The crew were deeply involved. In fact, Tom [Stafford] and the crew were one of the first ones to say, 'Boy, there's one heck of a lot of stuff we've got to do. There's stuff we've got to learn going down to the Moon and coming back, there's rendezvous.' Nobody had done a lunar rendezvous before. Was our tracking adequate? Did we have all our ducks in a row for lunar rendezvous? Because if you missed it, you were in deep trouble. It wasn't a thing where if you missed it, you could have another go in 90 minutes or so many revs later. Around the Moon, you either made it or you didn't. There were a lot of things that required a lot of practice."

Rose said more information, collected at lower altitudes, was needed about the mascons. As well, the LM could be flown to the vicinity of the Moon on Apollo 10, where the procedures on the flight out to the Moon, lunar orbit operations, and rendezvous could be rehearsed. "What Apollo 10 did for us, and a lot of people didn't realize this, was that it was flown as a carbon copy of Apollo 11. It was the same flight plan, it was going to the same landing site, same timing, same everything. And the whole point was that Apollo 10 would do everything up to the lunar module descent. We'd done it all before so that the 11 crew wouldn't have to go through a whole bunch of new procedures." As well, there was the question of training. The crew of Apollo 10 was going through intensive training for a mission that didn't include a landing. With a landing, the astronauts would have to start new training in landing procedures, including flights in the LLTV, and rehearsals for a walk on the Moon.

"So Bob Gilruth asked me to come out to his house, and we'd have a meeting at his house about it," Maynard said. "So George Low and Chris Kraft and Tom Stafford and Gilruth and me met at his house. And they had all heard this story before, but they wanted to go through it one more time. So I told them again, and they had to agree with what I said, and then they said, 'Well what do you think, Tom?' and what flabbergasted me was, I had just given him the opportunity to be hero of the century, and he says, 'Well, I think we ought not to land on Apollo 10. We should just go down to altitude and verify that. Yeah, we flew Apollo 9, but lunar orbit is different.' I would interject that it was easier, and, 'Well, yeah, it's three days away,' and I said, 'What difference does it make? If you've got any reason to come home, you come home.' 'Well, I think operationally that we ought to do everything except the final touchdown.' I thought the final touchdown was easy. When Tom said that, I said, 'Well, I must be missing something operationally.' But I said, 'I have a lot of respect for you guys, you Chris Kraft, and you Tom Stafford. If you are willing to give up the opportunity to be first man on the Moon, you must think it is pretty operationally important, so I bow to your superior wisdom.' So I accepted that and went on to do my best for Apollo 10."

Rose said Stafford was deeply familiar with the problems facing Apollo 10 from attending the flight operations panel meetings for his mission. "Tom was very unassuming, and he was also a super guy. He, as much as anybody, could understand the operational implications of what 10 was going to be about. He'd have liked to have landed on the Moon. Don't get him wrong - Tom wanted to go to the Moon as much as the next guy. But he recognized how vitally important it was to get all those operations down pat."

The decision to fly Apollo 10 as the F mission was made within a few days of Apollo 9's return to Earth. If anyone had wanted to overturn the decision, they would have to deal with the fact that Apollo 10's LM was too heavy to be used for a landing. Since it wasn't designed to make the final descent, the LM didn't carry the fuel necessary to land and return safely.

Apollo 10, with Stafford, lunar module pilot Gene Cernan and command module pilot John Young on board, lifted off from Cape Kennedy on May 18, 1969. On its way to the Moon, Apollo 10 sent the first colour TV broadcasts from space. Once in lunar orbit four days into the mission, Stafford and Cernan separated their LM Snoopy from the CSM Charlie Brown with Young aboard. Snoopy descended to 14.5 km above the Sea of Tranquillity, where Stafford and

The crew of Apollo 10: (l to r) Eugene Cernan, John Young, Thomas Stafford
Courtesy NASA.

Cernan took a closeup look at the main landing site for Apollo 11 and pronounced it smooth enough. Snoopy flew out of control for a brief but harrowing moment as its ascent and descent stages separated, but this was later traced to an out-of-position switch. Except for this and a minor engine problem, the LM proved itself in the lunar environment. Snoopy successfully returned to Charlie Brown, and the astronauts headed home. Said Rose: "It was not until the results of the [Apollo 10] F mission were analyzed that we realized the real influence of the lunar mascons on our trajectory and navigation programs and software. This flight was indeed an essential precursor to Apollo 11."

By the time Apollo 10 splashed down in the Pacific Ocean on May 26, Apollo 11 was already standing on Pad 39A at the Cape, but there was still plenty of work to do before it was launched.

•

Rose's flight operations panel, which had been working for years on the lunar landing mission, had held its 41st and final meeting in April to sign off on the flight operations plan for Apollo 11. Rose recalled that 128 people attended that meeting, twice the usual number. The panel, which had developed missions for launch windows in July, August and September, had debated, among other things, how long the crew of Apollo should remain in lunar orbit before they were rested enough to begin their descent to the surface, and whether the LM should land on the Moon and later dock with the CSM with its remaining ascent engine fuel still on board. Spacecraft returning to Earth had dumped fuel before splashdown, but the group decided to dispense with this practice for lunar operations.

In 1968, Apollo managers had also abandoned the plan outlined at the 1966 symposium to have two surface excursions during the first landing. Instead, the two astronauts would only make one Moon walk, during which they would gather samples, take photographs and deploy a stripped-down version of the Apollo lunar surface experiments package.

In the months leading up to Apollo 11, a design review board headed by Jim Chamberlin concentrated on the extra-vehicular mobility unit, the space suit and portable life-support system (PLSS) that the astronauts would depend on while they walked on the lunar surface. This group of two dozen engineers held regular meetings that worked through issues ranging from the addition of silicon coatings on gloves to make it easier to grasp tools and lunar samples, to the astronauts' balance with the 85 kg PLSS on their backs. Tests showed that if the astronaut leaned forward slightly, it would allow for a good balance in the one-sixth gravity of the Moon. Chamberlin's group also made changes to the suits to be worn by astronauts on the Moon when Neil Armstrong and Edwin "Buzz" Aldrin complained of fit problems after tests in altitude chambers.

Owen Maynard and John Hodge were caught up in the special preparations for the lunar landing as members of the Apollo site selection board charged with deciding where Apollo 11 and subsequent missions would land. For Apollo 11, five sites had been originally designated. The sites moved from east to west on the Moon, with the first two sites on the Sea of Tranquillity, the third on Sinus Medii or Central Bay, and the fourth and fifth on the Ocean of Storms. All these sites on the lunar "seas" had appeared promisingly flat in photos taken by Lunar Orbiters. Having more than one site for a landing allowed launches to be put off by only a day or two if the spacecraft and rocket weren't ready to launch at the appointed moment for the first site. Although each lunar day lasts two weeks, NASA decided that Apollo would land shortly after lunar sunrise to ensure that the astronauts had a good view of their landing site. If there was only one landing site, and a launch had to be postponed, the next launch attempt would have to wait for a month.

The site selection board had met several times, but one of the most crucial meetings took place at NASA Headquarters in Washington on June 3. The meeting was the culmination of years of work and haggling by scientists and engineers. Maynard told the meeting that site 1 had already been dropped since it would require a launch at the same time as site 2, also on the Sea of Tranquillity. Site 2, which was deemed to be a better place to land and the more interesting of the two sites for scientists, became the prime site for Apollo 11. Also available for Apollo 11 if the scheduled July 16 launch was postponed were site 3 in the Central Bay, and site 5 in the Ocean of Storms. Site 4 was eliminated because the complex geometry of lunar flight dictated that it wouldn't be easy to reach until late in the year. Maynard told the meeting that the crew of Apollo 11 was training for the three sites, but emphasizing site 2. Mission commander Armstrong and LM pilot Aldrin trained to land on the sites, while Michael Collins, who would remain in lunar orbit aboard the CSM, learned how to find the sites using a small telescope on his spacecraft and to pinpoint where the LM had landed. The meeting also included an extensive discussion of where the second landing would take place.

About the same time, a few weeks before the launch of Apollo 11, Maynard remembered attending a meeting of senior Apollo managers convened by Sam Phillips, the program director, where the main subject was the readiness of Apollo 11 to fly to the Moon. Phillips assigned everyone at the meeting a specific area to look at as part of a final effort to make sure that Apollo would succeed. Owen Maynard was asked to look into all aspects of the landing itself. He spent time working with the LM's manufacturer, Grumman, to make sure everything was right. One area of concern at that late date turned out to be the gold mylar foil insulation covering the lunar module legs. Recent tests had shown that the heat coming out of the descent engine heated the mylar more than had been anticipated, and the mylar could come loose and interfere with the landing radar, which told the astronauts how far above the surface they were as they descended.

"One of the very last things I did on the Apollo 11 mission was to work with Grumman personally on adding insulation to those legs. And the last thing I did was I went into the [launch adapter containing the LM atop the Saturn V] and inspected it to make sure that they hadn't done anything to preclude the deployment of the gear since we added the insulation while the gear was stowed. It's real easy to put a piece of tape in the wrong place." Maynard took his wife Helen and their three daughters Merrill, Beth and Annette down to the Cape for a small vacation, and Merrill and Beth threw rocks toward the Apollo 11 stack, which was safely beyond their reach. Because Maynard rarely visited a launch pad, he had to take a safety course on all the dangerous fuels present in and around the spacecraft. When he entered the adapter nearly 100 m above the surface of launch pad 39A, he took some tape with him. "I looked at the descent stage and decided that it could use a little tacking down here and there." Phillips, Maynard and the others responsible for Apollo 11 gave it a green light to attempt a launch on July 16 for a flight to site 2 in the Sea of Tranquillity.

•

While the Americans made their preparations to land astronauts on the Moon in the early months of 1969, the Soviet space program continued its pursuit of the Americans in case Apollo faltered or failed. On successive days in mid-January, the Soviets launched Soyuz 4 and 5, the first with one cosmonaut aboard, and the second with a crew of three. The following day, the two spacecraft docked and two members of the Soyuz 5 crew donned their space suits, left their spacecraft and transferred to Soyuz 4, in which they returned to Earth. Although it was not publicly known at the time, this was a vital test for the Soviet lunar program. A few days later, another attempt to send a robot Zond around the Moon failed at launch, and the Zond program was grounded until after Apollo 11. Then on February 21, the first launch of the 105 m tall N-1 rocket failed 70 seconds after launch when an oxidizer line to one of the 30 first-stage engines failed due to the rocket's heavy vibrations. Despite the successes of Apollo 9 and 10, the Soviet effort continued, and on the night of July 3, another N-1 with a test spacecraft aboard was launched. Within seconds, a metallic engine part entered an oxidizer pump, causing the rocket to explode, destroying itself and its launch pad. The damage from the catastrophe appeared on U.S. reconnaissance satellite photos, but remained a secret for nearly two decades.

Finally, three days before the Apollo 11 launch, a Proton rocket lifted a Luna probe into a path to the Moon. A cryptic announcement from the Soviet TASS news agency revealed little about Luna 15's mission, but rumours that proved to be accurate suggested that Luna 15 was designed to scoop up lunar soil and return it home before Apollo 11. An identical probe had failed at launch in June, and now the last Soviet chance to steal the American thunder was on its way to the Moon with just hours to spare.

The countdown for Apollo 11 began three days before the launch of Luna 15, and the flight of the Soviet probe only further focused the world's attention on the pad at Cape Kennedy. Newspapers and magazines published lunar supplements, and television and radio broadcast special shows dedicated to the lunar mission. The most popular sport that week was guessing what Neil Armstrong would say when he made the first step onto the lunar surface.

Close to a million people converged on the beaches, waterways and towns surrounding the Kennedy Space Center to watch the launch. Thousands of media were present, and VIPs who came to the launch included Charles Lindbergh, the first person to fly nonstop from New York to Paris 42 years before, and many of the people who had made Apollo 11 possible, including former president Johnson and former NASA administrator Webb. Two nights before the launch, the crew of Apollo 11 responded to reporters' questions from their prelaunch quarters in a nationally broadcast press conference. "Fear is not an unknown emotion to us. But we have no fear of launching out on this expedition," Armstrong said. "After a decade of planning and hard work, we're willing and ready to attempt to achieve our national goal."

Right at the opening of the launch window for site 2, 9:32 a.m. Wednesday, July 16, Apollo 11 set off from Cape Kennedy. Armstrong, Aldrin and Collins followed the now routine opening hours of their flight, riding the Saturn V into Earth orbit and then, on the second orbit, relighting the third stage to set them on course for the Moon. Then the CSM Columbia separated from the stage, turned around and docked with the LM Eagle, which it extracted from the now spent Saturn stage. The next day the Apollo 11 astronauts fired their service module engine for a midcourse correction and sent back a television broadcast from their spacecraft. President Richard Nixon declared a federal holiday for Monday, July 21, in honour of the planned Moon landing. And Luna 15 went into orbit around the Moon. The Soviets remained tight-lipped about Luna 15's mission, except to say that its path wouldn't interfere with Apollo 11.

Maynard settled into his flight routine, taking a shift in the spacecraft analysis (SPAN) room just outside the main mission operations control room. Maynard and other senior Apollo managers from NASA and the contractors, including Tom Kelly of Grumman, spent missions in this room, where they would be ready to deal with any problems that cropped up. SPAN was connected by telephone to the plants of the spacecraft contractors, where senior engineers were always ready during missions to deal with any problems or questions sent to them by SPAN. The managers in SPAN had access to monitors with the instrument readings coming down from the spacecraft. Sometimes the people in SPAN acted as an extra set of eyes for the flight controllers in the MOCR, looking at the incoming data. "If we wanted to know how something was going to behave in an unusual situation, and get it from the horse's mouth, we would have that [responsible engineer] come in," Maynard said. "The flight control guys had no way to get in touch with North American or Grumman, except through the [SPAN] operations manager." Because of their senior positions, the people in SPAN knew who was responsible for every major part of the spacecraft. SPAN worked with engineers in another room in a nearby building called the mission evaluation room, where subsystem engineers were connected with their contractor colleagues, ready to solve the problems identified in the MOCR or the SPAN room. During Apollo 11, as he did during most Apollo flights, David Ewart took shifts in the mission support room at the North American Rockwell plant at Downey, California, along with personnel from the contractor, waiting for calls from the SPAN room or other back rooms in Houston requesting information or help with problems.

Meanwhile, Dr. Bill Carpentier was aboard the aircraft carrier USS *Hornet* in the Pacific Ocean, where he had joined the members of the ship's crew in their final rehearsals to pick up Apollo 11. Carpentier had taken part in pre-flight medical examinations of the Apollo 11

Owen E. Maynard (with microphone), Thomas J. Kelly of Grumman (foreground) and Dale Myers of North American Rockwell (with eyepatch) in SPAN room during Apollo 11.
Courtesy NASA.

astronauts, so he would be sensitive to any changes he might find when the crew returned from space. In addition to preparing his tests aboard the quarantine trailer, Carpentier had to be ready to leap out of the recovery helicopter if necessary to assist an injured astronaut.

Apollo 11 was on the third day of its mission, and the astronauts visited Eagle for the first time in flight and took television viewers with them. The next day, Apollo 11 entered lunar orbit and the astronauts took their first closeup look at the site where they would attempt to land the following day.

On Sunday, July 20, the three astronauts donned their space suits and after final preparations, Armstrong and Aldrin floated into Eagle, closed the hatches and separated their spacecraft from Collins aboard Columbia. On the back side of the Moon, Eagle fired its descent engine for the first time to lower its orbit. Coming around the Moon, Eagle got a "go" to try a landing. On a command from Armstrong, the throttleable descent engine began firing again, beginning the 12-minute-long landing sequence that took the astronauts down from 14 km. During the final phase of the landing, Armstrong looked out the window to see where the computer was taking Eagle, and Aldrin called out information from displays on the control panel. Armstrong realized early on that Eagle was overshooting the planned landing point, and the tension of the moment was heightened by computer alarms caused by an overflow of data from landing and rendezvous radars. Armstrong saw that Eagle would land in a football-field sized crater full of boulders, so he took control and set Eagle down in a cloud of dust beyond the crater with the gages showing less than 30 seconds worth of fuel left. After Aldrin called out technical information about the landing, Armstrong announced: "Houston, Tranquillity Base here. The Eagle has landed." It was 3:18 p.m. Houston time.

Hundreds of millions of people were watching television or listening to radios throughout the dramatic descent. Although the landing was for many the climax of Apollo 11 and represented the achievement of Kennedy's goal, Maynard had a different view. "We had this almost 100 per cent unknown thing of these guys going out on the lunar surface ahead of us, and doing it not in accordance with the plan. And then they have to do this lunar liftoff. Never have done that before. And then they had to do this rendezvous in lunar orbit that had only been done once before. All of that was ahead of us yet. And so that lunar landing appeared to be a pretty insignificant thing. The mission wasn't successful until we returned them safely. Until we got them home."

On the Moon, the astronauts decided to depart from the flight plan, which had them sleeping before they walked on the Moon. Instead, once they had completed checks of Eagle's systems to make sure they were in good condition for the lunar liftoff scheduled for the next day, Armstrong and Aldrin donned their life-support backpacks, depressurized Eagle's cabin, and opened the front hatch. Armstrong crawled out. When he reached the top of the ladder that would take him to the surface, he pulled a lanyard that deployed an equipment storage package on Eagle's descent stage. Atop the package was a black-and-white television camera that broadcast back to Earth Armstrong's first steps on the Moon. "That's one small step for man, one giant leap for mankind," Armstrong said as he placed his left foot on the Moon at 9:56 p.m. Houston time. A few minutes later, Aldrin joined him on the Sea of Tranquillity, declaring the sight to be "magnificent desolation."

Astronaut Edwin "Buzz" Aldrin stands next to the leg of the Apollo 11 lunar module "Eagle" in the Sea Of Tranquillity. July 1969. Courtesy NASA.

During the 2 hours and 21 minutes they spent walking on the Moon, the two astronauts took photos, collected 22 kg of rocks and soil, and set up a seismometer and a laser reflector to allow scientists to measure lunar seismic activity and the precise distance to the Moon. As well, they deployed and then recovered a sheet of foil designed to capture particles of the solar wind, which never directly strikes Earth because of its magnetic field. Armstrong and Aldrin unveiled a commemorative plaque, raised the American flag, and spoke to President Nixon, while the world watched ghostly black-and-white television images that only added to the sense of wonder Earthbound viewers had of the moment.

Although the two Moon walkers were focused on carrying out their assigned tasks, they experienced the strange sensations that came with being on another world. Armstrong and Aldrin needed time to learn how to move in gravity that was a sixth of what they were used to. They had difficulty estimating distances because on the Moon, with a diameter only a quarter of the Earth's, the horizon is much closer than it is at home. The airless environment of the Moon made distant objects look even closer. The sky was black, except for a blindingly brilliant Sun and the beautiful blue-and-white globe with green accents that Armstrong and Aldrin had come from.

The people who watched the Moon walk on television all had their special feelings as this unprecedented spectacle took place. Bryan Erb, deputy manager of the LRL, was focused on the samples the astronauts were bringing back. Most of the lunar samples weren't picked up until late in the Moon walk, and while Armstrong and Aldrin were busy learning how to walk on the Moon or unfurling the U.S. flag, Erb admitted that he muttered at his television: "Stop jumping up and down, and pick up the damn rocks!"

As for Maynard, he was probably asleep during the Moon walk. "I was pretty well glued to the television set when they did the landing on the Moon," said Maynard's son Ross, 18 years old at the time, who remembers his father sleeping through the lunar surface activities. Ross filled his father in on what had happened while he had been asleep. "He went and had his cereal and didn't say much and got dressed and went back to do his shift. At the time it kind of impressed me that he had the discipline to know that he had to get a good night's sleep to be able to work on anything that came up during his shift. That always stuck in my mind and really impressed me that having worked on that for so many years, he just shut it all out and was able to get a good night's sleep while they were doing the thing that they had worked so hard for."

In the early hours of July 21, Armstrong and Aldrin completed their Moon walk. Weight remained a concern; to compensate for the lunar samples going up, the astronauts threw out their life-support backpacks once they had reconnected themselves to the LM's environmental control system. They then settled down for a fitful sleep, the first on another world.

After 21 hours and 36 minutes at Tranquillity Base, the ascent stage of Eagle blasted off on Armstrong's command. The ascent and rendezvous that had caused Apollo's designers and managers so much anxiety went off with only one minor hitch at the moment of docking. The 7-minute, 20-second ascent burn was nominal, and on the second orbit of the Moon after liftoff, Eagle rendezvoused and docked with Collins aboard Columbia. Two hours later, Armstrong, Aldrin and their haul of samples and photos were aboard Columbia, Eagle was cast off, and a few hours later, Columbia fired its big engine to head out of lunar orbit and back to Earth. The next two days were spent in the transearth coast, and the astronauts rested and sent two television broadcasts back home. In their last TV broadcast, which featured the astronauts' reflections on their flight, all three paid tribute to the people who made it possible. "This operation is somewhat like the periscope of a submarine," Collins said. "All you see is the three of us, but beneath the surface are thousands and thousands of others, and to all those I would like to say, thank you very much."

As Apollo returned home, the Soviets announced that Luna 15 had crashed on the Moon's Sea of Crises just before Eagle's ascent stage left the Moon. The Soviet announcement blandly claimed that Luna 15 had completed its unspecified mission, but in spite of this secrecy, it was clear that their last-ditch attempt to catch the U.S. in the Moon race had failed.

On Thursday, July 24, the crew of Apollo 11 awoke to prepare for their return to Earth. Aboard the *Hornet*, near Apollo 11's planned splashdown site 1,500 km southwest of Hawaii, Bill Carpentier was already in quarantine in the MQF trailer sitting in *Hornet's* hangar bay. With him was John Hirasaki, the recovery engineer assigned to take care of the trailer and of Columbia when it was brought aboard the *Hornet* and connected to the MQF. As he had done for previous recoveries in Gemini and for Apollo 7, Carpentier boarded the recovery helicopter, ready to care for the astronauts as soon as they splashed down. Also aboard the *Hornet* was President Nixon, who watched the splashdown from the bridge. After dropping the service module and plunging through Earth's atmosphere, Columbia splashed down safely nearly 20 km away from the *Hornet*, 8 days, 18 minutes and 18 seconds after leaving Cape Kennedy.

Aboard the helicopter designated Recovery 1, Carpentier waited in his wet suit as Navy divers entered the water and fastened a flotation collar around Columbia, then deployed a liferaft. One of the divers donned a biological isolation garment, opened the spacecraft hatch and passed in three BIGs for the astronauts. When the astronauts donned their BIGs, they opened the hatch and came out into the liferaft while the diver closed the hatch and cleaned the area around the hatch with disinfectant. After the astronauts washed the outsides of their BIGs with disinfectant, one by one they crawled into a basket slung from Recovery 1 and were reeled up to the door where Carpentier helped them in. Aldrin later wrote that Carpentier was "standing in the door, smiling, his hand extended in greeting .... We shook hands and he helped me out of the bulky Mae West."[10] Because of the noise of the helicopter and the masks of the BIGs, the only conversation on the ride back to the *Hornet* was a thumbs up from each astronaut indicating that they were all right. As part of the quarantine procedures, the crew and Carpentier rode in the back of Recovery 1 behind a line painted on the helicopter's floor. Once the helicopter landed on the Hornet and the helicopter had been moved from the flight deck to the hangar bay, the three astronauts in their BIGs emerged and walked 20 feet to the door of the MQF as much of the world watched on television. Carpentier followed, closing the door as he entered the MQF. Carpentier then took his biological swabs, blood samples and measurements of Armstrong, Collins, and Aldrin, after which each was free to shave and shower. Changed into flight suits, the three astronauts presented themselves at the window at the end of the MQF where President Nixon welcomed them back to Earth. "This is the greatest week in the history of the world since the Creation," the president declared in a piece of hyperbole that didn't immediately stand out in a week of superlatives. The ceremonies over, Carpentier began further medical tests on the astronauts.

"I had to get some blood drawn right away. I had to get blood pressure, pulse, etc., right away. We agreed to do certain things right away. The whole recovery business was a bit of a blur because I was so busy," Carpentier said in 1995. "I was by myself. I had three astronauts and I had to go through the whole medical program. It took me seven hours to get all the data. This was one of the most incredible adventures in the history of man, stepping out on the Moon for the first time, and I was there very close, and you would think I would remember more than I do. I had gone through so many rehearsals, it was so routine. I certainly knew it was an historic occasion, but I didn't think to remember details."

The blood samples and the swabs from the astronauts' skin, noses and throats that Carpentier obtained immediately after splashdown were flown to Houston with the lunar samples and photographs for testing to see if they had been exposed to any germs on the Moon. The post-landing samples were compared to similar samples taken from the crew before launch.

The safe recovery of the three astronauts meant that President Kennedy's goal was achieved, and when the astronauts stepped out of the helicopter, a raucous celebration began in mission control, which filled with VIPs and with engineers from around the Manned Spacecraft Center. Cigars were lit, flags unfurled, and Kennedy's 1961 challenge was flashed on the display

[10] Aldrin, Col. Edwin E., and Wayne Warga, *Return to Earth*, Random House, New York, 1973, p. 5.

Dr. William R. Carpentier follows the Apollo 11 crew to the Mobile Quarantine Facility aboard USS Hornet, *July 24, 1969. Dr. Carpentier is just stepping off the helicopter behind the crew. He and the MQF technician shared the MQF with the crew from then on until they got to the LRL in Houston.* Courtesy NASA.

board at the front of the control room. That evening, Houston was the scene of the biggest ever splashdown parties.

"I certainly did go to several parties," Maynard reflected. "I don't remember them that much. It was a long time ago, and I haven't recovered from this yet, resolving in my mind how sort of grateful and thankful I was to the people who worked for me. I was just thrilled for them. Landing a plane, flying with a navigator, leads to a sense of accomplishment. Then there's being

NASA officials join flight controllers in mission control to celebrate the successful conclusion of the Apollo 11 lunar landing mission. Identifiable in the picture, starting in foreground, are: Dr. Robert R. Gilruth, MSC director; George M. Low, manager, Apollo spacecraft program, MSC; Dr. Christopher C. Kraft Jr., MSC director of flight operation; U.S. Air Force Lt. Gen. Samuel C. Phillips (with glasses, looking downward), Apollo program director, Office of Manned Space Flight, NASA headquarters; and Dr. George E. Mueller (with glasses, looking toward left), associate administrator, Office of Manned Space Flight, NASA headquarters. Rod Rose is talking to Dr. Mueller. *Courtesy NASA.*

a husband and a father. Then there's a deal like Apollo. There's a vast array of people involved."

Inside the MQF, Carpentier and Hirasaki still had work to do. While Carpentier was conducting his medical examinations of the astronauts, Columbia was brought on board the *Hornet* and wheeled alongside the MQF. A plastic tunnel connected the MQF to Columbia.

"I guess what I recall the most was the Moon dust," Carpentier said. "Once they got the spacecraft connected to the MQF, I had to go in and get [biological] samples. I had to get samples preflight and now I had to get the same samples postflight. Anything that was cultured would have to match. I had to open the suit bag. I had to culture them in a couple of places on the suits, get swabs. As soon as I opened the suit bag, there was a cloud of Moon dust. It was very fine dust. It was on my hands, it was in my nose, in my eyes. Once you smell Moon dust, you don't forget it. It was a combination of wet ashes and gunpowder. After John Hirasaki got out the rock boxes, we had to package them, seal them and put them out. There was Moon dust on the table.

SPAN room during Apollo 11. Owen Maynard (with microphone headset) stands at the back. Thomas J. Kelly of Grumman, directly in front of Maynard, is on the phone. Courtesy NASA.

So the first thing I did when I got a break was scrape a bit of dust off and put it on a microscope slide and look at it under the microscope. Other than the crew, I was the first person to come in contact with the Moon. And I was the first person to look at it under a microscope. It was interesting because I had my blood taken along with the crew to make sure we weren't developing any strange condition."

While the astronauts' tests showed nothing unusual, Carpentier's own blood tests showed that two measures were slightly abnormal but no cause for alarm.

"The thing that happened to me that didn't happen to anybody else was that I had strange things happening in my blood. The same thing happened when I was in quarantine after Apollo 14. When I went into the spacecraft after Apollo 14 and opened up the suit bag, I had much more of a reaction than I did after Apollo 11. My nose began to itch, I started to sneeze and it bothered me in an allergic kind of way. The same thing happened. My basal cells went up.

"I went out in Apollo 15 and as soon as I opened the suit bag, I started to sneeze so bad I had to go out, take a deep breath and go back in and get the swab samples. I could hardly breathe when I got exposed to the Moon dust. The same thing happened. My basal cells went up. I think I am probably the only person in the history of mankind to have a documented allergy to Moon dust." None of the astronauts exposed to lunar dust reported a reaction except Harrison Schmitt, who reported a slight allergic reaction to lunar dust after his first Moon walk on Apollo 17, but the problem did not persist after his second and third walks.

The two rock boxes brought home on Apollo 11 were flown off the *Hornet* that day and arrived in Houston the next day, where scientists in the lunar receiving laboratory began their tests. The astronauts relaxed aboard the MQF, with Armstrong strumming a ukulele and the crew watching a replay of the televised coverage of their Moon walk. Carpentier delivered a report on the astronauts' condition to the media through the MQF window. "They're fine as far as I can tell," he said, adding that he only found a minor inflammation in one of Armstrong's ears. "Actually, it looks like I'm out of a job," Carpentier joked. Collins recalled that Carpentier served as bartender on the MQF. When the *Hornet* arrived at Pearl Harbor in Hawaii on July 27, the MQF was slowly trucked to Hickam Field, loaded aboard a transport aircraft, and flown overnight to Houston, where the astronauts and those locked in with them remained in quarantine until August 10. No lunar germs were found in the astronauts or lunar samples, and the quarantine in the LRL quickly settled into a tedium that had everyone waiting impatiently to get out.

With the debriefings done in quarantine, Armstrong, Aldrin and Collins began a gruelling round of parades, dinners and other ceremonies in the U.S. and, later that fall, in 23 countries around the world. Bill Carpentier accompanied them on a round-the-world trip that proved more difficult in some ways for the astronauts than the flight to the Moon because of the new and sudden pressures of their celebrity. Among the stops on their tour were London and Ottawa, the capital cities of the home countries of the Avro engineers.

11 Expeditions to the Moon

When most people in NASA were concentrating on Apollo 11, John Hodge was thinking about what Apollo should do after the first landing was achieved. In 1967, Hodge began to phase himself out of flight control, and that August he attended a meeting discussing where subsequent Apollo missions should go. About 150 scientists gathered on the campus of the University of California at Santa Cruz to discuss lunar exploration after the first landing, following on the work of similar conferences in 1962 and 1965 that set out general lunar exploration goals. Hodge, then still chief of the flight control division, spoke to the geologists and other scientists about Apollo operations on the lunar surface. The scientists at Santa Cruz spoke hopefully of ambitious flights involving dual Saturn V launches and automated driving or flying vehicles that could move between landing sites, collecting samples and taking photos of lunar seas, craters and rilles. But the soaring costs of the Vietnam War and domestic programs, coupled with disquiet in Congress over the Apollo fire, resulted in cuts to NASA's budget that dashed the hopes of the scientists for multi-vehicle expeditions to the Moon.

The ardour for further Apollo flights wasn't universally shared within NASA, either. Bob Gilruth and others were all too conscious of the great dangers that faced each and every Apollo crew, and they wanted to cut Apollo off after the first landing. But like the scientists, Hodge was looking to further missions. In May 1968, Hodge was put in charge of a lunar exploration working group that he had already been setting up for months. "I was beginning to worry that no one was looking at the second lunar landing. The whole organization was concentrating on the first, and it seemed to me that if you have 10 or 12 sets of hardware, you have to start worrying about the rest of the program."

George Low, the Apollo program manager, told Hodge that he wanted to concentrate on the first landing and directed him to set up the working group. "We did the planning studies for the extended lunar program, including extension of the [LM fuel] tanks, and things like that. And then we had conferences on where should we land, regular conferences." Some of the people who went with him to plan the advanced missions included Dennis Fielder, Hodge's fellow engineer from Avro, who had been instrumental in setting up the control and tracking for Mercury, Gemini and Apollo; Joe Loftus, who had worked closely with Maynard building Apollo and would design most of the modifications to the CSM and LM; Andre Meyer, who had studied the LM after working with Chamberlin on Gemini; and some of Hodge's colleagues from flight control. "When I got into this business, I was fairly familiar with what the scientists had hoped to do in the way of a lunar exploration program. And I think we were able to put together a program which met reasonably well the requirements of the scientists to do significant exploration, and the requirements from the technology standpoint to be sure that we could do the things that we were saying. We had to decide what orbits to put the sites in and what modifications we had to make to the spacecraft." By February 1969, Hodge's group had written specifications for a LM that could stay longer on the Moon, carry more equipment such as a lunar roving vehicle down to the Moon, and bring more lunar samples home. That spring, his group also drew up changes to be made to the service module to allow the placement of scientific instruments to photograph and measure the Moon.

"It was really just a question of straight engineering," Hodge said. "Everybody had lots of good ideas. Owen Maynard was involved in that because he was division chief in system engineering for the Apollo program office at the time, and his group did a lot of the studies that said you can increase the size of the tank, and that will give us some more fuel. When we did the calculations, it gave us enough leeway to put enough environmental control system consumables for a couple more days. There was room for a rover."

To make it all possible, the performance of the Saturn V was stretched by its designers at Marshall Space Flight Center so the heavier spacecraft could be lofted to the Moon. And honouring the deal made seven years earlier when Wernher von Braun agreed to fly to the Moon using lunar orbit rendezvous, Marshall was put in charge of developing the lunar roving vehicle. But these changes wouldn't be ready until the later flights to the Moon. At the time Apollo 11 returned home, NASA hoped to send nine more expeditions to the Moon. The initial flights following the Apollo 11 G mission, which ended Maynard's original A to G mission sequence, were designated as H missions, and the advanced missions Hodge was working on were known as J missions. H missions included longer stays on the Moon to allow for two Moon walks by the landing crews and emplacement of the full Apollo lunar surface experiments package (ALSEP) as envisioned by Bob Vale at the 1966 symposium.

•

Apollo 12, with commander Pete Conrad, LM pilot Alan Bean, and CSM pilot Richard Gordon, would have been ready to fly in September 1969 if Apollo 11 had fallen short. Now it was scheduled for a November flight to the Ocean of Storms, where scientists wanted to sample a "younger" lunar sea than the "old" Sea of Tranquillity. At the June 3 lunar landing site selection meeting attended by Hodge and Maynard, several sites were suggested for Apollo 12, including one next to Surveyor 3, a robot that had soft landed in the Ocean of Storms in 1967. Pieces cut from Surveyor could be examined to determine the effects of prolonged exposure to the harsh lunar environment. As well, Surveyor 3 would give Apollo 12 a fixed target to land near. The Surveyor 3 site was not as "young" as geologists would have liked, but it lay in one of the rays emanating from the crater Copernicus, allowing for sampling of rocks drilled out in that great impact. Although the June 3 meeting didn't decide on a site for Apollo 12, Apollo 11's off-the-mark landing caused Apollo management to make a pinpoint landing Apollo 12's major goal, and this ultimately led to the selection of the landing site next to Surveyor 3.

Lunar rocks undergoing examination in the LRL. Note gloved hands of operator in foreground wielding small brush to remove dust. Courtesy NASA.

Apollo 12, the first of the H missions, set off from Cape Kennedy on a rainy November 14. During the first stage burn, the vehicle was struck twice by lightning, but quick action by controllers and the crew restored spacecraft systems and saved the mission. Four and a half days later, Conrad and Bean left Gordon behind in the CSM Yankee Clipper and descended to the Moon in their LM, Intrepid. Mission planners responded to Apollo 11's targeting problem by giving Intrepid's computer information on its exact position in space just as it began its descent. The plan worked, and Conrad and Bean landed Intrepid right on target near Surveyor 3. In their first walk on the Moon, Conrad and Bean set up the ALSEP, took samples, and caused their colour TV camera to fail when Bean accidentally pointed it at the Sun. After a sleep period, the two astronauts took samples at various craters near their landing site and visited Surveyor 3, removing its television camera and lunar scoop for return home. After Conrad and Bean rejoined Gordon in Yankee Clipper, the crew of Apollo 12 returned to Earth and splashed down near the *Hornet* on November 24 with 34 kg of lunar samples. Quarantine procedures similar to those used on Apollo 11 were followed for Apollo 12, and the crew and samples were returned to the lunar receiving laboratory at the Manned Spacecraft Center.

The acting manager of the LRL at the time was Bryan Erb, who found the LRL to be one of the more difficult experiences of his career. From 1967 to 1968 he left Houston for a year while he attended the Massachusetts Institute of Technology on a Sloan Fellowship, which is given to up-and-coming managers in government and industry.

"After I finished that year, I stayed in my old area of structural mechanics for about three months. I was very much looking for new opportunities. Just then, the lab was being activated for the first sample return. I was recruited by the space sciences and applications directorate to come over as assistant manager of the LRL.

"What I brought to that job wasn't any particular understanding of lunar samples, or even laboratory operations, but good management practices. The laboratory was a blend made up of medical researchers who wanted to examine the astronauts, biological research people who wanted to see what the effects of the material might be on Earth life forms, people who were

The entrance to the crew quarters of the lunar receiving laboratory. Courtesy NASA.

concerned with containment in case the lunar material contained any nasty pathogens, and lunar researchers who believed there weren't any bad bugs on the Moon and just wanted to get their hands on lunar rocks and start doing their thing from a geochemical or geophysical point of view. It was an interesting mix of scientists from different disciplines with different motivations and different agendas, and somehow these things had to be pulled together in time to receive the rocks."

The LRL was constructed to accomplish a rigorous biological containment for the lunar samples, the crew, the CM, and the workers examining the samples. The building's air pressure was set slightly below outside air pressure, and exhaust air was passed through biological filters. People going in and out wore protective clothing and showered upon exit. In the end, the quarantine was judged fully effective. The lunar samples were tested on 35 plant species, 15 animal species and a variety of cell cultures to show that they would not be hazardous to Earth life forms. Researchers debated at great length the choice of species, how to expose them to the lunar material and the protocols for control testing. Many of the plant species thrived in the presence of lunar materials, absorbing significant quantities of helpful elements. No adverse effects were noted, supporting the conclusion that the Moon is sterile.

A good example of the complexity of the research work done in the LRL comes from Keith Richardson, a Canadian geologist working for NASA, who was involved in setting up the radiation testing facility for lunar rocks, which was built 15 m below the ground under the LRL. The lunar rocks were expected to have very low radiation levels, he said, "and so it was necessary to build a laboratory and equipment that would be very sensitive and not have any background radiation, so you'd be able to see the signal above the noise. So everything that went into that lab was analyzed to see that it had no detectable radioactivity in it. Actually, when it was put together, we had a great big rubber balloon to keep the water out. And we had something like five-foot-thick [1.5 m] concrete walls in the laboratory downstairs. Inside that was three feet [1 m] of olivine, that's magnesium silicate, an ultra-basic rock that has little or no radioactive elements in it. And then a steel liner inside that. That was to keep almost all the radiation out and not contribute to the background at all. And fairly fancy pieces of lead-shielded equipment. We had a neutrino shield and all sorts of shielding to stop all the radiation and surround the sample that went inside there."

Said Erb: "Unfortunately, [the LRL] had been underfunded for a year or two, and not adequate attention had been paid to the fact that it had to operate under containment conditions. NASA hadn't wanted to believe that it had to be taken seriously, and about the end of 1968 they got the message, and for the first part of 1969 we scrambled like hell to get the lab ready and make sure the containment was right and do simulations to verify it and satisfy everybody that the lab would function as advertised.

"Getting all these various scientific disciplines and containment constraints to march together was a challenge to whatever management skills I had. And I felt good that it all came off, and I'll take some credit for it, because the lab manager at that time was a quintessential example of one of the scientists who believed that it didn't have to be taken seriously. As long as he got his rocks he was happy. I had to do the job of getting the lab ready even though I was only the deputy manager."

When Erb became LRL manager in 1970, he was one of many people at NASA who faced the wrath of a scientific community that felt that NASA wasn't taking its needs seriously enough. The scientific discontent, which covered issues including the lack of scientists on Apollo crews, landing site selection, and the general priority given to science in Apollo, came to a head in late 1969 and early 1970. Bob Gilruth made a number of changes, including upgrading the role of scientists in directing missions and increasing the scientific training astronauts were given. "That was an interesting experience," said Erb. "In fact, I got frustrated with the scientific community, who really didn't like the discipline imposed on them by the laboratory operations. So I had my differences with some of the lunar scientists who just wanted to examine the lunar rocks and didn't want to be subject to all of the constraints that were necessary."

Erb moved from the LRL in late 1970 to a job in the area of remote sensing, which was just getting started at MSC. Erb's replacement at the LRL was another Avro veteran, Peter Armitage, who had spent much of 1969 and 1970 at Stanford University in Palo Alto, California, on a Sloan Fellowship.

"Landing and recovery was about to be disbanded, as the last of the water-landing spacecraft programs were coming to a close," Armitage said. "Chris Kraft asked me to consider joining the staff of one of two areas and to go talk to the directors involved. I decided that space sciences was closer to my experience base since that organization was responsible for flight experiments and the lunar receiving laboratory, which were programs I had experience with. I joined the staff as assistant director and proceeded to learn about the organization and the science programs. The next Apollo mission was Apollo 14. The LRL was in big trouble with the outside science community. The outside scientists were very vocal in their unhappiness with the way the LRL was being run. They felt that engineering and operational expediencies were detracting from the requirements for good science. We have all heard of the many stories that tell the difference between an engineer and a scientist. Both groups do excellent and essential work, but with a very different set of agendas and priorities. The leadership of the external lunar science group complained to Gilruth and said they were prepared to go to Congress if things were not changed."

When Armitage decided to take on the job of LRL manager, he was immediately introduced to the scientists making up the lunar science and planning team. "In my opening remarks to the team I said, 'What's a nice aeronautical engineer like me doing in a place like this?' The ice was broken, and as I ad-libbed an outline for getting the science community involved at the top level, everyone became happy. They seemed to accept me, and I never had a problem with the science community during the two Apollo missions in which I was manager of the LRL, Apollos 14 and 15. I had always been interested and intrigued with management psychology methods. I used to ask their advice even when I didn't really need it and generally go out of my way to show them that their science needs had priority. I closed down some LRL systems that were not meeting the science requirements and changed the organization to be more responsive to the concerns of the science community."

Some of the problems both Erb and Armitage had to deal with included handling of lunar samples in the vacuum which some researchers required. The gloves that were used to manipulate samples in a vacuum were very stiff and prone to breakage. New methods had to be developed to quickly isolate samples in a vacuum. After Apollo 11, most samples were handled in a controlled nitrogen atmosphere because of the problems with maintaining a vacuum and because samples exposed to the open air could be contaminated.

The first three lunar landings flew under strict quarantine rules, which complicated the work of scientists taking their first look at samples. For example, when the vacuum chamber gloves broke, the affected scientists were put in quarantine. And the use of lunar samples to test for biological effects was not popular with geologists. This biological work was reduced after the quarantine was relaxed, reducing the strain on Armitage and others running the LRL.

The scientists who did research on the lunar samples came from various parts of the world. Most of the top lunar scientists were from the U.S., but scores of scientists in many other countries, including Canada and Great Britain, did important work. Out of the nearly 140 different investigations done on Apollo 11 samples, 15 teams came from Great Britain and 6 from Canada. Armitage got at least one trip back to England related to his work at the LRL.

Although several Canadian scientists worked on lunar samples, most of them with the Geological Survey of Canada, the best-known of their number was a native of Simcoe, Ontario, named David Strangway. Aged 35 when Apollo 11 landed, Strangway was a professor of physics at his alma mater, the University of Toronto, after work at the University of Colorado in Boulder and at the Massachusetts Institute of Technology. As an expert on the magnetic properties of lunar samples returned by Apollo 11, Strangway became familiar to Canadians watching and reading media coverage of the first lunar landings. In 1970 he moved to NASA and became chief of the geophysics branch at MSC, where he was involved in selecting experiments and landing sites, handling lunar samples and training astronauts for upcoming Apollo missions. During lunar landing missions, Strangway worked in the scientific back room in the mission control center, and he was principal investigator on an electromagnetic sounding experiment flown aboard the Apollo 17 service module. After Apollo ended, he returned to the U of T in 1973 and eventually became acting university president. From 1985 to 1997 he was president of the University of British Columbia.

"My particular interest in the lunar samples had to do with two major areas," Strangway said. "One of them was the magnetic properties of the lunar materials, and the other had to do with the measurement of the electrical properties of the lunar materials. You can use magnetism to determine the composition of the metallic materials that are in the samples, such as iron. The other thing you can do is reconstruct ancient magnetic fields. So we were very interested in reconstructing the magnetic field at the time those lunar samples were formed." The pure metallic iron found in lunar samples differs from iron found naturally on Earth, which is usually compounded with oxygen, as in rust. "The second thing we found in terms of the magnetic fields and the preservation of them was that there were very clear indications that they were very weak. There were indeed significant magnetic fields present in the lunar materials. The Moon was formed about 4.5 or 4.6 billion years ago. What we found where we had samples of 3.3 billion or 4 billion years of age, they still had a significant memory of some magnetic field that we believe must have existed at that time. So that tells us that unlike today, when there is no magnetic field on the Moon of any significance, that during that period of the Moon's evolution, there was a magnetic field or possibly there was even a small core causing a small magnetic field."

The lighter areas of the Moon, known as the highlands, were formed about 4 billion years ago, and most of the rocks there are breccias, created from the impacts that left the Moon covered by craters, Strangway said. The darker parts of the Moon, known as seas, appeared between 3 and 4 billion years ago and are made up of volcanic basalts.

Another Canadian geologist from Toronto also worked at the Manned Spacecraft Center. A specialist in radioactivity in rocks, Keith Richardson trained at the U of T and Rice University in Houston. He was hired by NASA in 1964, and his work included helping train astronauts, in a series of field trips around the western U.S. and Iceland, for geological work on the Moon. Richardson worked on setting up the LRL and was a principal investigator on geological experiments on Apollo 11 and 12 before he returned to Canada in 1971 for a career at the Geological Survey of Canada.

•

Apollo 12 left behind the first ALSEP, which began sending back data from a set of experiments that included a seismometer to measure moonquakes and other devices that recorded information on the Moon's properties. Similar packages were left by all the Apollo crews that landed after Apollo 12.

An Avro engineer who joined a U.S. aerospace contractor helped design the radioisotope thermal generators (RTGs) that powered the ALSEP. Robert Cockfield, a native of Toronto who worked in flight test at Avro and remembers testing ejection seats in the CF-100 with Peter Armitage, went to the General Electric Space Division near Philadelphia in 1960. He started working in a military communications satellite project, but then was brought in to help the military with its troubled Discovery satellites. Discovery was the public name for a secret reconnaissance program now known as Corona. Discovery satellites suffered a series of failures when they tried to return to Earth re-entry capsules containing photographs of Soviet military targets, and Cockfield helped equip the satellites with cold-gas thrusters that spun up and stabilized them, contributing to their ultimate success.

The RTG used in the ALSEP was known as SNAP-27, for Space Nuclear Auxiliary Power. This generator converted the thermal energy created by the plutonium dioxide fuel into

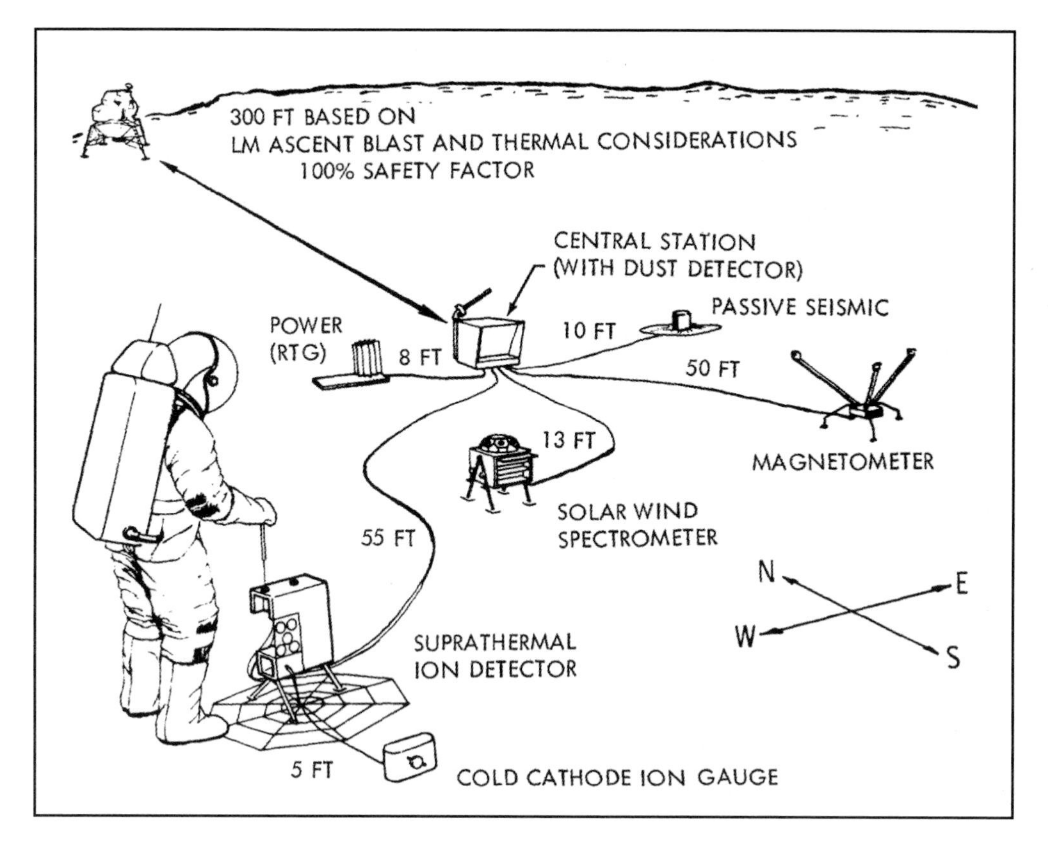

The ALSEP deployed by the crew of Apollo 12 in the Ocean Of Storms in 1969.
Courtesy NASA.

electrical power, which ran the ALSEP experiments through the lunar day and the long, dark lunar night, when solar power wasn't available. The team at GE that included Cockfield began work on SNAP-27 in 1965 under contract from the U.S. Atomic Energy Commission.

"One of my contributions to the design of SNAP-27 was the vent and filter assembly on the fuel capsule," Cockfield said. "A vent was required in the fuel capsule to allow the release of helium, a product of radioactive decay. The vent incorporated a filter with tiny laser-drilled holes, too small to allow particles of fuel to escape. Venting to the lunar atmosphere was delayed until the astronauts had departed."

Until the astronauts placed the nuclear fuel inside the RTG using a special tool, the plutonium was contained inside a graphite cask that could withstand explosions on the launch pad or re-entry at high speed. "Safety testing was a big part of the engineering development of SNAP-27, to ensure that no fuel would be released in the event of various hypothesized launch accidents. We tested fuel capsules by heating them up to operating temperature and then firing them with a gun to impact on a slab of granite. The fuel capsule was designed to withstand an incredible amount of deformation without rupturing," Cockfield said.

After they developed the RTGs for Apollo, Cockfield and his colleagues worked on RTGs that powered the Voyager spacecraft, which flew by Jupiter, Saturn, Uranus and Neptune in the 1970s and 1980s. His work has also included work on underwater habitats and wind turbine generators. By 1997 he was manager of Space Power Engineering, responsible for the

RTGs that were launched that year on the Cassini mission, which is due to arrive at Saturn in 2004. The former General Electric Space Division is now part of the aerospace giant Lockheed Martin.

•

Following the return of Apollo 12, changes were made at the Manned Spacecraft Center. As he had planned two years earlier, Owen Maynard left NASA early in 1970 and took a job in private industry at Raytheon in the Boston area under his old boss, former Apollo manager Joe Shea. Before he left NASA, Maynard visited Canada, where he gave several speeches and made media appearances to talk about Apollo 11. John Hodge, whose work with the lunar exploration working group had expanded into being manager of the advanced missions program office at MSC, left NASA in June 1970 for a new post in the U.S. Department of Transportation. Jim Chamberlin left NASA early in 1970 to take a job at McDonnell Douglas. At the same time, most of the top management of the Apollo program in Houston, Huntsville and Washington was changing, and new people managed the remaining Apollo flights.

•

The third Apollo Moon landing was set for an area called Fra Mauro, where geologists believed that astronauts would be able to find lunar samples representing various parts of the Moon's history. The new emphasis on science that infused Apollo included more training in geology. The crew of Apollo 13 – commander Jim Lovell, a veteran of two Gemini flights and Apollo 8, LM pilot Fred Haise and CSM pilot Ken Mattingly – underlined Apollo's new dedication to science by placing the motto "Ex luna, scientia" (From the Moon, science) on their crew patch.

By the time of Apollo 13's launch on Saturday, April 11, 1970, public interest had dimmed considerably since the flights of the previous year. Even the unprecedented last-minute substitution of backup CSM pilot Jack Swigert for Mattingly failed to stir much interest. The swap took place after Mattingly was exposed to a child with measles and doctors feared that he would become ill during the flight - fears that proved to be unfounded.

For its first three days, Apollo 13 followed the routine set out by the lunar flights that had proceeded it. On the evening of Monday, April 13, the crew sent down television pictures from inside the CM Odyssey and the LM Aquarius. A few minutes after the broadcast ended and the hatch to the LM was closed, Houston asked Swigert to perform the routine procedure of stirring the liquid oxygen and liquid hydrogen tanks in the service module. The tanks were linked to the fuel cells that supplied power to Odyssey, and the oxygen tanks also supplied air to breathe. Swigert began the procedure, and soon the crew heard – and felt – a bang as alarms went off aboard Odyssey. "Okay Houston, we've had a problem here," Swigert radioed to Houston. When the CAPCOM asked him to repeat what he'd said, Lovell said, "Houston, we've had a problem." The first indications were confusing, but soon the seriousness of the situation became clear. Power from the SM's fuel cells began to fail, and it became clear that Apollo 13 wouldn't land on the Moon. Then the crew noticed that the quantities in the SM oxygen tanks were falling to zero, and Lovell looked outside the window and saw something venting into space. The crew of Apollo 13 was now in a struggle to survive. They soon reopened the LM, which became their lifeboat, and powered down the CM.

Although they wouldn't learn the details until investigations after the flight, what had happened was that a series of mishaps and errors during ground tests had stripped away insulation from inside one of the two SM oxygen tanks and rendered a thermometer inoperative. When the fans inside the tank were switched on by the unsuspecting Swigert, teflon inside the tank caught fire, and the tank exploded, blowing the cover off that section of the SM and causing the contents of the other oxygen tank to leak into space through a broken line.

Apollo 13 was more than 340,000 km away from Earth, and the loss of the SM and its giant engine meant that turning around and heading home was out of the question. Apollo 13 would have to loop around the Moon and then come home, which meant that the LM had to support three astronauts for three and a half days when it was designed to support only two astronauts for a day and a half. Under the leadership of flight director Gene Kranz, flight controllers in Houston and experts around the U.S. went to work to save the crew of Apollo 13. Once Aquarius was up and running and Odyssey shut down, the astronauts fired Aquarius's descent engine to put the spacecraft on a course that would bring it back home quicker after it looped around the Moon. Fortunately, the crew of Apollo 9 had tried using the LM engines to maneuver the LM with a CSM attached, but much of what the Apollo 13 crew had to do had never been tried before. The service module was dead, but wasn't cast off because of the protection it gave to Odyssey's heatshield. The command module was powered down, a first in space, and mission controllers had to draw up a procedure to return it to life just before re-entry to Earth, a complex and difficult task because power was very limited. The lunar module was powered down to conserve water and power. While there was enough oxygen, there were not enough cartridges in the LM to scrub out the carbon dioxide exhaled by the astronauts, and the CM cartridges didn't fit the LM system. A team at mission control concocted a device to fit the CM cartridges into the LM system using flight plan covers and duct tape.

Carpentier, who was not assigned to Apollo 13's recovery, was in the staff support room of mission control when Odyssey's oxygen tank blew. "Being a flight surgeon in mission control is not an exciting job. It is a boring job, for the most part," he said. "They had to have some guidance on what happened when the carbon dioxide level hit a certain point, and they needed some kind of timelines on the problems that had to be solved. I was there kind of on the periphery giving advice when I could give it." Outside of this work, there was little he could do. "You could give some guidance when it came to physiology. But it was an engineering problem. It doesn't matter how much oxygen you have or how much of the carbon dioxide you're going to get rid of, or how much fuel there is. If you don't save power, you aren't going to make it. Shutting down the guidance system - that had never been tested before."

The Mission Operations Control Room during Apollo 13. Flight director Eugene Kranz watches Fred Haise on the TV monitor. Courtesy NASA.

The astronauts had to fire the LM's descent engine again to bring Apollo 13 back on the path to the selected splashdown point on Earth shortly after they passed behind the back of the Moon on Tuesday, the evening after the explosion. Then they had to endure long hours of cold and dampness in a powered-down spacecraft as they headed to a re-entry scheduled for the Friday morning. On Earth there was furious activity at mission control, the Manned Spacecraft Center and in contractor plants around the U.S. as engineers scrambled to help devise the fixes needed to bring Apollo 13 home.

On Thursday afternoon, with less than 24 hours to go before splashdown, an engineer from Grumman put in a call to the University of Toronto's institute of aerospace studies in Downsview, north of Toronto. Barry French was pulled out of a staff meeting and took the call, and soon he and others in a group including Ben Etkin, Irvine I. Glass, Philip A. Sullivan and Peter C. Hughes went to work on answering the question that came from the LM manufacturer. When Apollo 13 cast off Aquarius the next morning in the vicinity of Earth, the LM had to be separated farther than it normally would be to make sure there was no collision with the CM during re-entry. In a normal separation, the air was removed from the tunnel that connected the CM and the LM, and a set of explosive charges physically separated the ring holding the two craft together, pushing them apart. To separate the craft further in this special situation, NASA proposed to leave some oxygen in the tunnel, which would give the two craft an extra push. The U of T group was asked whether leaving some air in the tunnel would be safe given the pyrotechnics used in the separation, and whether the air would be effective in pushing the LM out of harm's way. "We did some rough calculations," Etkin said. The air pressure in the spacecraft was 258 mm of mercury, a third of the air pressure at sea level on Earth. The U of T group concluded that air at about half of the spacecraft's normal air pressure was the right pressure for the tunnel. They phoned their conclusion back to Grumman. "In the end, they told the astronauts to do pretty well what we said."

Odyssey's separation from Aquarius went according to plan, and the LM was destroyed on re-entry, except for a specially designed cask containing radioactive fuel for the ALSEP. The cask now rests at the bottom of one of the deepest trenches in the Pacific. Apollo 13 splashed down safely in the Pacific Ocean on Friday, April 17, and was recovered by the USS *Iwo Jima*, watched by a worldwide television audience that rivaled the viewership for the Apollo 11 Moon walk. The crew was chilled to the bone, had lost weight and Fred Haise was ill, but they survived. The astronauts and those who worked in mission control, including Carpentier and Rod Rose, were awarded the presidential Medal of Freedom, the highest civilian award in the U.S. Although Apollo 13 was perhaps mission control's greatest moment, luck had also played a part. Had the oxygen tank explosion taken place later in the mission, after the LM had begun operations at the Moon, the crew of Apollo 13 would have been doomed because there wouldn't have been enough consumables left in the LM.

•

Although President Nixon flew to Houston for ceremonies honouring the controllers and later went to Hawaii to welcome Lovell, Haise and Swigert home, space held a lower priority for him than it had for his two immediate predecessors. Budget cuts meant major reductions in post-Apollo programs, and before Apollo 13 had flown, plans for an Apollo 20 mission were cancelled. Later that year, two more Apollo missions were cancelled. NASA decided to fly one more H mission, Apollo 14, to Fra Mauro, and then fly three advanced J missions before Apollo ended.

Because of the need to investigate the causes of Apollo 13's near-catastrophe and then make changes to Apollo systems to prevent a repeat, Apollo 14 didn't lift off until January 31, 1971. The flight was commanded by Alan Shepard, the veteran of America's first space flight, who had recently returned to flight status after an inner ear problem that had grounded him was cured through surgery. Flying with him were two rookies, CSM pilot Stu Roosa and LM pilot Edgar Mitchell. After landing the LM Antares successfully at Fra Mauro, Shepard became the only one of the original seven astronauts to walk on the Moon. In two Moon walks, he and Mitchell set up an ALSEP and then, pulling a specially designed cart to carry tools and cameras, they explored near the rim of a large crater behind their landing site. Shepard wrapped up the second Moon walk by hitting two golf balls with "a genuine six iron" attached to the handle of a sampling tool. The two Moon walks lasted more than nine hours, and Antares' stay on the Moon was 33 hours. The three astronauts splashed down February 9 with 43 kg of lunar samples stowed aboard the CM Kitty Hawk.

Apollo 14 was the last of the lunar missions to be quarantined, and Carpentier was locked into the MQF aboard the USS *New Orleans* with Shepard, Roosa and Mitchell. Although parts of the quarantine procedure were relaxed for this flight, Carpentier said the experience was similar to Apollo 11 because there was a great deal of work to be done with the astronauts, the lunar samples and the spacecraft during the quarantine.

"There wasn't a lot of love lost between Al Shepard and the medical directorate," Carpentier said. "He had been grounded for a long time, and he was not happy with physicians in general and NASA physicians in particular. I was talking to one of the astronauts, and I said I was going to do Apollo 14. And he said, 'Don't do it. Quit your job, go back to Canada, but don't tangle with Al Shepard because he will chew you up and spit you out.' I was going to leave anyway, and I wanted to do that flight. I called him up and said, 'I'd like to have a few moments of your time to talk about the postflight medical testing.' He was great. He would say to whatever I would ask him, 'Whatever is possible.' I think that for us flight surgeons on an operational level, there was no problem getting along with the crews. But on the political level, there were a lot of problems."

Carpentier had had a similar experience with Walter Schirra, the only other Mercury veteran who flew in Apollo, on his Apollo 7 mission. Although Schirra did not always work well with mission control on that flight, and Carpentier was warned that he wouldn't cooperate in the postflight medical examinations, Schirra cooperated fully after Carpentier had explained the whole medical program to him before the flight.

After undergoing intensive training in geology, the crew of Apollo 15 - commander David Scott, CSM pilot Al Worden, and LM pilot Jim Irwin - set off from Cape Kennedy on July 26 in the first of the advanced J missions. Their target was a small valley known as the Marsh of Decay, nestled between the Apennine mountain range and Hadley Rille. The site, the first one that wasn't close to the lunar equator, was a challenge to land in but was geologically rewarding, and its beauty matched up to science fiction predictions of lunar landings. The LM Falcon landed on the Moon on July 30, and the next day Scott and Irwin unfolded and set up the lunar roving vehicle that took them farther from the LM than any crew before them. The battery-powered rover carried the two astronauts and their equipment up to 65 km at speeds up to 17 km per hour.

The rover was often out of sight of the LM, so a special communications unit was needed to keep the astronauts in touch with the Earth. The unit also permitted television transmissions from a colour TV camera mounted on the rover. Avro veteran Norm Farmer was part of the group that was drafted to build what he named the lunar communications relay unit. "I had a lot of objections because they thought the abbreviation LCRU could be mistaken for the crew. Anyway, it stuck," he said. The group designed the unit, which was built by RCA. The LCRU sat at the front of the rover, attached to a high-gain dish antenna, which the astronauts aimed at Earth using a special sight developed by Farmer's group. Since an astronaut in a space suit and a helmet couldn't use a normal boresight to aim the antenna, the group developed a sight that projected the Earth on a small screen when the antenna was correctly aimed. With this system, the rovers on the J missions sent back breathtaking television of the mountains, hills and craters of the Moon. The television camera was controlled from Houston, leaving the astronauts free to explore while controllers watched their activities. Before working on this system, Farmer had built test instrumentation for earlier Apollo spacecraft and boilerplates that provided important information on how the vehicles operated. This instrumentation included cameras that

Jim Irwin stands next to the first lunar rover during Apollo 15.
Courtesy NASA.

were aimed out the windows of the unmanned Apollo 4 and 6 vehicles, taking memorable photos during their high Earth orbit flights.

Scott and Irwin explored craters, mountains and the edge of Hadley Rille in their rover, drilled core samples under the surface, and set up another ALSEP in three Moon walks totalling nearly 19 hours. Among the 77 kg of samples they brought back was the so-called Genesis rock, made up of material believed to be from the Moon's ancient crust. After 67 hours on the surface, Falcon blasted off. For the first time the lunar liftoff was televised, using the television camera left behind with the rover. Worden, on board the CSM Endeavour, operated the scientific instrument module in the SM, which contained instruments that photographed the lunar surface and took measurements of the Moon. Endeavour also released a small subsatellite that would take further measurements. After Scott and Irwin rejoined Worden and the work in lunar orbit was done, Endeavour headed back to Earth. Nearly 320,000 km away from Earth, Worden left the spacecraft to retrieve photographic film cassettes from the service module, and on August 7 the flight of Apollo 15 came to a successful close. Carpentier was aboard the USS *Okinawa* for his last recovery operation - this time without the quarantine - before he left NASA. He also left the field of aviation medicine. His training in handling nuclear medicines, which allowed him to conduct full batteries of tests on the Apollo astronauts in quarantine, led him to a new career in nuclear medicine at Scott and White Clinic in Temple, Texas. He was director of the nuclear medicine department there for 12 years. His wife Willie, a nurse who earned a Ph.D., also conducts research at the clinic, and their son Brad is an anesthesiologist there. Their other son, John, works in procurement at the Johnson Space Center.

•

The extensive geologic training for Apollo's lunar landings included field trips to various parts of the U.S. and some foreign sites. David Strangway, who was involved in many of these field trips, described what happened. "We asked them, 'What would you sample, what would you pick up, what photographs would you take, what choices would you make to describe the nature of the [formation]?' There were other trips that we went on, craters in Nevada and other places, trying to understand the dynamics of what happened in the cratering process. What kinds of samples should be taken, what features should be looked for, observations that should be made so that people could understand the nature of that crater."

Canada joined the list of countries visited by astronauts seeking to learn about the Moon when, in July 1970, the crew of Apollo 15 and eleven other astronauts were taken to the Canadian Forces testing ground at Suffield in southern Alberta to witness a 450-tonne TNT explosion that created an 86-metre-wide crater. Examining the crater following the blast helped the astronauts know what to expect in and around impact craters. Better known are the visits of the landing crews of Apollo 16 and 17 to Sudbury, Ontario. The oval-shaped basin containing the nickel that is the basis of Sudbury's wealth was originally a crater caused by the impact of a meteor about 1.7 million years ago. There are many fractures, faults and shatter cones in the area, like those caused by similar violent events on the Moon. And while impact breccias - rocks associated with impacts - are rare on most parts of the Earth, they are plentiful in the Sudbury area and on the Moon. Accompanied by instructors, Apollo 16 astronauts John Young and Charles Duke came to Sudbury in July 1971, and Gene Cernan and Jack Schmitt of Apollo 17 visited in May 1972.

For the heavily scientific J missions, simulations were aimed at training both the astronauts and the scientists. Strangway recalled simulations involving astronauts driving an earthbound version of the lunar rover across a simulated Moonscape, while scientists watched via television from their back room in Houston. "They would be down there running a mockup of the lunar rover around, and the medical guys said we have 10 more minutes of science. What do you want to do in that time? So there were all kinds of simulations of that kind, which were very

Astronauts John Young and Charles Duke undergo geological field training near Sudbury, Ontario, in July 1971. Courtesy NASA.

interesting exercises and actually forced you to look at your priorities and to think of what you wanted to do in that mission so that you could get the maximum return after whatever crisis came up. What it did was force us to deal with the competing interests, the physicists, the chemists, the geologists, the astronomers, all of these people who wanted the maximum information return. What it really did was force us to exercise our minds as to what was really important. I think it had a real impact on the actual design and the actual layout of what happened in the missions."

Apollo 16 left Earth on April 16, 1972, bound for the highlands of the Moon. Ken Mattingly, who had been bounced from Apollo 13 at the last minute, flew Apollo 16's CSM, nicknamed Casper, while commander Young and LM pilot Duke landed their LM, Orion, on the Plains of Descartes between two large craters. The mission went smoothly except for a problem with the SM's large engine that caused mission control to delay Orion's landing for a few hours. Once on the Plains of Descartes, Young and Duke found impact breccias and saw shatter cones almost everywhere their rover took them. This confounded scientists who had expected the area to be rich in volcanic, rather than impact, formations. On their third Moon walk at North Ray Crater, Young picked up a rock and recalled his trip to Canada. "It looks like a Sudbury breccia, and that's the truth," he said as he described the rock. The flight concluded on April 27 with the crew bringing home 94 kg of samples.

•

Only one more expedition to the Moon remained, and with it came the special dangers of such a flight. Most of these arose from the fact that a return home from the Moon required at least three days, but there were other dangers that were little appreciated by the public. Although space lies outside the Earth's atmosphere, there is still "weather" to contend with due to the stream of particles from the Sun known as the solar wind. The Earth is protected by its magnetic field and the huge radiation belts named after their discoverer, Dr. James A. Van Allen, that follow the magnetic field. But the Moon lies outside these belts, leaving the astronauts open to danger from solar radiation. Much of the time this radiation is at safe levels for space travellers, but occasional storms on the Sun sometimes result in dangerous, and in rare cases lethal, levels of solar radiation at the Moon. This danger received almost no publicity during the Apollo program, but doomed the fictitious Apollo 18 mission in James Michener's best-selling 1982 novel *Space*. In a real situation the CSM would provide shelter from most of the radiation, but astronauts in the LM or walking on the Moon's surface were exposed to any solar radiation that might strike them.

NASA constantly monitored the Sun and its radiation during the four years that the nine Apollo crews flew to the Moon or its vicinity, and Rod Rose chaired the radiation constraints panel that set safe limits for radiation exposure and settled radiation issues between and during flights. The panel included space physicists, physicians and two astronauts - Bill Anders, who was a nuclear physicist, and Joe Kerwin, a doctor. It set the maximum allowable dose as the amount of radiation that would double the incidence of leukemia in a white male subject between 25 and 39, about 400 rem full body dose.

"It turned out that with the Sun rotating, the solar particle events that were coming up on the east limb were the ones that we were concerned with, because that's where the spiral of matter ejected from the Sun would come into the Earth-Moon area," Rose explained. "If you had a solar particle event going around the west limb of the Sun, it really didn't bother us much unless it was sustained when it came back around the east limb of the Sun, because the ejecta from the west limb would go out into deep space. If you had a major event, the guys came up with a program on the computer such that by taking four to six hours of data, they could project what the peak flux was going to be in the Earth-Moon environment, and therefore they could work out what the dose was going to be for the crew, whether they were in suits on the lunar surface, in the LM or in the CSM around the Moon or in free space. We set up a whole set of limits and procedures. For example, if the crew were on the Moon and an event occurred, we could have enough time to alert the crew, get them back into the lunar module, set up and get ready, take off, rendezvous with the CSM, put the CSM in what we called the hatch down mode, and that put the thick section of the command module facing outward away from the Moon, using the Moon as a shield, and that way you could cut the free space dose down to about a third."

Astronauts on the Moon would have about 16 hours' warning of any major solar event, during which they would have to get from the Moon's surface back to the safer confines of the CSM. Although the Sun goes through an 11-year cycle that includes periods of low and high activity, Rose said a major solar event could occur at any time.

"Purely fortuitously, we flew Apollo between each of these major solar events. In an actual Apollo mission, we never had to do anything from the radiation point of view. We did have one [event] that if it did occur at the wrong time, we would have had to take some action. I think it was an event where if the crew had been in their suits on the surface and we hadn't have done anything, they would have exceeded the maximum dose. But I hasten to add, the dose would not kill the crew." Indeed, the maximum allowable dose was well below the level where radiation would have caused vomiting, the mildest direct immediate effect of a radiation overdose.

Another former Avro engineer spent much of his career at NASA dealing with another hazard of space travel: meteoroids and micrometeoroids. Burt Cour-Palais had worked on Apollo heatshield designs early in the program, but soon he was working on protecting spacecraft and astronauts from meteoroids. Every day the Earth is bombarded by thousands of micrometeoroids and meteoroids, which are classified as meteors when they enter the atmosphere, where almost all of them burn up harmlessly. Only a few of these objects reach the Earth's surface, but they can pose a danger to spacecraft, and Cour-Palais, the Apollo subsystem manager for meteoroid protection, was responsible for developing standards for meteoroid shielding for Apollo spacecraft and space suits. His work included testing various materials and structures to see how well they protected against meteoroid strikes. A special cannon that fired small projectiles was set up at Ellington Field near Houston, and later Cour-Palais and his colleagues set up a more sophisticated test facility on the Manned Spacecraft Center site.

Gene Cernan and the lunar rover during the flight of Apollo 17.
Courtesy NASA.

Meteoroids were a concern even in Mercury, and Leslie St. Leger, another former Avro engineer, had done meteoroid studies in Mercury and early meteoroid protection work for Apollo. St. Leger drew up a proposal for an unpiloted Mercury capsule to research meteoroid conditions in space for two weeks. This plan didn't fly, but the last three of the nine Saturn I rockets carried Pegasus satellites into orbit in 1965. Pegasus was made up of the second stage of the Saturn fitted with large wings containing sensors that recorded meteoroid strikes. As well, returning spacecraft were scrutinized for evidence of meteoroid damage. As Cour-Palais found, the best place to look for it was the windows, since they provided flat surfaces that were relatively unaffected by re-entry forces. Research on meteoroids showed that having a bumper, an outer layer of metal separated from the next layer, provided the best protection against meteoroids, because the bumper would dissipate the meteoroid's energy before it struck the next layer.

Cour-Palais was born in India in 1925, the son of a British railway worker, and was educated at St. Francis Xavier College in Calcutta and the College of Aeronautical Engineering in London, England. He worked at Vickers-Armstrong Ltd. and Bristol Aircraft Ltd. and on his own as an aircraft structures analyst before joining Avro in 1957, where he continued his work as a structures engineer. He joined STG and NASA in January 1960.

Although micrometeoroid strikes of spacecraft are routine events, no meteoroid event has yet endangered astronauts. Cour-Palais said he had a scare when Apollo 13's oxygen tank exploded. "We had gone to see the launch, and we were on the way back when we heard the bad news and got home," he said. "I tried to convince everybody that it was a micrometeoroid."

•

The last and perhaps greatest of the six Apollo expeditions to the Moon's surface left Cape Kennedy shortly after midnight on December 7, 1972. Hundreds of thousands of people gathered around the Kennedy Space Center to witness the only night launch of a Saturn V, and the light show from Apollo 17 didn't disappoint. The crew included commander Gene Cernan, CSM pilot Ron Evans, and LM pilot Harrison "Jack" Schmitt, the only geologist in the astronaut corps. Cernan and Schmitt headed for the valley known as Taurus-Littrow, a site that scientists hoped would provide a variety of new and old lunar material to further illuminate the history of the Moon. The two astronauts landed their LM Challenger at Taurus-Littrow on December 11, while Evans worked on his orbital experiments aboard the CSM America. Cernan and Evans spent 74 hours on the Moon, including 22 hours of Moon walks, during which they collected 110 kg of lunar samples. They set up another ALSEP and drove their rover 36 km in all, ranging as far as 7.37 km away from the LM. Their excursions took them up two massifs that bordered the valley, bringing breathtaking views. On December 14, Challenger lifted off, ending the first round of human exploration of the Moon, and five days later, after some further work in lunar orbit and the return to Earth, Apollo 17 splashed down in the Pacific on December 19, 1972, bringing Apollo to an end.

The program had seen 11 piloted spacecraft fly, including nine that flew around the Moon, eight in lunar orbit, and six that landed 12 astronauts on the Moon's surface.

Although Apollo 17 was the last American spacecraft to visit the vicinity of the Moon until the Clementine probe entered lunar orbit in 1994, the first era of lunar exploration didn't end until later in the decade, mainly due to continuing Soviet efforts. After the N-1 disaster of July 1969, the Soviets tried twice more to launch the N-1, once in 1971 and again just two weeks before the liftoff of Apollo 17. Both failed, as their predecessors had done, while the first stage was firing. The N-1 lunar program was suspended in 1974 and terminated two years later, ending any plans to put Soviet cosmonauts on the Moon. The circumlunar program was wound down, also without ever flying a cosmonaut. Two weeks after Apollo 11's return, Zond 7 set off on a successful flight around the Moon, and in 1970, Zond 8 ended the series with a loop around the Moon and a splashdown in the Indian Ocean. That year was also notable for two other Soviet lunar firsts. In September, Luna 16 succeeded where Luna 15 had failed. It landed in the Sea of Fertility and returned a small soil sample to Earth. Two months later, Luna 17 soft landed a six-wheeled robot car called Lunokhod, which drove for 11 days on the Sea of Rains, taking photos and analyzing soil. The Soviets sent seven more Luna probes to the Moon, including two orbiting spacecraft, another Lunokhod vehicle, two sample return missions that failed, and two that succeeded. Luna 24, which returned samples drilled from the Sea of Crises in August 1976, was the last Soviet spacecraft to visit the Moon. On September 30, 1977, ALSEPs left behind by Apollo crews were switched off in an economy move, ending the first era of lunar exploration with a whimper rather than a bang.

Apollo had cost $25 billion in uninflated U.S. dollars, or $80 billion in 1990 U.S. dollars. The war in Vietnam, which helped bring Apollo to what many saw as a premature end and curtailed NASA's ambitious post-Apollo plans, cost seven times as much. A month after Apollo 17 returned home, former president Lyndon Johnson died on his Texas ranch, and soon the Manned Spacecraft Center he helped set up was renamed the Johnson Space Center in his honour. Also in 1973, the name Cape Canaveral, one of the oldest geographical names in America, was restored to the spit of land in Florida that contains the launch pads for American astronauts and many of its satellites. The space center where astronauts depart the Earth still bears the name of John F. Kennedy.

The samples and results from Apollo, Ranger, Lunar Orbiter, Surveyor and Luna spacecraft gave astrogeologists and other scientists information that they are continuing to digest. In 1984, a meeting of lunar scientists effectively brought the long argument over the origin of the Moon to an end when most agreed that the Moon was created when the young Earth was struck by an object the size of Mars. Matter cast off in the collision from the Earth and the other object formed the Moon.

The results from Apollo's research into lunar history have taught scientists a great deal about the early history of the Earth as well as that of the Moon, according to David Strangway. Before Apollo, the Earth's early history was obscured because of the Earth's atmosphere and the forces that continue to change the Earth's surface. "The lunar samples that are 3 or 4 billion years old look as fresh as lavas that came out last week on the Earth. They are absolutely unaltered from the point of view of their chemical composition.

"What became very clear is that the rate of impacts taking place in the solar system was very non-uniform. It was very high up until 4 billion years ago, then slowed down immensely between 4 and 3.3 billion years ago. What this suggests is if you look at the period between the formation of the solar system, 4.6 billion years ago, and then the end of this most intense period of bombardment, 4 billion years ago, the solar system itself probably underwent an intense set of bombardment activity. And therefore, there was a whole piece of the Earth's history that was probably the same as the early history of the Moon, but there's no surface left of that age anymore, so we didn't even know about that. We learned a lot about the Earth's early history in a way that we hadn't been able to record."

The findings from Apollo drove home to the scientific community the importance of impacts in the histories of both the Moon and the Earth. Apollo forms one link in the chain of discoveries that led scientists to conclude that catastrophic impacts by asteroids and comets have altered Earth's history on several occasions, one of them causing the extinction of the dinosaurs. Today, more scientists are working on the search for objects that could strike the Earth and wipe out the human race.

The legacy of Apollo includes many technological advances, particularly in the field of computers. The photographs brought home by the astronauts taught humans to look at their home planet in a new way. Thanks to Apollo's scientific returns, humans have a better idea of our planet's history and a better chance to preserve our future.

12 Building the Space Shuttle

Long before Apollo 17 closed the Apollo program, the question of what would follow the lunar landings had become the biggest issue inside NASA. At the beginning of 1969, Apollo assembly lines were already being wound down, and the Apollo Applications Program, NASA's ambitious plan to continue human space exploration using Apollo hardware, had been reduced to a minimal effort. If NASA didn't act quickly, it would soon be out of the human spaceflight business.

As John Hodge wrapped up his work with the exploration working group, he handed the job of making the J missions a reality to the Apollo program office and the Apollo lunar exploration office. In October 1968, Hodge's working group became the advanced missions program office, and Hodge was named manager. The office was charged with looking beyond Apollo to programs such as a permanent space station, a reusable spacecraft, and trips to Mars.

"By that time, [Robert] Gilruth was beginning to say, 'What are we going to do after Apollo?'" Hodge recalled that Gilruth wanted a space station and commissioned studies to back up this idea. "The conclusion that I came to, personally, was that we should build a shuttle. You really couldn't do a space station program without a reliable transportation system. Our conclusion was that it had to be reusable. I remember the presentation we made to Gilruth and the staff at Houston. Dennis [Fielder] made the presentation. We were very excited about the idea of building what eventually became the shuttle. And everybody was interested."

One reason for the interest was the extremely high cost of sending satellites, spacecraft and payloads into Earth orbit or beyond. Each kilogram put into orbit cost thousands of dollars because the launch vehicle was destroyed following each use. It was like throwing away the jetliner after every flight. Faced for the first time with serious economic and political pressures to cut costs, NASA looked to a reusable spacecraft as the key to space travel after Apollo.

As Hodge recalls it, Gilruth and Max Faget wanted to press on with a space station before the shuttle, so they didn't react positively to Hodge and Fielder's presentation. Gilruth was not always the avuncular figure he is remembered as today, Hodge said, and there was an argument. Whatever happened that day, Faget appeared to change his mind within weeks.

In February 1969, MSC and Marshall Space Flight Center jointly awarded contracts for a study of what they called an integral launch and re-entry vehicle (ILRV), a spacecraft that would be completely reusable. Four contractors – McDonnell Douglas, North American Rockwell, Lockheed and the Convair division of General Dynamics – conducted design studies on the ILRV, which included a piloted launch vehicle that would return to its launch site, and a piloted orbiter vehicle. Martin Marietta also conducted an ILRV study using its own funds.

The idea of a winged vehicle that could fly into space, return home and then be reused is almost as old as the airplane itself. The first serious design for such a vehicle emerged in the 1930s from the minds of Eugen Sanger, a Viennese engineer, and his mathematician wife Irene Bredt, in the form of the Silverbird, a winged vehicle with a rocket engine. During the war they proposed a transatlantic bomber of similar design to the Luftwaffe. As the space age began, the U.S. Air Force pressed for its own piloted orbital spacecraft, a glider known as the X-20 Dyna-Soar that would be launched atop a Titan rocket. The project was cancelled in late 1963. Despite this setback, the concept of reusable winged space vehicles continued to advance through the design and test flights of lifting bodies - wingless aircraft that designers hoped would lead to vehicles that could fly at high speeds from space but allow their pilots time to choose a suitable landing site. As well, many designers in the United States, Europe and the Soviet Union proposed reusable space vehicles.

Although all these ideas led to what would become the shuttle, the studies on the ILRV marked the real beginning of the shuttle effort at NASA. Following the pattern established at the beginning of Apollo, engineers at MSC began at almost the same time to do their own in-house study of ILRV. Under the direction of Faget, NASA's top spacecraft designer and MSC's director of engineering and development, a design group for ILRV was set up in March and April of 1969. At the meeting where he announced this group, Faget showed a model of the proposed vehicle he had made in his workshop at home, and in a July memo he described the vehicle being developed as a two-stage fully reusable craft. Both the orbiter and the booster had fixed, straight wings, and the orbiter rode piggyback on the upper forward fuselage of the booster. The booster would fire its engines at liftoff, and the orbiter would not use its engines until the booster's work was done. The vehicle would stand 57 m high at launch and weigh 613,000 kg. The orbiter would be 53 m long, weigh 143,200 kg at launch and carry 3,175 kg of payload to orbit, plus up to 10 passengers.

Working under Faget was Jim Chamberlin, then design and analysis manager in Faget's directorate. Although Faget was in charge of this "skunk works" effort that went on in secret in a secured area, Chamberlin was given supervision of the day-to-day work. A design review board set up to review the in-house work at MSC included Faget as chairman and Chamberlin as alternate chairman.

Early in the process, Faget directed Chamberlin to do wind tunnel tests examining heat transfer to the vehicle during the early phases of re-entry, because there was little data on this phase of flight. He also wanted Chamberlin to look at gimballing, or steering, the main rocket engines of the vehicle together rather than separately, and to draw up re-entry profiles taking into consideration the tolerance of crew and passengers to deceleration during re-entry.

Working with Chamberlin in Faget's group was Leslie St. Leger, who also came from Avro. "We were put in a separate building," St. Leger recalled. "We were doing design studies on the shuttle. Faget's idea was a simple, big-winged shuttle. The designs from the contractors had delta wings, swept back. The one we looked at was straight winged." A straight-winged shuttle would land at a lower speed than a delta-winged vehicle, Faget reasoned, and so it could land at almost any major airport. At the time, NASA was also looking at an orbiter with a payload capacity in excess of 10,000 kg, but the smaller capacity was kept to the fore because NASA hoped to use the shuttle mainly to deliver astronauts and equipment to a space station in Earth orbit. If the space shuttle became a stand-alone program, the larger payload capacity would be needed.

Peter Armitage, then near the end of his decade in the landing and recovery division, asked Chamberlin if he could help with his shuttle work. "Jim asked me if we would design and test a 1/10th scale shuttle and air drop it to prove the flight dynamics as it transitioned from supersonic to subsonic flight. The project was a natural follow-on to the air drop work we did on the Gemini paraglider. The project was successful and the 1/10th scale shuttle was air dropped at the New Mexico White Sands Test Range."

The question of whether the shuttle would fly, along with the future of the other plans drawn up by Hodge and his group, was up to a higher authority. In January 1969, Richard Nixon became president and his Republican administration began a review of space activities. In his first month in office, Nixon formed the president's space task group, chaired by Vice-President Spiro T. Agnew. The group reported in September, recommending three options for future space activities. The options included a human expedition to Mars in the 1980s, a space station in lunar orbit leading to bases on the lunar surface, and a 50- to 100-person space station in Earth orbit, serviced by a reusable space shuttle. The most modest option included only the Earth-orbiting space station and the shuttle. The trip to Mars and other features of the most ambitious option reflected the work being done by Hodge's advanced projects office. In a memo that August on NASA's planning for the coming decade, Hodge said NASA's planning "has focused upon a manned landing on Mars in the early 1980s." As a result of the success of Apollo 11, President Nixon "has expressed his belief in pursuing an aggressive manned space program and a bold program in that it sets the manned exploration of Mars as a goal in the early 1980s."

In fact, Nixon was silent on the options laid before him, and soon it became clear that he would not fully endorse even the least ambitious of the three options. At the same time, NASA's budget continued to face cutbacks. NASA was left scrambling to find a program that would pass muster with the Nixon White House and the Congress, which were preoccupied with funding the war in Vietnam and dealing with domestic problems. Both the president and Congress were reacting to public opinion surveys that showed Americans believed that the space race had gone far enough. When Nixon took office, NASA's annual budget had lost nearly a quarter of the funding it enjoyed at its high point in 1965, and the Johnson administration had not promoted follow-on programs to Apollo in Congress as part of an effort to protect funding for the Moon program. At its peak, NASA funding had exceeded 4 per cent of the U.S. federal budget. By the end of the 1970s, NASA's budget fell below 1 per cent of the federal budget, the level it had been at before Kennedy launched the Moon landing effort. It has remained at or near that level ever since. The public wanted more money spent on earthly concerns, and the politicians obliged them.

In early 1970, Hodge could see that many of his ambitious dreams would not be coming to fruition in the near future, so in June he left NASA for a new post in the U.S. Department of Transportation. In January 1970, Jim Chamberlin also left NASA to go to work for McDonnell Douglas, which was bidding for the shuttle contract. At the time Chamberlin left NASA, St. Leger remembers that the "skunk works" he directed was disbanded. The design he had worked on with Faget remained NASA's prime design for the shuttle in 1970. At that time it was nicknamed the DC-3 in honour of the sturdy aircraft of the 1930s that serves even today as an airliner and transport plane.

•

As Hodge, Chamberlin and Owen Maynard were leaving NASA, another Avro veteran was joining. Late in 1969, Bob Lindley left McDonnell Douglas in St. Louis and came to Washington, where he became NASA's special assistant to the associate administrator for manned space flight.

"About that time, George Mueller, who was running manned space flight at headquarters, wanted somebody to do something with the shuttle," Lindley said. "I'd contributed to a tailless delta on the Arrow. And I'd done the preliminary design on the Vulcan when I was very young and wild. So I knew about tailless deltas, which they wanted the shuttle to be, and I knew something about space because I'd been involved in Gemini. So I went to NASA headquarters and worked on, in theory, the shuttle. Now that was an interesting experience because George Mueller had moved me and got me there. The second day I was there, he resigned. Chuck Mathews came in, and he was holding the fort until they got a new associate administrator. So I said, 'Chuck, what am I supposed to do?' And he said, 'We're thinking about it.' I said, 'I get bored, what can I do?' He said, 'Go to meetings.' I said, 'We all have meetings here. I've been to meetings. I don't like it.' He said, 'I've got a job for you.'"

The job was dealing with a request from the office of management and budget (OMB), the U.S. government agency that was in the process of slicing major portions from NASA's budget even before it got to Congress. OMB wanted an economic study of the shuttle's benefits because it believed that more proof was needed to back up NASA's claim that the shuttle would save money. Lindley recalls that he began this work at first by himself with the help of an official from OMB. But Lindley was not an economist.

"So I went and placed a contract with an outfit called Mathematica. We got that great study going. My slant on it went like this: Here we've got this wonderful shuttle. What's it for? The word was, we were going to put payloads into orbit, but for an order of magnitude less than what it costs now. It would cost one-tenth of what it costs now. You could take our annual budget and ask what percentage of that goes into putting things into orbit, and what percentage goes into the things that are going into orbit? Oh, about 10 per cent goes into putting things into orbit. So we're going to drop 10 per cent to 1 per cent. It's not going to revolutionize the world, is it? So we concluded that the only way to sell this thing was to prove that the payloads would cost less."

Since Gemini ended in 1966, Lindley had worked at McDonnell Douglas on projects such as Skylab, which was the space station that grew out of the Apollo Applications Program, and McDonnell proposals for various military aircraft, including the swept-wing F-111, which went to another contractor. Lindley became a vice-president of the company, a position he said he did not enjoy.

When Lindley joined NASA, the primary shuttle design at NASA was still Faget's straight-winged vehicle rather than the tailless delta that Lindley was familiar with. But in 1970 and 1971, the fiscal and political pressures on NASA, combined with some engineering realities, eliminated Faget's DC-3 concept from the shuttle sweepstakes.

As it became clearer that NASA would have only one major human space project - the shuttle - and even that was in danger, NASA persuaded the Department of Defense and the U.S. Air Force to get on board. This meant a vastly enlarged payload capacity was needed so that the shuttle could carry reconnaissance satellites and other defence payloads into space. As well, the Air Force wanted the shuttle to have a large crossrange landing capability because it planned to launch the shuttle into a polar orbit where it would deploy a satellite and then return at the end of one orbit. Because the Earth's rotation would move the landing site more than 1,600 km away from the shuttle's track, the shuttle would have to be able to glide off its path during re-entry to return to the launch site or to a designated airfield. At the same time, analysis revealed serious engineering problems with the DC-3 concept. This meant that the straight, stubby wings of the DC-3 concept were out, although both it and the delta-winged concept remained on the table for most of 1970 and 1971.

At the same time as he commissioned the Mathematica report in early 1970, Lindley continued to press the economic case for building a fully reusable shuttle vehicle. Lindley's reasoning, which impressed officials at the OMB, was that the shuttle would lower payload costs because it would permit the use of larger, heavier and more standardized components.

Basing his estimates on a fully reusable shuttle system, Lindley in 1971 projected that the shuttle would cut the "direct cost of space activity" in half. He added that the costs of developing this system would be justified if the space program operated at a level of 40 shuttle flights a year.[11]

As the year wore on, it became clear that another major change to the shuttle would result from the fiscal pressures facing NASA. The OMB, supported by President Nixon and Congress, made clear that NASA's budget would remain flat for much of the 1970s. This made it impossible for NASA to build a fully reusable shuttle. The Mathematica report in 1971 supported the use of a partially reusable stage-and-a-half vehicle, and Lindley moved to build support for this concept within NASA and among the contractors. This partially reusable vehicle would include recoverable boosters that parachuted back to Earth, and a fuel tank that would supply engines on the orbiter during the boost phase and then be cast off and destroyed on re-entry. Unlike the DC-3 concept, the orbiter and booster engines would all start burning at liftoff. The partially reusable shuttle would have much lower development costs in the 1970s, but it had higher operating costs in the 1980s and beyond, due to the need to build new tanks and refurbish boosters for each flight.

The Mathematica report is blamed by some shuttle critics for helping to sell the claim that the shuttle would dramatically lower launch costs, a claim that was not borne out when the shuttle flew. In fact, the report made the point that the only way to make a major dent in launch costs would be to fly the shuttle much more frequently than it has in fact flown. So NASA can be blamed for being overly optimistic in projecting the need for the shuttle. But Mathematica's analysis did endorse from an economic standpoint the use of a partially reusable shuttle. While it is hard to say what would have become of the shuttle had the fully reusable design been adopted, the Mathematica report was certainly instrumental in leading NASA to its final shuttle configuration, with all its drawbacks, economic and otherwise. The strength of the concept was that it could be sold politically.

[11] Since the shuttle program began, the shuttle has never flown more than nine times in a year.

In 1970 and 1971, NASA and leading contractors created many different shuttle configurations, including shuttles launched atop Titan and Saturn rockets, and even a concept (described in Chapter 7) known variously as Big Gemini or Big G. But by the end of the year, NASA's new administrator, James Fletcher, went to the Nixon White House with a delta-winged orbiter attached to a disposable external tank and two boosters. Pending a decision on whether the two boosters would be solid- or liquid-fuelled, the configuration was that of today's space shuttle.[12]

•

On January 5, 1972, Fletcher and his deputy George Low met Nixon at his western White House in San Clemente, California, to put the finishing touch on their discussions of the shuttle. Less than an hour later, the men from NASA emerged with a two-page statement from the president endorsing what was called the space transportation system and saying it would "take the astronomical costs out of astronautics" and be ready to fly by 1978. In the face of the strong opposition the shuttle had encountered from many sources, Nixon stated his belief that American astronauts must continue to fly. The reasons included national security and prestige, and the need to breathe new life into an aerospace industry that was then suffering from one of its worst downturns. Many of the affected contractors were just a short drive from San Clemente, and Nixon needed California to win the presidential election later that year.

During this time, Jim Chamberlin, who had gone to McDonnell Douglas and moved to St. Louis, was part of the McDonnell Douglas/Martin Marietta consortium bidding for the shuttle contract. In 1970, NASA awarded this consortium and North American Rockwell contracts to study fully reusable shuttles. Each team produced two designs, one with straight wings on the orbiter for low crossrange capability, and another with delta wings similar to today's shuttle orbiter. McDonnell Douglas handled the orbiter designs in the proposal, and the orbiters were

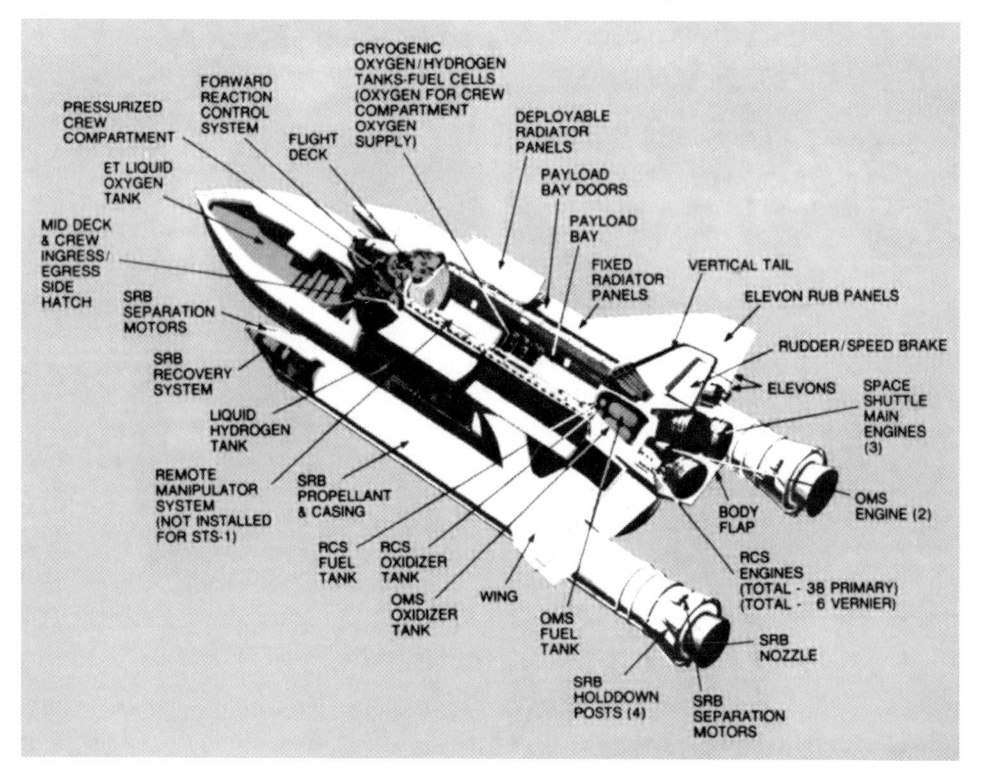

Schematic of space shuttle. Courtesy NASA.

[12] The decision to use solid rocket boosters was made in March 1972, in the belief that solid boosters, which had never been used before for human spacecraft, would save money.

equipped with jet engines in addition to rocket engines, which would make the orbiters much more than gliders when they re-entered Earth's atmosphere. These contractors were believed to have the inside track for the final contract, but other contractors were granted NASA contracts to look at alternative shuttle concepts.

In October 1971, with presidential approval near for the shuttle, NASA issued a request for proposals for the contracts for final design and building of the shuttle. McDonnell Douglas/Martin Marietta, North American Rockwell, Lockheed and a Boeing/Grumman consortium produced proposals for this competition.

Chamberlin was one of the people responsible for design integrity in the McDonnell Douglas shuttle bid, tying together the various parts of the complex proposal. John Yardley, who was in charge of the McDonnell Douglas bid, said that before Chamberlin had come on board, the firm was having problems with the configuration of the tanks and other components of the shuttle in its existing proposal. "He was good with these ideas and analyzing why they wouldn't work and what would. So he was very helpful. And we put Jim to work on how we think it would be, and he came up, really, in my opinion, with the configuration that finally won. And that isn't documented at all. But anyhow, I think Jim came up with the best one. And I pitched it myself to Fletcher, and they bought it. Now they bought it, but we didn't win."

On July 26, 1972, NASA administrator Fletcher announced that North American Rockwell had won the contract to build the shuttle orbiter, narrowly beating the Grumman/Boeing proposal. The McDonnell Douglas/Martin Marietta proposal placed a distant third. The contract award became the subject of controversy, as did the contract for the space shuttle main engine work, which went to a North American Rockwell subsidiary, Rocketdyne.

Yardley said his company's bid failed because McDonnell and Douglas had just merged. "It was not a very friendly thing. [NASA] said, 'You know, you just can't run this thing with half Douglas and half McDonnell people 2,000 miles apart and expect this thing to really work.' And they were right. So that's the reason we lost."

The man in charge of North American's winning bid was John W. Sandford, who had spent most of the 1960s working at the firm after two years at Avro Canada. Born in Yeovil, Somerset, he apprenticed at Westland Helicopters and studied at Cranfield before coming to Canada and working on special projects at Avro. When North American won the Apollo contract, he joined the company and worked on the Saturn V second stage, on nuclear rocket studies and on North American's early preparations for shuttle. Sandford said North American had the advantage of experience on both the Apollo spacecraft and the Saturn rocket, which made the firm stronger for all aspects of its proposal. "The amount of intellectual work that goes into proposals and marketing is just amazing," Sandford said. "You're bidding for $10 billion." All the bidders kept close tabs on what the White House and Congress wanted as they prepared their proposals, he added. When North American Rockwell won the shuttle contract, the company held "the biggest, wildest party you'd ever seen" because this win meant an end to the layoffs that came as Apollo wound down. After his work on the shuttle bid, Sandford moved to other parts of the company and since has led several aerospace firms, including Gulfstream Aerospace, Rolls Royce North America, and de Havilland Canada.

The aerospace firms that lost to North American Rockwell soon won contracts to build various parts of the shuttle system. Grumman built the orbiter wings, Martin Marietta was contracted to build the external tanks for the shuttle, and Thiokol Chemical Corporation built the solid rocket boosters. McDonnell Douglas built the three orbital maneuvering system pods on each orbiter.

After the contract decision, Chamberlin worked as technical director for advanced space programs at McDonnell Douglas in St. Louis. Later on, NASA awarded a support contract for shuttle development work to McDonnell Douglas. Under this contract, which was the first of its type for the firm, McDonnell Douglas employees performed much of the detailed design, development, simulation and test work for the shuttle that in earlier programs had been done by

NASA employees. Chamberlin was named to help head up this effort as McDonnell Douglas's technical director for space shuttle engineering in Houston, and he and his wife moved back to Houston in August 1974. Chamberlin remained with McDonnell Douglas until he died, following a heart attack, at age 65 on March 8, 1981, just a month before the first flight of the space shuttle. Later that year, NASA posthumously awarded Chamberlin its Exceptional Engineering Achievement Medal for his "outstanding achievements in providing technical leadership and technical integration management to the McDonnell Douglas Technical Services Company system engineering operations support to the Space Shuttle Program. His engineering excellence contributed to the development of the space shuttle and to the first space shuttle orbital flight."

•

While NASA and aerospace firms were gearing up to build the shuttle, NASA wound up Apollo in 1972, and in 1973 it flew three crews to the Skylab space station using Apollo hardware. In July 1975, NASA flew its last expendable human spacecraft in the Apollo-Soyuz Test Program, where an Apollo CSM docked with a Soviet Soyuz spacecraft. Apollo was equipped with a docking module that permitted astronauts and cosmonauts to move between the different atmospheres in the two spacecraft. NASA hoped the joint flight heralded a new era of cooperation between the two superpowers, but the end of the short period of détente soon after the joint flight halted cooperative space efforts for nearly two decades. The splashdown of Apollo also began a long hiatus in American human spaceflight. At the time, the shuttle was scheduled to fly in 1979, but engineering and technical problems put that target back by two years.

The first orbiter was a prototype not capable of flying in space. But the craft, named Enterprise at the behest of fans of the *Star Trek* television program, rolled out of the shuttle plant at Palmdale, California, in September 1976. The following summer it flew atop a 747 airliner modified to carry shuttle orbiters. In a series of test flights, the shuttle's gliding capabilities were examined when Enterprise separated from the 747 and glided to a landing at Edwards Air Force Base.

The space shuttle main engines became a major challenge during the shuttle's development. These reusable and throttleable engines represented a major increase in performance over previous rocket engines, and as a result they had many problems during development that pushed back the date of the first flight.

The first flightworthy orbiter, Columbia, didn't roll out of the Palmdale plant until March 1979, and on its flight to Cape Canaveral atop the 747, another major shuttle problem was plainly evident. Many of the 31,000 tiles on the shuttle, which were designed to protect it from the heat of re-entry, detached from Columbia's aluminum skin. To meet the Defense Department's crossrange gliding requirement, the shuttle required heavier shielding than a design without this requirement would have needed. The reusable tiles, made by Lockheed out of lightweight silica, were a major advance on the ablative heatshields used in Mercury, Gemini and Apollo, but they were fragile and proved to be difficult to attach to the shuttle. Work on this problem delayed Columbia's first flight by several months.

•

During this time, many of the Avro engineers remaining at NASA were busy getting the shuttle ready to fly.

Tom Chambers, who helped develop onboard computer equipment for Apollo, did similar work for the shuttle, where he served as deputy chief and later chief of shuttle avionics integration division. Computers were changing so much by the 1970s, and the shuttle was so complex, that its systems bore little resemblance to Apollo. The shuttle is controlled by five computers, including a spare. The number, Chambers explained, was chosen to ensure that the shuttle could continue to operate with two failures. The shuttle starts as a rocket vehicle, then operates as a spacecraft, and finally becomes a glider. Its systems are so complicated that they can't work without the assistance of computers, so a total computer failure cannot be allowed. The job of setting up and testing the shuttle's software was an around-the-clock operation,

Chambers said. "You can't really tell what's going to happen with everything until you put it all together. We were facing problems every day. Some minor, some enormous. It all made the software content grow. When you made the software content grow, you had to increase the memory capability in the computer." But the memory could only grow so far, and decisions had to be made about what would go in the shuttle computers.

"The problem on the shuttle was that we had so many contractors involved. The associate contractor on the software was a subcontractor to Rockwell. The problem was that no contractor had overall responsibility, so the civil service people had to handle discipline and technical decision-making. The other thing we did on this job was tie in the Cape people. They would come up and tie in their equipment. We would tie in a piece of the solid rocket with the Marshall people, and then the Cape people would come up with their stuff. It was difficult sometimes to keep them all lined up. It was like a League of Nations." Chambers officially retired from NASA in 1980, but he was hired back to make sure the software was ready for the first flight in 1981. Even though the computers on the shuttle are 1970s machines that are primitive by today's standards, they are still used because they are highly reliable and can take the shaking, acceleration and radiation of space flight. Instead of replacing them and starting from scratch on the software, the shuttle computers are supplemented by laptops carried by the crews.

Chambers, a Yorkshireman who served in the Royal Navy for more than a decade during and after the war, was trained in electrical engineering and worked on missiles at de Havilland in England before coming to Avro Canada in 1957. At Avro he worked on flight simulation, and once at NASA he worked on the Mercury attitude control system before moving to the Apollo computer.

Thomas V. Chambers receives award from Maxime A. Faget in 1975.
Courtesy NASA.

Two other members of the Avro group who contributed to the shuttle in the 1970s were Bob Vale and Leonard Packham. After working on the Apollo lunar surface experiments package, Vale returned in 1967 to the structures and mechanics division at MSC, where he worked on design details for Apollo and shuttle. In 1971 he became chief of the division, a position he held for the rest of the 1970s. He died in 1980 at the age of 57, not long after he retired from NASA. Packham continued his work on communications equipment at MSC until he retired from NASA in 1974. He died in Houston in 1979, also at the age of 57.

Dick Carley was located at NASA Headquarters, where he helped design systems in the shuttle, particularly the shuttle's avionics and control systems. Carley worked in the Gemini program until it ended in late 1966, then worked briefly in Apollo, where he said some of his ideas for Apollo's control systems were picked up. But soon he moved to NASA's Electronics Research Center near Boston, where he worked until it was shut down in 1970. Carley said that the researchers at the center were trying to promote the use of satellites, and one use that was mooted was satellites to help with air traffic control for the military. In these discussions, Carley said he pushed for such satellites to track aircraft not only by latitude and longitude, but also by altitude. The satellites that came out of these discussions are now famous as the Global Positioning System (GPS) that helps soldiers and civilians determine their exact location, be it in the air or on the ground. Carley said his major contribution to the shuttle was insisting on the inclusion of what he called the "fail-op/fail-op" capability that led to decisions such as the one to put five computers on board. This operational capability meant that systems are redundant and that one failure can't knock out both the primary and secondary systems. As well, it means that both primary and backup systems work in a similar fashion because, he said, if they operate differently, "you've got two different ways of doing [things], and it's a big mess." With fail-op/fail-op, the operators can "maintain consistent procedures even if two single point failures occur." Carley worked on shuttle systems at headquarters until he left NASA in 1981.

Dave Ewart continued to represent NASA at the Downey, California, plant of North American Rockwell, which became known as Rockwell International. "From 1973 to 1984 I was head of the systems engineering resident office in Downey. We went through a very similar phase as what we had in Apollo, supporting the development and the design and the certification and qualification testing of the systems for the shuttle orbiter." Before Enterprise was built, the contractor produced a structural test article to allow engine testing, but soon work began on Enterprise and then Columbia. Different sections of the shuttle structure were shipped from all over the U.S. to the Rockwell plant at Downey and tested before being sent on to the plant at Palmdale, where the parts were put together and tested again. Ewart said Columbia faced the most complex testing done to that time. "And the testing would be supported by the crew, the astronauts. Whenever they could, they would send down the crew that would be flying it to support the final integrated test. Between Columbia and the next one, Challenger, there was a weight reduction effort so that Challenger and subsequent vehicles were capable of launching heavier payloads. But basically the system stayed pretty much the same."

Les St. Leger worked for a couple of years as a liaison between JSC and the Marshall Space Flight Center on the solid rocket motors that boost the shuttle in the initial two minutes of its flight. "NASA needed someone to make sure there was no conflict between managing and designing it. We just made sure that what they were doing met up with the overall requirements. We had to make sure that requirements weren't overlooked. At the end of that, I got a job in the structural test department. That's where we did structural tests on the various components of vehicles including the space shuttle, although the contractors did a lot of that. We did tests for anyone who wanted them. About the only thing we did test from the direct shuttle was the leading edge of the wing under the auspices of the contractor." St. Leger worked in the structures test laboratory at JSC until he retired in 1989.

Rod Rose worked on preparing flight operations plans for shuttle, similar to what he had done for Apollo. But because the shuttle was seen as an operational vehicle flying standardized

David D. Ewart (left) receives NASA Exceptional Service Medal from Johnson Space Center director Gerald D. Griffin in 1985. Courtesy NASA.

missions, Rose and his colleagues worked on standardizing flight plans. "A lot of stuff became standardized, for example the network support, and then all you'd cover would be the exceptions or changes with the way we'd handle it. The same with some of the flights. You'd try to get the flight into standard phases and you'd try to do things by exception rather than start from scratch."

Rose was also handed the job of chairing the communications and data systems integration panel. "What we did was cover each phase of the mission - preflight checkout on the pad, liftoff, each stage, and we're going through every piece of data flow, what path it flowed down." The panel would review whether the communication facilities were adequate for each task and for all the users." While the shuttle was being built, Rose sat on the program manager's level 2 board, the orbiter control board and other review boards. "We'd review every change that was going to go on the program. I put in my two cents worth, operations point of view. We used to meet every Friday, and sometimes they would go on for it seemed like forever. I was involved in both the planning aspects and getting the hardware up to scratch from the point of view of putting in the operations or mission support input." Another task Rose had was negotiating with U.S. Air Force range safety officials to determine how the range safety rules would work with the shuttle. This was critical because if a vehicle strays off course, the range safety officer can destroy it to protect populated areas.

When the shuttle began flying, Rose had the enviable job of certifying shuttle records with the Fédération Aéronautique Internationale, the Paris-based group responsible for deciding such records. This job involved annual trips to FAI meetings in Paris. With the exception of the Soviet Buran shuttle, which only flew once, the U.S. shuttle is in a class of its own:

"aerospacecraft." As well, Rose traveled to Australia to discuss potential impact points for the shuttle's external tank with officials of that country, which was under some of the tank's possible suborbital paths to destruction. Rose retired from NASA in 1984 after helping the shuttle program get underway and seeing it through its first dozen flights. He worked at Rockwell for a time until he and his wife moved to their retirement home in the Texas hill country.

Morris Jenkins came to the shuttle program after short stints working in Apollo Applications (Skylab) and with Hodge and Fielder on future planning, where he concentrated on plans to go to Mars. He returned to the mission planning and analysis directorate, where he had worked during Apollo, and concentrated on trajectories for the shuttle. "For some time, I was the technical assistant to the division chief and later became the chief of the ascent analysis branch." Jenkins and his group prepared launch profiles for the shuttle, and Jenkins' area of special interest included aborts that might be necessary if the shuttle engines failed during launch. "Early in ascent there is a possible return to launch site abort mode and this required a tremendous amount of analysis," he said. "It entailed a complicated set of maneuvers, and not surprisingly, every astronaut who tried it in the simulator hated it. But very early in the ascent, it is the only abort option." This work brought him together with Rod Rose and his work on range safety. Like Rose, Jenkins retired from NASA in 1984.

To help save precious funds and to bring America's allies on board the post-Apollo space program, NASA invited the European nations and Canada to work on the shuttle. Canada agreed to build a robot arm for the shuttle, and the European Space Agency built a laboratory module known as Spacelab that rode in the shuttle's cargo bay. Spacelab was a flexible system that allowed scientific equipment to be mounted inside a pressurized laboratory or on pallets exposed to space, permitting a full variety of scientific experimentation to take place on shuttle flights.

Bob Lindley spent much of a year in 1976 and 1977 in Paris, working as NASA's deputy associate administrator for space flight (European affairs). Spacelab was going through some difficulties at the time, and Lindley was brought to Europe to provide management advice to the European governments and aerospace firms working on the project. Douglas Lord, NASA's Spacelab director, credits Lindley with getting the major German contractor to strengthen its relationship with other members of the consortium building the lab. Spacelab went on to be a major success, carrying a variety of scientific payloads in 22 shuttle flights from 1983 to 1998.

One by one, the problems that delayed the first flight of the shuttle were solved. The shuttle's final configuration is an orbiter 37.2 m long, weighing 70 tonnes empty. The delta-winged craft has three main engines that fire during the boost into orbit, drawing fuel from a 47-metre-high expendable external tank, which separates from the shuttle as it reaches orbit. The boost is assisted by two 45.5 m long solid rockets that are jettisoned after two minutes of flight and descend with parachutes. These boosters are recovered from the sea and refurbished for reuse. The shuttle's payload bay, which carries satellites and modules such as Spacelab into orbit, is 4.5 m wide and 18.3 m long, and could carry 27,000 kg into orbit.

•

On April 12, 1981, by coincidence 20 years to the day after Yuri Gagarin began human spaceflight, Gemini and Apollo veteran John Young and first-time space flyer Robert Crippen ended America's six-year-long space flight drought when they took Columbia into space for a 50-hour, 36-orbit shakedown flight.

As the flight's purpose was simply to prove that the shuttle worked, the only equipment in Columbia's payload bay was a development flight instrumentation (DFI) pallet, which weighed more than two tonnes and recorded detailed information on conditions in and around the orbiter, and information on the operation of every orbiter system. Norm Farmer, who had retired from NASA not long before Columbia's first flight, worked on this flight instrumentation as he had in Apollo. "The DFI that flew on the shuttle was a huge thing, it was probably the biggest system in the world that's ever gone on a flight vehicle. There were 4,000 channels on that thing.

The first space shuttle is launched, carrying astronauts John Young and Robert Crippen.
Courtesy NASA.

Once you take it out, you have a thing called the scar effect. That scar effect is that which remains in the vehicle. You have cable runs and all sorts of things submerged in the wings and whatever. You have to design it to minimize this scar effect." Unlike previous spacecraft, where the scar effect wasn't an issue because each spacecraft only flew once, the task of designing easy-to-remove test equipment was critical in the reusable shuttle.

Columbia, which was equipped with ejection seats for its initial tests, flew two-man crews in four test flights in 1981 and 1982. The second flight included the test of the Canadian-built robot arm, and when the fourth flight landed at Edwards Air Force Base on July 4, 1982, an elaborate ceremony took place to mark the end of the test flights. President Ronald Reagan was on hand to welcome the crew of STS-4 home and declare the shuttle system operational. When Columbia made its fifth flight into space that November on its first "operational" flight, the ejection seats for the two pilots were no longer armed because this time there was a crew of four. The crew of STS-5 launched two communications satellites, including a Canadian satellite, from its payload bay.

The shuttle's sixth flight in April 1983 marked some departures for the shuttle program. STS-6 was the first flight of Challenger and included the first space walk of the shuttle program. Its mission was to launch the first of a series of satellites that revolutionized the tracking of spacecraft, the Tracking and Data Relay Satellite (TDRS). A network of at least two TDRS satellites allows the shuttle to send its signals up to geosynchronous orbit instead of down to tracking stations. As a result, the shuttle has continuous contact with the ground rather than spotty coverage through tracking stations, and TDRS costs far less to operate than the stations. As well, many civilian and military satellites can also use TDRS. This represented the coming to fruition of a proposal made years before by Dennis Fielder, the Avro veteran who had been instrumental in setting up the network of tracking stations that TDRS was replacing.

Among the people working for contractors on TDRS was one of the former Avro engineers who had helped Fielder put together NASA's network of tracking stations for Mercury, Gemini and Apollo. George Harris had left NASA nearly 15 years before and gone to Darmstadt, Germany, where he helped build the first European space control center for the European Space Research Organization, one of the forerunner organizations of the European Space Agency. Harris remained with ESRO and ESA until 1975, when he and his wife returned to the United States. During his time in Europe, Harris was involved in controlling European satellites and on at least one occasion, rescuing one in trouble in orbit. In the U.S., Harris got work in private industry developing a center in South Dakota for processing data from Landsat satellites, which were making images of the Earth that would be used by scientists, prospectors and others.

Harris had been working on TDRS since 1979, setting up the control center in Las Cruces, New Mexico, and making sure that the system was up to NASA's standards. "I did that, and then I was told, 'You're now going to be the man pulling together this launch.' Just to set the stage, we had a TDRS that had never flown before, with an [inertial upper stage] which it was going to be attached to, to bring it to geosynchronous orbit. It had never flown a commercial mission before. We had a brand new control center, and a control team that had never flown a mission before. We had a ground station that had never tracked a target before, and we had a brand new shuttle that had never flown before, the Challenger. So my job was to pull it together. So I put a plan in motion and procedures together to try to get the Air Force Control Center at Palo Alto, Houston, and Las Cruces, pull them all together, and put together a mission management structure."

Once Challenger got into orbit, astronaut Story Musgrave deployed the TDRS satellite. When TDRS and the IUS, a solid-fuelled, two-stage booster developed for the Air Force, were safely out of the shuttle's way, the IUS began to fire to lift TDRS from low Earth orbit to geosynchronous orbit, 36,000 km above the Earth. Harris was on the control loop with the shuttle flight director in Houston and an Air Force flight director in Palo Alto.

"The first burn of the IUS circularizes beautifully. The second burn has a burn-through of the gimbal seal, and the whole thing goes unstable and begins tumbling with the TDRS still attached to the IUS. We are on battery time. We have 10 minutes of IUS left, and [the Houston flight director] is on the loop with me and he's saying, 'Okay, TDRS flight, what do you think?' I've already asked the Air Force to bring up their command stations. I asked the big eye-in-the-sky to look at it and see that we're still attached, and they said, 'Yeah, you still are attached.' Well, I better tell him that we just about lost it. Out of the clear blue sky, somebody had the brilliant idea that we had a computer lockout, couldn't get a command in because the TDRS accelerometers, sensing the tumbling, were telling the computer that we're still burning. So we had to get an override in. We got the override in, followed immediately by separation. And all of a sudden my attitude control guy said, 'We've got one valid frame of data that says we've separated.' It was just before loss of signal. I got on the line and said, 'Houston flight, we've separated.' I made the call on one frame of data. We had separated."

If Harris's team had not ordered the TDRS to separate from the IUS, the satellite would have been lost. Quick thinking on the part of the team had saved TDRS. But the satellite had to stop tumbling, which it did using its thrusters, and then deploy its solar panels and antennas. TDRS was in an orbit 35,000 km by 21,000 km, and controllers in Houston and Las Cruces laboured to bring TDRS into the proper orbit, using TDRS's thrusters in 27 maneuvers over 70 days. TDRS was then usable. There was a long delay before the second TDRS was launched while the problems with the IUS were investigated and fixed. "We got it up there just in time, because the United States was very heavily obligated to provide TDRS for the European Spacelab on STS-9," Harris said. "For that, NASA very kindly awarded the public service medal to me. My company was very good to me and gave us a lovely cash bonus."

About the same time TDRS was put in its proper orbit, Challenger blasted off again on STS-7. This flight was notable in that four of the five crewmembers came from the first group of astronauts chosen specifically to fly on the shuttle. All previous seats on the shuttle had been occupied by astronauts selected in the 1960s. The 1978 intake of 35 astronaut trainees included, for the first time, women and members of visible minorities. The first of the women, Sally Ride, flew aboard STS-7, and America's first black astronaut, Guion Bluford, flew on the next flight, STS-8. As 1983 closed out, veteran John Young commanded STS-9, the first flight of Spacelab and the first flight to include a non-American astronaut, Ulf Merbold, a German from the European Space Agency.

During 1984 and 1985 the shuttle made another 14 flights that included satellite launches, satellite rescues, research flights, classified missions for the Department of Defense and tests of astronaut maneuvering units. Guest astronauts, called payload specialists, who took part in these missions included scientists flown on behalf of private industry, a U.S. senator, and non-American astronauts, including more Europeans, a Canadian, a Mexican, and a Saudi prince. The shuttles Discovery and Atlantis also made their first flights during this time, giving America a fleet of four space shuttles.

Ambitious plans were in place for 1986. A set of ambitious shuttle missions were on the manifest, including deployments of the Hubble Space Telescope and the Galileo probe to Jupiter. The first shuttle mission to be put into a polar orbit was scheduled to fly from the shuttle's new launch site at Vandenberg Air Force Base in California. On January 12, the first mission of the year blasted off successfully from Cape Canaveral. Columbia carried a commercial satellite, a set of experiments and a crew of seven, including a U.S. congressman.

The second mission of 1986 was to deploy the second TDRS satellite, and its crew of seven included a schoolteacher, Christa McAuliffe, who would conduct lessons from outer space. On January 28, Challenger blasted off, but just 73 seconds later the shuttle and its boosters disintegrated. The crew – commander Dick Scobee, pilot Mike Smith, mission specialists Ellison Onizuka, Judith Resnick, and Ron McNair, commercial payload specialist Gregory Jarvis, and McAuliffe - died in the disaster. Investigations revealed that Challenger was doomed because it was launched on an unseasonably cold day that, among other things, rigidized rubber seals that lay between segments of its solid rocket boosters. The boosters were assembled and shipped from Utah to Cape Canaveral in segments, and the rubber seals were designed to block the burning fuel from escaping from between the segments. Even though NASA knew that the seals had nearly failed on several of the 24 shuttle missions that preceded Challenger's final launch, nothing had been done. When Challenger lifted off, one of the seals failed, and the flames eventually broke through to the external tank, causing structural failure. The crewmembers were the first in-flight fatalities in the U.S. space program.

•

By the time of the Challenger disaster, only four of the Avro engineers were still on staff at the Johnson Space Center, and only one, David Ewart, was directly involved in the shuttle program. Ewart was still working in the Los Angeles area, representing NASA at the Rockwell plant where shuttle orbiters were repaired and refurbished. In 1985 he became NASA's resident manager at Rockwell after serving for a year as deputy manager. When Ewart became manager, the shuttle fleet was built and it appeared that work would be winding down. But the loss of Challenger changed all that.

"There was an interval before the decision was made to build one more vehicle to replace Challenger, and that of course was Endeavour. We had unexpectedly a new lease on life at Palmdale and Downey to build Endeavour. At the same time, there were enhancements to extend the [shuttle's] stay on orbit, so they contacted Rockwell to build a pallet to carry additional liquid oxygen and liquid hydrogen. It allows us to get more mileage out of the orbiter, and it allows us to extend our stay up on orbit to 16 or more days." As well, Rockwell was put to work modifying the orbiters during the shuttle stand-down from 1986 to 1988 to increase safety features as a result of safety reviews following the Challenger disaster. "We also were able to use Palmdale as a major modification site, bringing orbiters back to Palmdale and running a total structural inspection, modifying it as required to accommodate future missions, such as taking the airlock out from the inside of the crew module to the payload bay, and any other changes required to support the extended duration missions." Later on, in the 1990s, Ewart supervised development of docking equipment to allow the shuttle to dock with the Russian space station Mir and the International Space Station (ISS).

But the major job was building Endeavour. In 1987, Rockwell won a contract to build the fifth shuttle, a task made easier by the fact that parts were on hand from an earlier time when NASA had hoped to fly a fleet of five orbiters. On April 25, 1991, Endeavour rolled out of the plant at Palmdale, with Ewart signing acceptance on behalf of NASA. Endeavour was transferred by road to Edwards Air Force Base and flown to Cape Canaveral. A year later, on May 7, 1992, Endeavour and a crew of seven blasted off on a dramatic mission to rescue a stranded communications satellite. Dave Ewart, the last of the Avro engineers working for Johnson Space Center, retired in 1994 after nine years as resident manager at Rockwell and 30 years representing NASA at Rockwell and its predecessors.

The shuttles resumed flying in the fall of 1988. Up to the turn of the century, the four shuttles in the fleet returned to space more than 75 times, deploying satellites, carrying scientific experiments and full laboratories, and docking with Mir and the ISS. In the wake of the loss of Challenger, plans to launch shuttles into polar orbit from California were abandoned, and most commercial and defence satellite deployment work was curtailed. Ironically, just as the defence role of the shuttle was being phased out in 1988, the Soviet Union launched its own shuttle, which was developed to match the military capabilities built into the U.S. version. The Soviet Buran shuttle, which had a similar design to its U.S. counterpart, only flew once, without a crew, before being abandoned for economic reasons. Although NASA is working on new vehicles to succeed the shuttle, it appears to be destined to serve for many more years as America's workhorse vehicle for human space flight.

•

Many of the Avro engineers in Houston who had been involved in Apollo stayed with NASA but moved to other pursuits within the agency. After Apollo 11, NASA shifted its focus from the Moon to goals that served more earthly purposes, including remote sensing, using satellite sensors to image the Earth to search for minerals, check the health of crops, and other purposes. When he left the lunar receiving laboratory in 1970, Bryan Erb joined in the change. "My social conscience came to the fore, and I began to question whether or not the lunar program was all that important in the overall scheme of things. I'd always been somewhat interested in remote sensing, and the opportunity came along to join the remote sensing division, and I moved over to head an area looking at applications of remote sensing. At first we were looking at some local applications, forestry, agriculture, rangeland, wetlands, that sort of thing. And then in about '74 we formulated what was called the Large Area Crop Inventory Program. What we were trying to do was make estimates of wheat production around the world. The prime source of data was the Landsat satellite, and aircraft observations were used to calibrate the interpretations of the satellite data. The goal was to develop a technology that would allow us to use satellite data to estimate the areas planted with different crops, and use meteorological data to estimate the likely yields, and then combine these two estimates to predict total wheat production.

"The ironic thing is that though we had Department of Agriculture people working with us, the Department of Agriculture never picked up the technology. To me, that was a lesson in the merits or demerits of technology push. If you don't understand what your customer really wants to do, offering the best tools in the world, the customer is not going to agree with much if he isn't prepared to pick it up and staff it. Over the last 10 years the Department of Agriculture has come a long way and begun to use these technologies.

"Then we tried to expand that technology into other crops and what we call global

Bryan Erb (far right) and colleagues explaining Earth resources work to Shirley Temple Black, then U.S. Ambassador to Ghana, during a mission to Accra to present a short course on remote sensing, 1975. Courtesy Bryan Erb.

habitability. Can we with remote sensing techniques really understand enough about the biosphere so that we can assess productivity of various ecosystems and whether the world will remain habitable? My feeling is that there has been much too much emphasis on the satellite side of the technology and not nearly enough on how you use the information. It's been sort of easy and fun and technically snazzy to sell a satellite. So you look at the data and the pretty pictures, but how do you turn this raw data into something that a land use manager or a crop agronomist can make use of? There's some big gaps there, and a lot of work to be done in that area."

In 1984, JSC had the most competent remote sensing capability in NASA, but because of a reorganization at NASA headquarters and the press of shuttle and space station work at JSC, the remote sensing work was terminated. Erb was handed the task of shutting down JSC's effort and sending the data to the Goddard Space Flight Center. "There again, almost a rerun - not quite - of the Arrow experience." Erb then took early retirement from NASA in 1985, but continued his work on remote sensing and other advanced technologies as a consultant.

Peter Armitage, when he left the LRL, returned to the space sciences directorate at JSC, where he worked on the Skylab science program, including physical science and Earth resources experiments. The Earth resources work was similar to that being done by Erb using imagery from the Landsat satellite. Armitage worked as assistant director, handling management work for a parade of scientists who took the top position at what became in 1978 the space and life sciences directorate. Finally he retired from NASA in May 1986.

"I was very happy to be leaving NASA even though NASA had been a wonderful opportunity for me. It was time to get on and do something new. In the aftermath of the shuttle disaster in January 1986, the NASA that we knew was changing and was not a particularly endearing environment to work in. The engineering management that I really respected, Bob Gilruth, Chris Kraft, Bob Thompson, and Max Faget, had all retired from NASA, and the place was never the same as it was in the 'old days.'"

•

Many forces of change swept through NASA once the first Moon landing had been achieved. While many, such as layoffs, cutbacks and more bureaucratization, were negative, a change for the better was the arrival of large numbers of women engineers. The wives of two Avro engineers joined the ranks of skilled people working in Houston on the space program. One of them was Dona Erb, Bryan Erb's wife, who in 1972 became a computer programmer working on the space program at Lockheed, where she helped program the software for the shuttle Enterprise's approach and landing tests in 1977. She also tested and installed upgrades to JSC's UNIVAC software systems and became supervisor of the data distribution center in JSC's orbiter data reduction complex.

In 1978, Erb left Lockheed for the MITRE Corporation, a not-for-profit firm that serves government. She spent 15 years there, working on the mainframe systems at JSC, which then included IBM, UNIVAC and Control Data System computers. "The idea was to tune the systems to enable them to do more work so as to avoid the expense of buying more machines. Processor cycles and disk storage space could be wasted in a variety of ways that could be fixed with some detective work. One year, I saved JSC from having to buy yet another mainframe, for a saving of over a million dollars." This accomplishment was recognized by NASA with what was then known as the Manned Flight Awareness Award.

Dona Erb at her computer while working for the MITRE Corp., explaining a hypermedia prototype she developed for space station on-board training. The image on screen shows a joint of the Canadarm 2, the test case for the training effort. Courtesy Dona Erb.

As the space station became a major program during her later years at MITRE, Erb worked with the mission operations training division to help them prepare to train astronauts for long-duration missions. For shuttle missions, which usually last two weeks or less, each crew is trained for months and sometimes years to prepare for every move on orbit, following the tradition set in Mercury, Gemini and Apollo. But astronauts spending weeks and months aboard the station cannot rehearse or memorize everything they will have to do, so some form of "just-in-time" training is needed for those on board the station. Erb said she had developed a prototype using HyperTalk, an Apple Computer, Inc. language, to demonstrate computer-based training material that could be used in lieu of the hundred-plus pounds of documentation carried on board. It predated the creation of the similar point-and-click links that allow internet users to move from one page to another on the World Wide Web. "I wrote a simple training program as an illustration of what an astronaut might do to resolve a problem with a stuck joint on the space station arm. The idea was so fascinating at that pre-internet time that JSC had a video made of my demonstration to further disseminate the concept. An on-line hypermedia manual certainly beats the current method of carrying on board over 100 pounds of the paper flight data file and having to follow charts that skip from page to distant page to trace information on a particular aspect of the space vehicle." The ideas developed by Dona Erb are in use to allow astronauts to prepare for their work on long-duration flights on the ISS.

Since leaving MITRE in 1993, Erb has worked as a computer consultant with her own firm, Erb Software, Inc.

Prior to entering the computer business in 1972, Erb had worked as a teacher and was active in her community, like many other women during that time. Born Dona Marie German, she grew up in the same Calgary neighbourhood as her future husband, and while she attended school, she also worked summers in her older brother's law office. When she told her brother of her interest in becoming a lawyer herself, he asked, "But who would hire you?" Although he hired a woman for his firm a decade later, she reconsidered and entered the education faculty at the University of Alberta in Edmonton, where she majored in mathematics and physics. During that time she was dating Bryan Erb, and during her second year, when her future husband went to England to study at Cranfield, she studied for a year as an exchange student in Connecticut. Dona was still a year away from graduation when Bryan finished his studies in England, but instead of taking work and asking her to move, he returned to Edmonton and earned another degree. They married shortly after her graduation in 1955. While her husband worked for Avro and NASA, Erb taught school near their homes in Ontario, Virginia and Texas. Like many of the women whose husbands came to the Clear Lake area south of Houston to work at NASA, Erb helped create a community from empty scrubland when she wasn't teaching or raising her children. Years later, Erb put her computer skills to work for her community when she developed software for an early desktop computer to help administer her local church.

Bryan Erb spent a year as a Sloan Fellow at MIT in 1967 and 1968, and the family moved with him, but when they returned to Houston, Dona Erb decided on a change in direction. Instead of picking up her community activities, she studied administration and computer science at the University of Houston, completing masters degrees in 1970 and 1972. She had some concern about the move from academia to industry, having heard it was cutthroat, but she found industry more cooperative than the academic world, with its competition for students, money and grants. "Women were still very much in the minority in the engineering community. There were the nude pictures and 'jokes' that were probably intended to intimidate. While it was not a general problem in the JSC community, the time came when such displays and behaviour became illegal in the workplace and it was a relief. Mostly I worked with great guys who kept things on a high plane."

Another who followed a similar path was Joan Jenkins, wife of trajectory expert Morris Jenkins. She had studied mathematics in England, and after their children left for college she took a master's degree in education from the University of Houston. Teaching didn't agree with her, so Jenkins took her training in mathematics to McDonnell Douglas. There she worked for a decade in support of JSC's mission planning and analysis branch, She and her colleagues produced math models in support of engineering problems, including orbit and trajectory determination models to be used in conjunction with ground tracking.

•

Some of the Avro group continued their NASA careers outside of Houston. Bob Lindley worked at the Goddard Space Flight Center from 1972 to 1979, with time out for his work in Europe on Spacelab. At Goddard, Lindley was director of project management and worked on 27 satellite projects - including weather satellites, astronomy satellites such as the International Ultraviolet Explorer and the Orbiting Astronomy Observatory - and on the Delta launch vehicle.

Also at Goddard throughout this time was Tec Roberts, who from 1972 to 1980 was director of networks. This put him in charge of tracking virtually all of NASA's satellites and spacecraft. "What began as a diverse and diffuse tracking system for specific space missions has now grown into a single, unified network serving manned flights, such as space shuttle, scientific and applications [satellites] as well as commercial and foreign payloads," Roberts said in 1983. "We not only made great technical strides, we also learned how to manage large international projects involving other governments, private industry and academia. Much of the technology which NASA programs helped foster has found its way into the commercial sector where it has created new business opportunities." Roberts left Goddard in 1984 and died in 1988.

John Shoosmith returned to the Langley Research Center in 1964 after five years with STG and MSC. NASA sent him to the University of Virginia, where he earned a Ph.D., and for much of his career he was head of the scientific computer applications branch in the computing division at Langley. Most of his work involved running the computers at Langley to support researchers there with their aeronautics and space applications research efforts. During much of this time, Langley had Control Data computers, unlike the rest of NASA, which relied on IBM mainframe computers. Eventually Langley acquired Cray supercomputers.

"I have seen the most fantastic changes in computing equipment in my career. I could never have imagined them," Shoosmith said in 1996. "We have gone through almost 12 orders of magnitude in increasing speeds. At the very beginning of my career, computers were only on the order of 10 operations per second. Ten to the one. By the end of this century, in fact within the next year or so, we will have computers with the capability of a trillion operations per second. Ten to the 12th. You've gone all the way from 10 to 10 to the 12th in one career. That's an incredible change. When we started out with computers, they failed very regularly. Today they are very reliable.

"There are a lot of good and exciting things going on at Langley today. We have got involved in understanding the atmosphere, using measurements from satellites. There are experimental sensors and lasers which are giving us a better sense of the behaviour of the atmosphere and the impact of human activity on the atmosphere. There will be a new generation of supersonic aircraft in the United States, and Langley is working on that. Planes can be flown automatically. It won't be too long before they're essentially run by computers. There will still be pilots in charge. Aircraft flown that way will be a lot safer. Langley's in the forefront of that work."

Shoosmith retired from NASA on April 1, 1995. He was the last of the engineers from Avro to retire from NASA, having worked there for 36 years. He has continued to work as an adjunct professor of applied science at the College of William and Mary in Williamsburg, Virginia.

Like Shoosmith, few of the Avro engineers actually stopped working when they retired from NASA. Many continued to work on the challenges of space. One challenge in particular beckons as a new generation of engineers and researchers moves into the world's space programs. It predates the space age and has been the subject of great efforts by both NASA and the Russian space program. As the new century opens, both space programs are working together to make space a home for humans as well as a place to visit.

13 The Troubled Ascent of Space Stations

When Jim Chamberlin, Owen Maynard and John Hodge considered missions to follow Mercury into space in the early 1960s, they and their NASA colleagues probably spent more time thinking about permanently inhabited space stations in Earth orbit than about going to the Moon. Even when President Kennedy set his lunar goal, many people suggested that the Moon vehicle be assembled and launched from a space station, which would form as much a part of the space infrastructure as the launching pads at Cape Canaveral and mission control at Houston. Yet the space station was passed over at the time, and it has taken the passage of a generation to make the dreams of a space station a reality.

One reason is that the concept of a space station has always been controversial. Perhaps the only space exploration issue that can start an argument faster than space stations is the question of human versus robotic space exploration. Proponents of space stations argue that Earth orbit is a good platform for research and an equally good transfer point for destinations beyond because of the major velocities required simply to get into Earth orbit. Once in Earth orbit, one is "halfway to anywhere." Opponents use a seagoing analogy, comparing space stations to building a large floating city a couple of miles offshore instead of investing the money in better ships to cross the oceans.

The first proposal for a space station was Edward Everett Hale's 1869 article in the *Atlantic Monthly*, "The Brick Moon." Hale's spherical satellite was made of bricks, had a crew of 37 in an orbit 6,000 km high, and would serve as a navigational aid. Pioneer space theorists picked up on this idea, with Hermann Oberth coining the phrase "space station" in 1923. It was Oberth's protégé, Wernher von Braun, who popularized the idea of a space station in his famous articles in *Collier's* magazine in 1952. Von Braun's concept, brought to life by the paintings of Chesley Bonestell, called for a gigantic wheel in the sky generating artificial gravity and serving as a way station for voyagers to the Moon and Mars. This idea was further engrained in the public imagination by Stanley Kubrick's 1968 film *2001: A Space Odyssey*. Prophetically, the space station depicted in that film was still under construction.

The critics were also active. A 1964 book, *The Space Encyclopaedia*, stated bluntly that "Such stations will probably never be needed," and set out six reasons to back up this contention, including the availability of robots to do observation work, and the claim that building such stations would be "ludicrously uneconomic."

When NASA was being set up, the space station and lunar landings were seen as major post-Mercury goals. Indeed, many experts saw a space station in Earth orbit as a necessary stepping stone to the Moon. As Maynard pointed out, the original concepts for Apollo in 1960 included not only the command and service modules, but also a mission module that could be put to many uses, including extended stays in Earth orbit. Following President Kennedy's decision to land astronauts on the Moon by 1970, the postponement of the space station became official when lunar orbit rendezvous was chosen as Apollo's route to the Moon. Earth orbit rendezvous might have involved a space station if it had been chosen.

Throughout the time of Apollo, many engineers at NASA, with encouragement from the agency, worked to keep the idea of the space station alive for the day when the Moon landing was achieved. One of the major space station concepts during this time came in early 1963 from Maynard, who with Rene Berglund, the head of the space station office at Houston; Willard Taub; David Brown, another member of the Avro group; Edward Olling; and Robert Mason proposed a Y-shaped space station that could be launched in one piece. This station concept was inspired by the work of H. Kurt Strass, a key figure in the early days of Apollo, and was designed under the supervision of Charles Mathews, who was about to move to the Gemini program.

The design of this station was based on the assumption that much of the work in orbit would best be done using artificial gravity. This idea was also reflected in the gigantic rotating

PATENT NO. → PLEASE WRITE IN WHITE SPACE ONLY. VOID IF FOLDED, STAPLED OR OTHERWISE MUTILATED

3300162

DO NOT ATTACH ADDITIONAL THICKNESS OF PAPER TO THIS COUPON
Will be accepted by the U.S. PATENT OFFICE in lieu of 50¢ in payment for copies of U.S. Patents

Address Coupons Only
Box 9
Patent Office
Washington, D.C. 20231
Form PO-194b

INVALID IF SEAL CANCELLED

CANCELLED PATENT OFFICE OCT

Do Not Write In This Box
400288

Coupon No.
0258294
Series of 1969

Jan. 24, 1967 O. E. MAYNARD ETAL **3,300,162**

RADIAL MODULE SPACE STATION

Filed Jan. 20, 1964 3 Sheets-Sheet 1

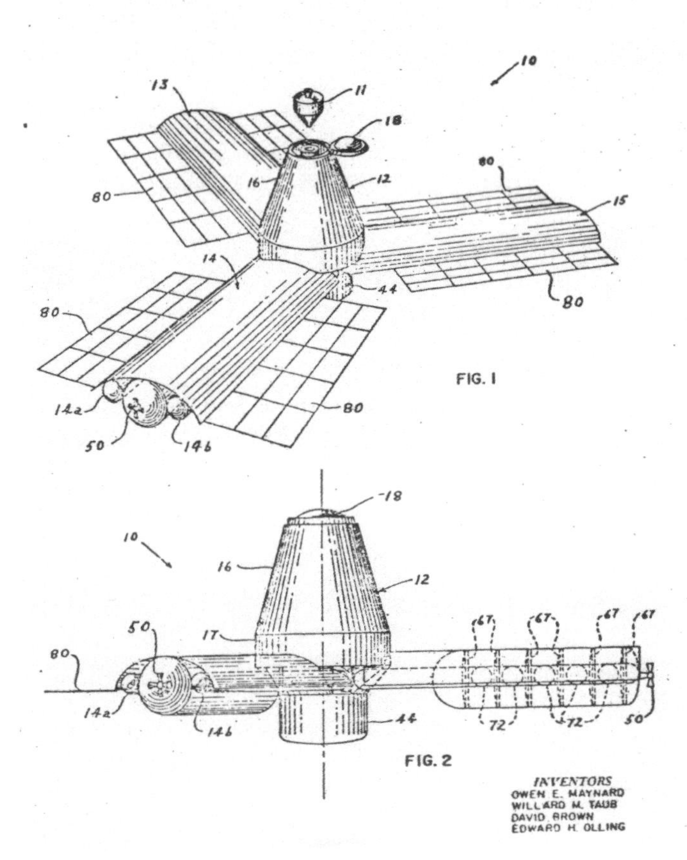

Copy of Maynard sketches of space station concepts.
Courtesy Owen Maynard.

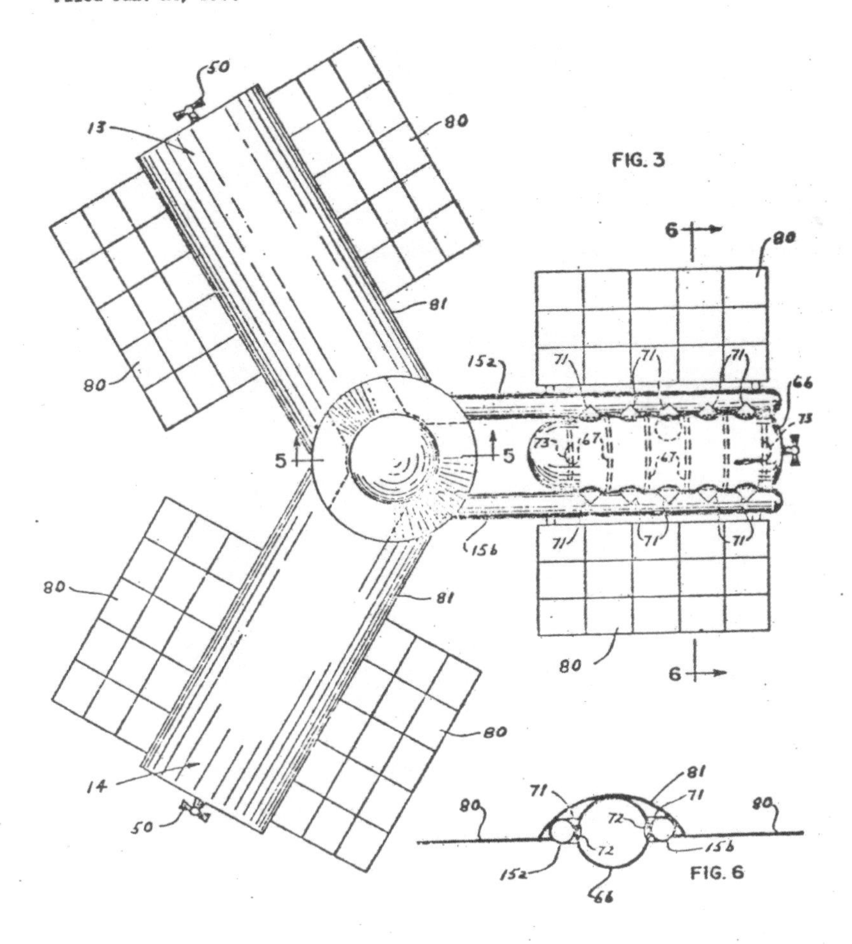

Copy of Maynard sketches of space station concepts.
Courtesy Owen Maynard.

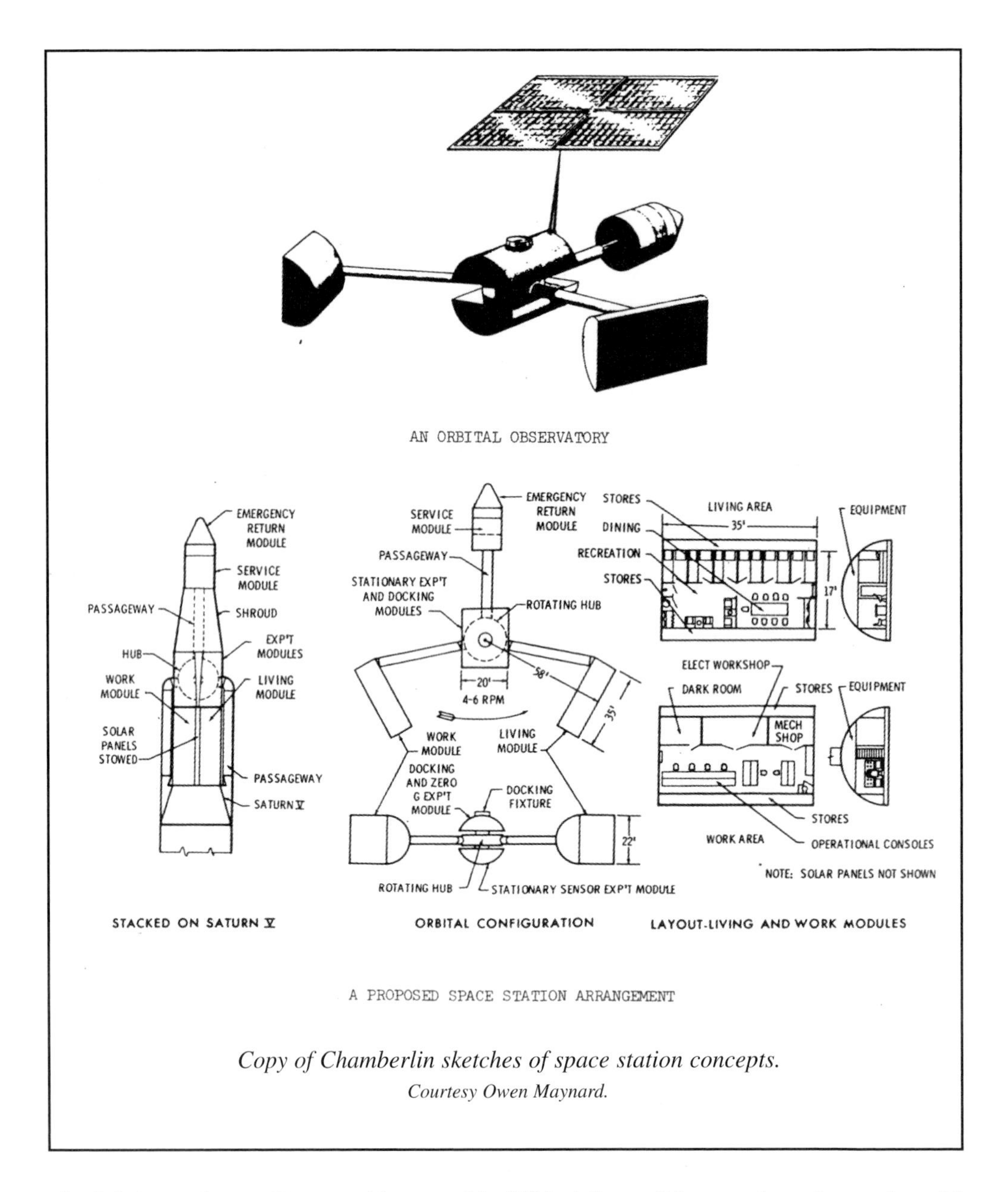

Copy of Chamberlin sketches of space station concepts.
Courtesy Owen Maynard.

wheel drawn up by von Braun and later used in *2001: A Space Odyssey*, where the rotation of the station created artificial gravity for the comfort of the station's inhabitants. The station designed by Maynard and his colleagues included a large, conical, nonrotating hub section where spacecraft could dock and experiments could be conducted in weightlessness. Attached to the hub were three radial sections, which at launch would be folded under the hub atop an advanced Saturn launch vehicle. Once the station was placed in orbit, the three sections would unfold like the ribs of an umbrella and spin around the hub at three revolutions per minute, creating a force equivalent to a quarter of Earth's gravity at the end of each radial section. The station would be powered by solar panels and weigh about 100 tonnes. Maynard and his colleagues applied for a patent for the station in 1964 and were granted one three years later.

In 1966, Chamberlin tried his hand at designing a space station. In a paper for an American Institute of Aeronautics and Astronautics meeting that year, Chamberlin also proposed a Y-shaped station with a nonrotating hub and three modules with artificial gravity at the end of the spokes. Designed to be launched on a Saturn V, this "space observatory" included a conical emergency return module sitting atop the rocket at launch. Below this module and the cylindrical hub, were two half-cylinders that would deploy as working and living modules. All three modules were connected to the hub by long spokes, and as in the Maynard concept, movement from one module to another would take place through the hub. Rotating at between 4 and 6 rpm, the station was designed to provide artificial gravity of between .31 and .71 Earth gravity.

Chamberlin's paper discusses, in far more detail than the earlier proposal, human factors for the crew of six to nine astronauts, including a comparison to the experience of Antarctic explorers. He proposed sending data back to Earth via satellites in geosynchronous orbit rather than directly to ground tracking stations and, going a step farther than the Maynard team, he discussed in detail the actual purpose of such a station. "The fact that a space environment provides many extraordinary capabilities for observing both the Earth and the universe has already been demonstrated extensively in the unmanned satellite program," Chamberlin wrote.

By late 1963, the major action on space stations had moved to the military front. The U.S. Air Force, having sustained the cancellation of its Dyna-Soar space plane program in late 1963, shifted its space focus to the Manned Orbiting Laboratory program. MOL was far simpler than any of the NASA plans of the time. A modified Gemini spacecraft attached to a cylindrical laboratory module would allow military astronauts to conduct reconnaissance from orbit, as well as perform experiments with military applications. Military astronauts were put into training and work was underway to build MOL, but the program was cancelled in June 1969. Many of the MOL astronauts moved to NASA, where they eventually flew aboard the space shuttle. One of the reasons for the cancellation of MOL was the new capability of reconnaissance satellites, which began sending back high-resolution photos of military targets.

With Apollo under development, it was inevitable that the new hardware would be factored into discussions of space stations. Langley Research Center designers proposed a Manned Orbital Research Laboratory using Apollo hardware and a large weightless laboratory space. In 1965, the Apollo Applications Program was set up at MSC, with the intention of using Apollo spacecraft for many different missions in Earth orbit and on the Moon. Over time, AAP became the primary victim of cutbacks that started in 1967 in the wake of the escalation of the Vietnam War and congressional disquiet over the Apollo fire. By 1968, AAP was a space station program, and the most popular idea involved using the spent second stage of a Saturn 1B rocket as an orbital workshop. Astronauts would clean out the "wet" workshop and move equipment in, turning the rocket stage into a home in space. A solar telescope, fitted to a modified lunar module ascent stage, would be attached to a docking adapter at the top of the workshop. The next year it became clear that not all the Saturn V rockets that had been built would be used for Apollo Moon landings, and support grew for a "dry" workshop. The bottom two stages of the Saturn V could loft the fully outfitted space station into orbit. Despite the protests of both NASA and the Air Force, Congress saw AAP and MOL as competitors, and MOL's cancellation in 1969 opened the way to keep the AAP station alive. By the time AAP was renamed Skylab in 1970, the "dry" workshop design was adopted.

Because much of Skylab was directed by the Marshall Space Flight Center in Huntsville, Alabama, the Avro engineers had little involvement in the program. After Gemini ended, Bob Lindley worked at McDonnell on preliminary concepts for the airlock module on Skylab, which used Gemini hardware, including a spare Gemini hatch as the passageway from the station outside. Dennis Fielder, who got out of tracking networks in 1966, went to Huntsville to help with mission planning in the early days of the program before returning to Houston to work with John Hodge on more advanced concepts. David Ewart, who spent most of his career representing the Manned Spacecraft Center at North American Rockwell in the Los Angeles area, took two years off from this duty to represent MSC at the McDonnell Douglas plant at Huntington Beach, California, where the Skylab orbital workshop was being built from the shell of a Saturn V third stage. Ewart was responsible for integration and testing of all the experiments aboard Skylab that came from MSC, and on issues involving the crews. "It was a good experience to be involved in Skylab," Ewart said. "It was very different from Apollo in terms of the accommodations in orbit. They learned quite a bit on Skylab. That was our first space station, and the accommodations were luxurious in comparison to Apollo." Many items that were considered essential for Skylab were shown not to be necessary, such as the "floors," which astronauts could attach their feet to using special shoes, and he said this experience is being put to use in designing today's space station.

The Skylab space station flew atop the last Saturn V launched from Cape Canaveral on May 14, 1973, and the first crew was ready to fly to the station the next day aboard an Apollo CSM and Saturn IB. But it soon became clear that something had gone wrong at launch. Two large solar panel arrays on the Skylab workshop had not deployed, which meant that the station was without half its power supply. Its meteoroid shield, which also provided temperature control for the station, was lost, too. The launch of the first crew was delayed while tools and sunshades were designed to solve these problems. In the meantime, flight controllers had to struggle to save Skylab from becoming uninhabitable due to its lack of power and surplus of heat. Finally, on May 25, the first crew of Pete Conrad, Paul Weitz and Joseph Kerwin lifted off. They found one of the two solar power arrays completely lost, and the other held in a closed position by an errant strap. Their initial attempt to free the stuck solar array failed, but the crew docked with the station and entered it the next day, wearing gas masks because of fears that the heat inside the unshielded station had caused materials to emit dangerous gases. The crew unfurled a sunshield through a scientific airlock, bringing Skylab's temperature problems under control. Nearly two weeks into the flight, Conrad and Kerwin freed the stuck solar array in a dramatic space walk. The three astronauts returned home after spending four weeks in space, at that time the record for duration in space. Five weeks later the second crew headed to Skylab, where they spent 59 days in space. In November the third and last crew began 84 days in space aboard Skylab. When they splashed

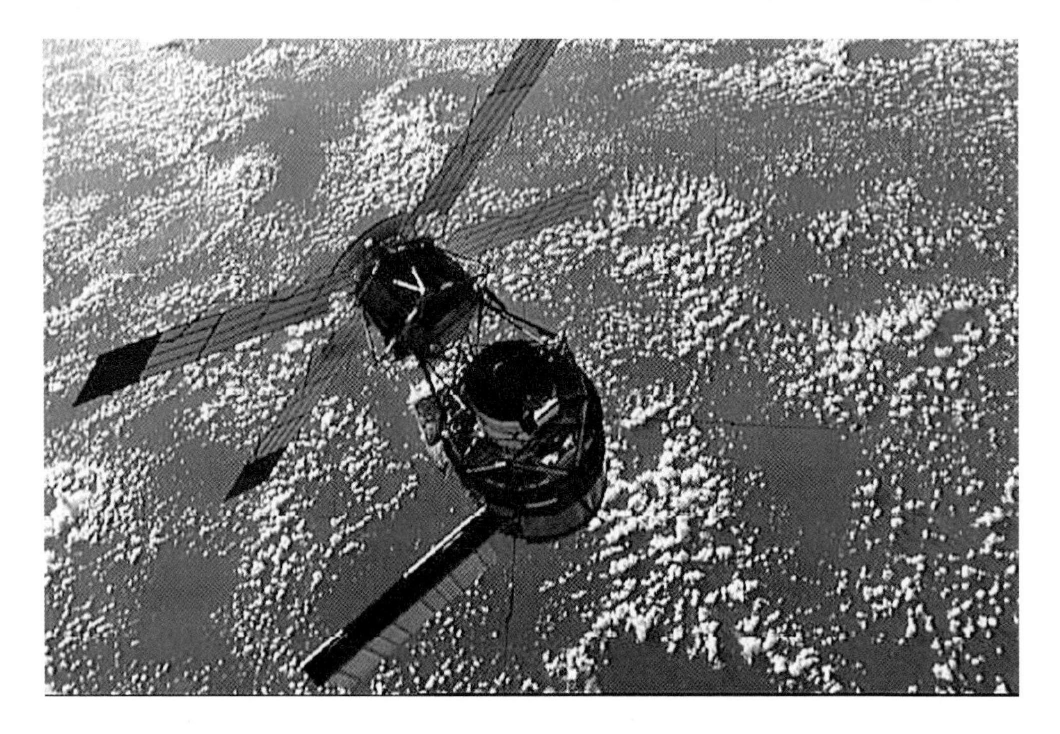

The Skylab space station in Earth orbit.
Courtesy NASA.

down on February 8, 1974, Skylab was abandoned to drift in orbit until its destruction on re-entry in July 1979. While the station was operational, Skylab astronauts conducted a series of biological and medical experiments, carried out extensive Earth resources observations with their cameras, and recorded solar activity with the station's solar telescope. Most importantly, the astronauts of Skylab overcame adversity and showed the value of humans aboard spacecraft.

When Skylab was conceived, NASA had hoped to follow it with more permanent space stations. Hodge, Fielder and the others working at MSC's advanced missions program office in 1968 and 1969 were planning both space stations and reusable shuttle craft to take astronauts there. In 1969, Hodge's group was looking at a space station launched by a Saturn V with a crew of 12 that would use shuttles to move between the station and Earth. The station, like those in earlier NASA proposals, had different sections with artificial gravity and weightlessness. But by 1970 it had become clear that President Nixon did not support NASA's ambitious plans. The new NASA administrator, James Fletcher, directed his agency to focus on the space shuttle rather than a space station.

Before Fletcher abandoned them, NASA's space station plans underwent a change that has never been reversed. After the station concept proposed by Hodge's group in 1969 was dropped and NASA's focus moved to the shuttle, NASA proposed what was called a Phase B modular space station. The station was to be made up of can-shaped modules carried into orbit by the shuttle and then assembled. For the first time there was no provision for artificial gravity, and NASA never again proposed a station with artificial gravity. Back in the 1960s, Chamberlin and Maynard expressed reservations about artificial gravity. "A recently compiled list of uses for a space station and the experiments that could be performed on board indicates that few applications or experiments have a requirement for artificial gravity," Maynard and Berglund noted in an article on their space station proposal. "There has been much controversy about whether artificial gravity is required for a space station, and what research is necessary to prove this," Chamberlin wrote. Although he admitted that a zero-G station is less complicated than one with artificial gravity, Chamberlin said having to live in zero-G "imposes a substantial additional burden on the crew." The long-term experience of weightlessness by astronauts aboard Skylab persuaded NASA that astronauts could get by for a few weeks or months aboard a space station without artificial gravity.

•

NASA was not the only place where space stations were being designed and launched. The Soviet space program turned to space stations when it fell behind in the race to the Moon. In early 1969, when two Soyuz spacecraft docked and two cosmonauts moved from one to the other in a space walk, the Soviet media laid claim to having the first space station, a statement that was laughed off. But on April 19, 1971, the Soviets did launch the first space station, which was named Salyut, a "salute" to the 10th anniversary a few days earlier of Yuri Gagarin's flight into space. Five days later, three cosmonauts aboard the Soyuz 10 spacecraft docked with Salyut. Due to a fault in the docking mechanism, the cosmonauts could not move into the station, and they quickly returned to Earth. The experience of Soyuz 10 was just a foretaste of the difficulties the Soviets would face with space stations, but their determination was more than a match for these setbacks.

On June 6, cosmonauts Georgi Dobrovolsky, Vladislav Volkov and Viktor Patsayev lifted off aboard Soyuz 11, and this time their docking and transfer into Salyut were successful. In a record-breaking flight that lasted 24 days, the cosmonauts conducted research and sent back television broadcasts from the station. Just before Soyuz 11 re-entered the Earth's atmosphere, a valve opened accidentally. The three cosmonauts, who weren't wearing space suits, were exposed to the vacuum of space, with fatal consequences. The Soviets made changes to their spacecraft as a result, and insisted that all crews wear space suits during ascent and descent. And they pressed on with space stations.

During this time, the Soviets actually had two space station programs – one civilian and the other military. The two programs used different space stations built at different places, although every successful station launch carried the Salyut name. The first Salyut was a civilian vehicle, and an unpublicized attempt to launch another civilian station in 1972 failed. Salyut 2, launched in 1973 a month before Skylab, was the first military station, but it tumbled out of control before any cosmonauts could visit it. The same week Skylab was launched, another civilian Salyut failed almost immediately after entering orbit. With these back-to-back failures, the Soviets were forced to watch while Skylab eclipsed all the hard-won records of the first Salyut. But when the Skylab program ended, the field of space stations was abandoned to the Soviet Union. In June 1974 another military station, Salyut 3, was launched. One crew spent two weeks aboard the station taking reconnaissance photos of Earth, and a second crew failed to dock with the station. Salyut 4, launched on December 26, 1974, hosted two crews in 1975. They stayed in space for 29 and 62 days respectively. The third and last military space station was launched in June 1976 and hosted two crews for flights lasting 49 and 17 days. By the time the mission of Salyut 5 ended in 1977, the Soviets were still behind Skylab. American astronauts owned the record for longest stay in space, and the astronauts of Skylab had ventured outside the station, something the Soviets had not done on any of their stations. And although Skylab's title as the most spacious space station still stood at the end of the century, the lifetimes of Skylab and its Soviet counterparts to 1977 were all limited by the fuel and supplies they were launched into space with.

The Soviet Union began to move ahead of the U.S. in the space station race in September 1977 with the launch of Salyut 6. This station marked a major advance over its predecessors in that it had two docking ports – one at each end of the station – and could be refuelled. The first crew to board the station stayed 96 days, breaking the record for longest time in space previously held by the final Skylab crew. Ever since, the duration record has belonged to Russian cosmonauts. And a new robot tanker spacecraft named Progress visited Salyut 6, performing the first refuelling operation in space. The twin docking ports on Salyut 6 allowed Soviet cosmonauts to remain on board the station beyond the lifetime of their Soyuz ferry craft. Crews flew fresh spacecraft to the station and, after a brief stay, returned home aboard the spacecraft that brought the long-duration crew to Salyut. The Soviets used these short-duration missions for diplomatic purposes by including guest cosmonauts from Soviet satellite countries and, later, from other nations. Salyut 6 flew until 1982, when it was replaced by Salyut 7, which had a similar design. During the time of Salyut 6 and 7, Soviet cosmonauts racked up ever longer stays in space and overcame obstacles including launch pad failures, docking problems, hazardous landings, and at least one medical emergency in space. One crew docked with an inert Salyut 7 and repaired it after a major systems failure when no crew was on board. Cosmonauts confined to the small stations for long stays often experienced depression and other psychological problems. As well, the Soviets learned that a great deal of the cosmonauts' time was taken up in maintenance of their station and its systems.

With Salyut 7 nearing the end of its lifetime, the Soviet Union launched a new station called Mir on February 20, 1986. The station was similar to Salyut 6 and 7, except that the docking unit at one end could accept five docked vehicles. The first crew to visit Mir was also the last crew to visit Salyut 7. After their stay on Salyut 7, the two cosmonauts brought equipment from the old station to Mir before they returned home. In 1987, the first expansion module – Kvant – was attached to Mir, and by the middle of 1990, two more expansion modules, Kvant 2 and Kristall, had joined the Mir base block. Crews aboard Mir continued the work of their predecessors, achieving the first stays longer than a year and welcoming more visits by guest cosmonauts. A Japanese television network paid for one guest cosmonaut, while another was the first person from Great Britain to fly into space, Helen Sharman. By the time the Soviet Union collapsed in 1991, to be replaced by a non-communist Russia and other independent countries, the Russian space program had a clear lead in space stations over the United States. And although the Soviet Union was gone, Mir still had nine years of life ahead as 1992 dawned.

•

Back in the United States, the end of missions to Skylab in 1974 meant that space station activity was restricted to studies while NASA concentrated on building the shuttle. By 1981 the shuttle was flying and NASA was looking for a new development effort. At the same time, a new Republican administration headed by President Ronald Reagan was taking office in Washington, D.C. Reagan chose as his NASA administrator a top executive from General Dynamics, James M. Beggs. The new administrator had worked for NASA in the 1960s, when he had been deputy associate administrator responsible for advanced research at NASA headquarters. Not surprisingly, during this time he had become friends with the man who was in charge of advanced programs at the space center in Houston, John Hodge. When Beggs was being confirmed by Congress for his new job in 1981, he stated that NASA needed a new goal: a space station with permanent human inhabitants. "When he came back to NASA, I called him up and said, 'Hey, is there another job?'" Hodge recalled. "He had been pushing space station in his congressional hearings, and he asked me if I would like to start up whatever space station was going to be. I took the chance and went back to NASA." Hodge was returning to NASA after nearly 12 years away, this time to continue with some unfinished work from his days at the advanced missions program office in Houston.

In the years between his two stints at NASA, Hodge had kept himself as focused on the future as he had been while at the space agency, though he was working on earthbound concerns. "I had decided to make a change in career. It was the first time I had done it without something dramatic happening," Hodge said of his departure from NASA in 1970. "This time, I had decided I either wanted to go into the environmental movement or transportation, both of which were key problems for the country at that time." When Hodge joined the newly established department of transportation in June 1970 as director of transportation systems concepts, the Transportation Systems Center was being set up at Cambridge, Massachusetts, on the site of a former NASA facility, the Electronics Research Center, which was established in 1964 to support Apollo and closed in 1970 as NASA retrenched.

"I had the job of sort of looking to the future. What directions technically and politically and all that kind of thing could we take the department to be useful in the field of transportation? It was a tremendously broad subject and I had a wonderful time. We did a lot of studies on the future of the air traffic control system. We did a lot of stuff on high-speed railroads, looking at passenger situations, and came to the conclusion that there was nowhere in the United States where railroad passenger traffic made any sense at all. It would never be profitable. Even the northeast corridor, which is probably the biggest one of the lot, just didn't make a great deal of sense economically. Then there was always the big fight between the ascendance of the automobile versus public transportation. We came up with all kinds of theories about what made up the city. What was the interaction between the city and the transportation system? The conclusion I came to there was if you wanted a city to maintain its strength, you had to keep the people in the city, not out of it. As soon as the public transportation system allowed people to leave the city, half of them would leave the city, and leave the poor inside the city. That was when we came up with one of these rapid transit systems that gave you great mobility inside the city but made access to the city very difficult. Those ideas got floated around. We ended up building a local personal rapid transit system which everybody said was a total eyesore but it was successful, actually. It's in Morgantown, West Virginia."

After four years at the department of transportation, Hodge was restless again when he realized that few of his ideas could be realized because politics usually outweighed technical merit in setting up transportation networks.

"What I was looking for was another institutional arrangement which would allow you to do things independently of either the political or profit motive, and that's what the Urban Transportation Development Corporation was offering." UTDC was a crown corporation for the government of Ontario, so Hodge, by then an American citizen, left his latest adopted home for a new job in Toronto, which he had left 15 years before when the Arrow was cancelled. He and his family moved back to Oakville, where he had lived while at Avro and where his wife Audrey had relatives. UTDC was planning to build a personal rapid transit system at Ontario Place, the new amusement and recreation complex in Toronto on the shore of Lake Ontario. Hodge stayed at UTDC for only two years. "I think UTDC as an experiment didn't work, because politics did come into it, and the technical solutions that they came up with were so far removed from what I thought was necessary that it wasn't a useful exercise for me personally, and I didn't think it was going to work very well.

"So we came back to Washington. I worked some more for the department of transportation. Really doing R&D policy for about four or five years. During that period, Jim Beggs was deputy secretary for transportation." Hodge had spent a total of 10 years at the department, interrupted by two years in Canada. Although he had stayed far away from NASA during these years, Hodge was now returning to the space program, this time at NASA Headquarters in Washington.

In May 1982, Beggs announced the formation of a space station task force headed by Hodge, who reported to Philip Culbertson, an associate deputy administrator. According to Hodge, Beggs believed NASA needed to build a space station, but the administrator didn't have much of an idea of what the station would do. "I don't know what the hell a space station is, and what it's supposed to be doing, so why don't you go and tell me?" Hodge recalls Beggs telling him. "So we had about a dozen people to start with. But that was a wonderful group we had. No one thought we had a hope in hell of selling the program."

Beggs and his deputy administrator, Hans Mark, had already been working to get President Reagan, who had demonstrated a strong interest in the space program both as governor of California and as president, to commit to a space station. Hodge attacked the problem of getting approval for the station by turning the usual procedures on their head. He believed that one of the problems afflicting the shuttle was that NASA had designed the vehicle before fully canvassing potential customers about what they would use it for. Costs for the shuttle increased with each new use that had to be built into the vehicle after the initial design work was done. Over the objections and against the instincts of many NASA engineers, who are always anxious to design things, Hodge accepted advice that he focus the task force on finding out what the potential users wanted in a space station. He threw out two proposed designs, one from Johnson Space Center and the other from Marshall, and for several months prohibited members of the task force from drawing any pictures of a space station. By consulting potential customers first, the task force built political and economic support for the station. Hodge decided to spend three years defining the station's mission before proceeding to design and development. "This strategy avoids a premature emphasis upon configuration and design. It forces us to think through clearly and at the beginning what any future space station would be expected to do. It is also the only way I know to protect against the potential of costly slips in schedule and performance," he said at the time. But during this period of definition, the station would be more vulnerable to attack from its many opponents than it would be once construction began.

Hodge, his deputy Robert Freitag, and the others in the task force employed a horizontal management technique and worked hard to get experts from all NASA's centers participating in the task force's four working groups. This was a difficult job at first, because often employees' major loyalty was to their center rather than to NASA as an institution. And while the task force produced books with station configurations for its internal use, its major means of communicating its decisions with outsiders was with viewgraphs, usually restricted to 20 per presentation. The viewgraphs created, in the words of historian Howard E. McCurdy, a "small, elusive target for opponents of the plan." Just as importantly, he explained, they did not include material that could be exploited by opponents, such as detailed plans, alternatives to the main plan, or information on differences within NASA over the plan.[13]

[13] McCurdy, Howard E., *The Space Station Decision: Incremental Politics and Technological Choice*, Johns Hopkins University Press, Baltimore, 1990, pp. 97-98.

The work of the task force was supplemented by eight study contracts given to major aerospace contractors, and by late 1983, Hodge was able to tell a Senate subcommittee about design conclusions reached by the studies and the task force. "To briefly summarize, the studies recommended an initial station consisting of a manned base with an unmanned platform at an orbital inclination of 28.5 degrees, an orbital maneuvering vehicle operating in conjunction with the station, and a second unmanned platform serviced by the shuttle but derived from the space station design, operating at a near-polar or 90 degree inclination." In other words, the station would operate in the standard equatorial orbit used by most spacecraft launched from Cape Canaveral, and it would have two robot platforms to satisfy requirements for experiments that did not require or were even harmed by the presence of humans. For example, some materials-processing experiments and telescopes are disrupted by the movements of astronauts around a station. The equatorial orbit for the station and the robot platform would be fine for materials processing, medical experiments and astronomical observations, while the polar platform would be mainly used for remote sensing and other Earth observations. The station would also be used as a satellite service depot - where astronauts could pull in satellites for upgrading, servicing, storage and repair - and for manufacturing new alloys and medications that could not be made on Earth. Although this potential use was not explicitly acknowledged, NASA hoped to use the station in the 21st century as a launch platform for the first human trip to Mars. Canada, the European Space Agency, and Japan would be asked to join the station, although the form of their participation wasn't spelled out.

Hodge stressed "operational autonomy" in his design concepts for a station that could operate independent of ground control. The former flight director realized that the cost of maintaining the ground control infrastructure would drive the station's costs to a prohibitive level. "What happened to me was I suddenly realized that we would have to be able to do this more cheaply, or nationally we wouldn't be able to afford it," he said later. "These things have been cancelled because of the cost. And it was true of the Arrow. The transportation stuff [in the 1970s] didn't go because of costs. So I became almost messianic about the need to get costs down. I was absolutely convinced that we could do so, but not in the context of the bureaucracy that we had."

Although Hodge and his colleagues were doing a good job building a consensus within NASA for the form of the station proposal, they had to win over many outsiders before Reagan would approve it.

Mindful of the military's role in helping NASA win approval for the shuttle in 1972 and of the high priority given military interests in the Reagan administration, NASA hoped to win military participation in the space station. But the military wasn't supporting NASA, in part because it feared that the station would divert NASA's attention from the shuttle, which was being used to loft military payloads into space. As well, Reagan had boosted funding for the military in general and military space programs in particular as part of his strategy to re-engage the Soviets in the Cold War after the period of détente in the 1970s. The symbol of this new military emphasis on space was Reagan's controversial decision in March 1983 to undertake the strategic defense initiative, known to its opponents as Star Wars, which was an effort to construct a "shield" against incoming nuclear warheads. This new program was getting started in the middle of NASA's lobbying for a space station, at the same time as Reagan was facing fiscal problems due to tax cuts he had made and a recession in the early years of his administration that put pressure on social programs. In this atmosphere, the U.S. Defense Department opposed the space station for fear it would represent a new source of competition for funding.

Reagan's science advisors, and the office of management and budget, were against the station. In the fall of 1982, the senior interagency group for space, known as SIG (Space), was formed to produce a report on the space station that the president could use as the basis for a decision. The group included representatives of NASA, the military, other security agencies, the state department, and the commerce department. John Hodge was named as chair, and he worked hard to bring the group to a consensus in favour of the station. Charged with producing a report by the summer of 1983, the group with all its divisions could only produce a report that laid out a number of alternatives, something clearly not suitable to present to the president. A sympathetic White House aide enlisted NASA help to draft another, more favourable report on the station, but when SIG (Space) met in August with top level representatives to make its recommendations to the president, it was deadlocked.

Beggs and Mark decided to remove all military involvement from the space station and make their pitch directly to Reagan with the help of several supportive officials in the White House. To escape the influence of the station's military opponents and to emphasize the potential commercial benefits of the station, NASA chose to bring its proposal before the president at a meeting of the cabinet council on commerce and trade. On December 1, 1983, a group of NASA officials, led by Beggs and including Hodge, made their pitch for the station to Reagan, using a model of the proposed space station and a set of viewgraphs to emphasize the need for America to reassert its strength in space. The estimated price tag was $8 billion. Although Reagan didn't tip his hand at the meeting, on January 25, 1984, he used his annual state of the union address to Congress to announce his decision. "I am directing NASA to develop a permanently manned space station, and to do it within a decade," Reagan said at the same spot where John F. Kennedy had issued his Moon landing challenge 23 years earlier. Said Hodge: "When the president made his announcement, we had a big party that night. The state of the union message, that was kind of everybody's dream."

Hodge and his team were instrumental in pulling the agency together for the station and in winning the president's support for the program. But they had also worked to win congressional interest in the space station at the same time. As everyone knew, the station still faced a fight in Congress. The initial skirmish took several months in 1984 and ended with an appropriations bill in early July that included funding for the station. In April, Beggs announced the formation of an interim office of space station at NASA headquarters, headed by Culbertson, with Hodge as his deputy. The office was made permanent on August 1, and Hodge became deputy associate administrator for space station.

In those early months, Beggs and his team at NASA worked to secure international support for the station from ESA, Japan and Canada. In 1985 ESA and Japan agreed to build scientific research modules for the station, and Canada signed on to build an advanced remote manipulator system that would be used for assembly and maintenance of the station. ESA also was planning to participate in the robot polar platform.

The space station team also began work on defining what the station would look like and, more importantly, who in NASA would do what. By late 1985 NASA was beginning to undergo a series of changes triggered in part by the departure of Beggs as administrator. Culbertson left his space station job, and as 1986 began, Hodge was acting associate administrator for the station. Soon NASA was reeling from the impact of the Challenger disaster, but for the time being, work on the station continued on schedule. In May, Hodge announced that the station would be built in a "dual keel" configuration, with two 110 m vertical keels to which experimental and living modules, and solar panels, would be attached. The first element of the station would be launched in 1993, and crews of up to eight astronauts would be able to start living on the station the next year. Hodge explained that this design was put together with the help of nearly 400 people at JSC in Houston working as a "skunk works" for three months. "Of course, the whole concept of the space station is that with time you can add things, but in an incremental way so you don't have to budget for it all at the same time."

Later that year, a new associate administrator for space station was appointed, and in a reorganization of NASA stemming in part from the effects of the Challenger disaster, Hodge was given a reduced role. In fact, infighting within NASA was growing. Every center in NASA was fighting for a part of the station, and members of Congress didn't hesitate to go to bat for their local NASA center or contractor. As Hodge explained: "I made the bureaucratic mistake of trying too many things at the same time. New ideas in management, technical management. By this

Artist's concept of space station Freedom.
Courtesy NASA.

time, NASA had become stodgy, very bureaucratic. Everybody knew what was right. Very few people were around who remembered that when we started, we didn't even know what was right. Innovation, delegation and things like that were the order of the day. It had become heavily bureaucratic." Many people at NASA had joined in the glory days of Apollo and then survived the nearly two decades of austerity and retrenchment that followed. Hodge got caught in the crossfire with the centers as he worked to set the operational parameters of the station. His idea to run the station autonomously of ground control proved unpopular, and Hodge faced "sort of a revolt," with his critics at JSC accusing him of carrying a grudge over disputes there in the 1960s. He left NASA for the last time in 1987 and became a consultant. "I thought that I had done my thing there, and got the program solved, and so it was time to do something else."

Since then, Hodge said he has enjoyed years of "picking and choosing who I'll work for." During that time, he has worked on assessing where the U.S. is going in space. "I think that the future depends on the ability to commercialize space and that it's time for the government to stand back and get out of the way, and encourage industry to do things on its own." Among other things, this means that NASA should get out of operating systems like the shuttle and stick to research and development of new systems.

•

The years that followed Hodge's departure were difficult for the station, which in 1988 was named Freedom in honour of America's Cold War struggle with the Soviet Union. It quickly became apparent that the $8 billion price tag announced by NASA in 1984 didn't cover the cost of the station, and this gave heart to the station's critics both within and outside Congress. Meanwhile, infighting over design elements and responsibility for those elements continued unabated. Billions were spent in repeatedly redesigning the station, and no metal was being bent. Although the shuttle program recovered from the Challenger disaster, NASA again came under fire in 1990 when the Hubble Space Telescope was found to have a defective mirror after it had been put in orbit by the shuttle. Meanwhile, the problems with Freedom continued to accumulate. In 1992, President George Bush named a tough aerospace executive from New York to take over NASA. Daniel Goldin was determined to shake up NASA and reshape it to fit his "faster, better, cheaper" philosophy. That year, the Republican Bush lost his bid for re-election to Democrat Bill Clinton. A new administration with no stake in the station, a changing NASA, and growing opposition in Congress combined to make 1993 a year of crisis and transformation for Freedom. As well, major changes were shaking the world outside the United States. The Soviet Union collapsed in 1991, two years after the opening of the Berlin Wall, and a new Russian leadership was eager to work with America and to preserve its own space program, one of the few areas in which it led the world.

President Clinton, who retained Goldin as NASA administrator, quickly ordered the space agency to sharply reduce the size of the space station. While the station was being refashioned, a bid in Congress to kill the station fell short by only one vote. NASA came up with a redesign for the station, by now renamed Alpha, but Goldin and the Clinton administration had one more major change in store for the station. In late 1993, a series of contacts Goldin and other officials such as Vice-President Al Gore had made with Russia resulted in an agreement to make Russia a partner in the station. The station, with the participation of the U.S., Russia, the 11 members of ESA, Canada, Japan and Brazil, became the International Space Station. The success in late 1993 of a shuttle mission to repair the Hubble telescope restored public and political confidence in NASA, and Goldin moved to stabilize the ISS by putting the Johnson Space Center firmly in charge of the program. As well, Boeing was made lead contractor for the American portions of the ISS.

As part of its agreements with Russia, NASA agreed to a joint program of flights to the Mir space station. In February 1995 the shuttle Discovery rendezvoused with Mir, and that June, nearly 20 years after the Apollo-Soyuz docking, which to that point had been the only joint flight of American and Russian spacecraft, the shuttle Atlantis docked with Mir and delivered a new crew to the station. Shuttles docked with Mir eight more times over the next three years, bringing supplies to keep the station going. As well, seven U.S. astronauts flew long-duration missions aboard Mir, the longest being Shannon Lucid's 188 days in 1996. But the economic problems plaguing Russia in general and the aging Mir station in particular plainly affected the shuttle-Mir program. American aid helped Russia complete the construction of Mir by adding two modules, Spektr and Priroda, to the station in 1995 and 1996. The Americans who flew aboard Mir noted that the station was seriously cluttered, and cosmonauts had to spend what seemed to be an inordinate amount of time just keeping the station operating. In early 1997 a fire broke out aboard the station when a carbon dioxide cleaning device failed, placing two cosmonauts and American astronaut Jerry Linenger in danger. An even worse crisis afflicted Mir following a botched docking operation on June 25, 1997. A Progress cargo ship collided with the Spektr module and opened a hole in it. Two cosmonauts and British-born U.S. astronaut Michael Foale had to move quickly to cut cables and seal Spektr off from the rest of the station before the air leaked out of Mir. The crew then had to struggle to restore power and control to the station. Mir's problems were eventually brought under control, and the shuttle-Mir program continued as planned for another year. Mir continued flying until its controlled plunge into the Pacific Ocean on March 23, 2001, ending 15 years of operation.

Mir's crises were highlighted by critics to point out the dangers of cooperating with Russia on the ISS. Russia was facing severe economic problems that were nowhere near being solved, one of them being a serious lack of money for government-supported activities such as the space program. In spite of delays in getting Russian components ready to fly, the first component of ISS to lift off was the Russian Zarya module, which was launched from Baikonur on November 20, 1998. In December, shuttle astronauts began the long process of actually building ISS, using the shuttle's Canadian robot arm to join the U.S. Unity node to Zarya. After the long-delayed Russian Zvezda module joined the station in July 2000, the first permanent

Space station Mir in Earth orbit.
Courtesy Rocket & Space Corporation Energia.

occupants of the ISS moved into the station on November 2, 2000, and construction of ISS moved into high gear.

•

While most people think of space as a limitless void, it is actually a dynamic and sometimes dangerous environment. Burt Cour-Palais, because of his work on meteoroid impacts on spacecraft, knew better than most people that space presents its own dangers. He and his colleagues had called for bumpers - meteoroid shields - to be installed on spacecraft to dissipate the impact of meteoroids, but proper bumpers weren't provided for until Skylab. Unfortunately, the bumpers on Skylab fell off during launch because they were installed incorrectly. The resulting thermal and power problems endangered the station until the astronauts made repairs. Cour-Palais believes his success in convincing NASA to include the troublesome bumpers on Skylab contributed to career setbacks he suffered later at NASA. He had to endure a pay cut and nearly lost his job. During the period after Skylab, he worked in the environmental effects office at JSC, researching another environmental question, the effect of shuttle exhaust in the lower atmosphere.

But he maintained his interest in meteoroids and impacts on spacecraft. Another researcher at JSC, Donald J. Kessler, suspected that many of the impacts that Cour-Palais and his team had recorded did not come from meteoroids but from orbital debris. First in an internal NASA paper in 1977, and a year later in a scientific journal, Kessler and Cour-Palais sounded the first serious warning about orbital debris. They said orbital debris would soon become a bigger danger to spacecraft than meteoroids and warned that as the rate of collisions between objects in orbit grew, the danger from debris would rapidly increase.

Burton G. Cour-Palais, Lake Country, Texas, November 8, 1999.
Photo by Chris Gainor.

Cour-Palais and Kessler's work soon put orbital debris on the map, both within NASA and in the scientific community in general. Orbital debris includes not only unused satellites and rocket stages still in orbit, but also pieces of equipment, fragments of exploded rockets and spacecraft, and even bits of paint that have flaked off space vehicles. The smallest fragments could threaten or destroy other spacecraft in low Earth orbit. Since the shuttle began flying, NASA has had to foot expensive bills for window replacements caused by collisions with paint flecks. Until the end of the 1970s, Cour-Palais said, he and Kessler were unpopular for bringing up this issue. "When you come up with a problem, you must have the solution at the same time," he said. But many people at NASA soon came around to see the danger of orbital debris.

By then, Dennis Fielder had placed Cour-Palais in the technical planning office at JSC, where he could resume his work on impact collisions, and by 1983, Cour-Palais, Kessler and other experts on meteoroids and orbital debris were reunited in the space science branch, again working full time on this issue. Cour-Palais remained there until his retirement from NASA in May 1989.

For the next five years he continued his work on impacts at McDonnell Douglas, preparing protection systems for what would become the International Space Station. Cour-Palais and Jeanne Lee Crews invented what they call the multi-shock shield, a variation on the standard meteoroid bumper that is named after its creator, astronomer Fred Whipple. "It was a simple shield, but it wasn't good enough. The problem with the space station is the fact that you have to take it up in the shuttle bay. Its size was limited, and the habitation modules have to fit within the confines of the payload bay. You had to have living space, and so you couldn't put a very big bumper on it. That bumper Jeanne and I put together involved multiple shields and was much lighter than the equivalent Whipple bumper. The beauty of our multi-shock shield that we got the patent for is that you can make it out of ceramic cloth or aluminum. Four layers of ceramic cloth or five layers of aluminum. Taking it up there was a problem. Now the idea is to add it after you get there."

The shields that will be put on ISS and other vehicles will probably be variations on the multi-shock shield. The work of Cour-Palais and his colleagues on shields also attracted the interest of the Defense Department, which wanted to build spacecraft that could withstand attack.

In 1994, Cour-Palais retired with his wife Audrey to a home in the Texas hill country, but he still continues his work on a part-time basis at the nearby Southwest Research Institute in San Antonio. Most tests of high-speed impacts have used spherical projectiles, he said, and he is looking into impacts caused by non-spherical projectiles.

•

Artist's concept of the International Space Station.
Courtesy NASA.

The International Space Station is quite different from the station that John Hodge and his team designed in the mid 1980s. The modules and solar panels will be attached to a single truss 108 m long. The station will have a crew of six or seven and fly 400 km high in an orbit inclined 51.6 degrees to the equator, which is easily accessible to spacefarers from both the U.S. and Russia but more expensive to launch to. The free-flying robot platforms are long gone, as are plans to use the station as a servicing depot for satellites. More than 450 tonnes of modules and materials are due to be lofted aboard 37 shuttle flights and nine rocket launches over five years as the station is assembled. The task of putting the station together will require 144 space walks lasting more than 1,700 hours, twice the amount of time spent on U.S. space walks in the 33 years preceding the launch of ISS. Cost estimates in 1998 put the construction cost of the station at around $50 billion, and over the station's lifetime the cost could go past $100 billion. The cost, a figure comparable to the price tag for Apollo, has been seized by ISS critics, who believe that the money would be better spent mounting a human expedition to Mars.

Much of those high operating costs are due to the requirement for expensive round-the-clock ground control capabilities, something Hodge tried to design out of the station. When asked what he now thinks of the space station he did so much to make a reality, Hodge said: "I try not to. There is hardly anything in the program now that vaguely resembles what we were trying to accomplish. Low-cost common systems, a growth path, large amounts of power, large crew capabilities. All of those things have disappeared. Advanced communications systems. Advanced power systems. Virtually none of them is what we had. The big thing was to get the costs down."

Bruce Aikenhead was just a few days back in Canada after leaving NASA when John Glenn made his historic Mercury orbital flight aboard Friendship 7 in February 1962. "I stayed home from work and was glued to the TV." He had just started a new job, but he called his office and said he was too emotionally involved in the Glenn flight to tear himself away from the television. Aikenhead had started work at NASA the same day as the Mercury astronauts, and since then he had worked to train them for their flights. But Aikenhead's decision to return to Canada was made almost at the moment in September 1961 that NASA decided to move the Space Task Group from Virginia to Houston, Texas. He quickly wrote to the last employer he had before Avro to see if it had a job for him. He flew to Montreal in October and accepted a job with that employer, CAE Industries, then and now known as a leading manufacturer of aircraft flight simulators. Aikenhead said he didn't want to start his new job until Glenn had succeeded in reaching Mercury's goal of putting an American into orbit. But Glenn's flight suffered from a number of postponements. "Finally, after stalling a couple of times, they said, 'Are you coming back or aren't you?' So I said I'd come back in February." He moved back to Montreal and found himself watching the flight on TV instead of from a tracking station, where he would have been had he stayed with NASA. Aikenhead and Stanley Cohn, who also left NASA and returned to Canada at the same time, found themselves in demand for a time as speakers on their experiences with the Mercury program. Before long they turned to their new lives in Canada.

Cohn joined IBM Canada in Toronto as scientific services manager. When he started his new job in Toronto, he was working with a familiar computer: the IBM 704 that Avro had leased in the late 1950s, which had reverted back to IBM after the cancellation of the Avro Arrow. Cohn remained with IBM in Toronto and Ottawa until 1968, when he moved to the University of Toronto, where he taught and did research until his retirement in 1989. Stan Galezowski, another who came back to Canada early in 1962, was, like Cohn, a computer specialist. The skills needed

Bruce A. Aikenhead in 1960 at the Langley Research Center.
Courtesy NASA.

for computers were evolving rapidly at that time, and Galezowski decided to change careers as well as homes. He moved to Arnprior, near Ottawa, where he operated a Canadian Tire store. He continued with the auto parts and hardware chain until he retired.

While several of the Canadians in the Avro-NASA group chose to come home, or, in the case of Fred Matthews, stay in the U.S. but move closer to home, the story was different for the others. Most of the Avro engineers who hailed from the British Isles chose to stay in the U.S. The two notable exceptions both left NASA after only a few months of working there in 1959. Frank Chalmers, the lone Scot in the Avro-NASA group, returned to Toronto, where he joined his wife's family printing business. He worked there until he died in 1975 at the age of 49.

The only ex-Avro engineer who is known to have returned to England, Joseph Farbridge, went to work at Hawker-Siddeley in England after he left NASA in the summer of 1959. Farbridge worked there until 1962, when he joined a tourism business operated by his father-in-law. But after two years he wanted to resume his work in aeronautics, so he returned to Toronto, where he took a job with de Havilland Aircraft of Canada. Farbridge remained with de Havilland until his death in 1993. At the time, he was manager for research and development. Among other things, Farbridge worked on projects such as short takeoff and landing aircraft, including advanced "augmenter wing" research on the de Havilland Buffalo. There he became part of a team under another Avro veteran, Don Whittley, who had gone directly to de Havilland Canada when Avro Canada was wound down after the Arrow cancellation.

While most of the British engineers who worked for Avro and NASA maintained their ties with their home country, the fact that only one is known to have moved back home, and then only for five years, is a graphic illustration of the decline of the aerospace industry in Britain. Once the home of arguably the top aeronautical industry in the world, Britain has continued to support an industry that still produces leading military aircraft, such as the Harrier Jump Jet. In the 1960s, British industry, supplemented by several experts from Avro Canada such as Jim Floyd, worked in cooperation with the French to build the Concorde supersonic airliner. Since that time, Britain has left the field of building large airliners, and now, like Canada, maintains an industry that manufactures smaller passenger aircraft.

Britain was also the home of pioneering space theorists such as Arthur C. Clarke and Ralph A. Smith. In the 1960s and 1970s, Britain possessed the ability to launch its own satellites. This capability was used only once, when a three-stage British Black Arrow rocket launched the Prospero satellite into orbit on October 28, 1971. Since then, the British government has maintained only a minimal British participation in space through the European Space Agency, leaving projects such as Spacelab and the Ariane family of booster rockets to the French, Germans and Italians. The British Blue Streak rocket had been the first stage of the ill-fated European Launcher Development Organization's Europa rocket, and the failure of this effort in the 1960s soured the British government on cooperative ventures. The lack of a British space agency and the reluctance of both Labour and Conservative governments to support a British space program has left the United Kingdom well behind other countries such as Canada in the field of space exploration.

•

The cancellations of the Avro Jetliner and the Avro Arrow prove that the Canadian government was just as capable as the British of making shortsighted decisions. Unlike Britain, however, Canada has a strong record as a spacefaring nation. And although Aikenhead didn't know it when he returned to his home country in 1962, he would spend most of his career working in Canada's space program.

Although his parents were both from Ontario, Bruce Alexander Aikenhead was born in Didsbury, Alberta, on September 22, 1923. His family soon returned to Ontario, settling in London, where Aikenhead grew up. During the war, Aikenhead was a radio mechanic in the Royal Canadian Air Force, attached to the Royal Air Force in Britain and later in India. After the war, Aikenhead took his degree in math and physics at the University of Western Ontario, and then got work in 1950 at Electrohome in Kitchener, where he did design and production engineering for radios and televisions. After five years at Electrohome, Aikenhead was restless and wrote for career advice from a university friend who had helped found a firm in Montreal called Canadian Aviation Electronics. CAE started by making aircraft flight simulators under license, and then got the go-ahead from the Canadian government to make simulators for the Avro CF-100. "I got a very brief reply back from him, saying 'Why don't you come here?' I had never thought of flight simulators. He suggested I get in touch with a classmate who, unbeknownst to me, was working there, and I got a glowing report. So I moved to Montreal in 1955, and I worked on a CF-100 simulator, and then we got a contract with Canadian Pacific Airlines," which he also worked on.

After three years at CAE, Aikenhead and his wife Helen wanted to move back to Ontario. In August 1958, Aikenhead started work at Avro under Dick Carley. He was soon working on simulators for the Avro Arrow. After six months at Avro, which ended with the cancellation of the Arrow, Aikenhead was off to NASA for three years before returning to Montreal.

One unfortunate aspect of life in Virginia livened up the Aikenhead home after they moved back to Canada in 1962. "The base in Virginia was segregated, so when we returned to Canada we thought there must be something people can do on a one-to-one level to improve human relations," Helen Aikenhead said. So their house became a home away from home for international students from McGill University.

Aikenhead and the others who returned to Canada had the grim satisfaction of watching the Diefenbaker government self-destruct in 1962 and 1963. Reduced to a minority government in the 1962 election, Prime Minister John Diefenbaker was soon forced to deal with one of the implications of his government's decision to cancel the Arrow. The government had decided to rely on the American Bomarc surface-to-air missile instead of the Arrow to defend southern Ontario and Quebec from Soviet bomber attack, but the Bomarc was of limited use in this role and was nearly useless without nuclear arms. In the face of the early 1960s ban-the-bomb movement and the concerns of some in the cabinet over American control of nuclear arms, Diefenbaker repeatedly hesitated to bring cabinet to a decision on equipping the Bomarc squadrons with nuclear arms. In early 1963, Diefenbaker's cabinet publicly split over the issue, and his government was defeated in Parliament. Lester B. Pearson's Liberals won the April 1963 election after they reversed position and came out in favour of nuclear arms. Pearson quickly agreed to nuclear arms for the Bomarcs, CF-104 Starfighters, CF-101 Voodoos and Honest John missiles of the Canadian Forces.

In the early 1960s the Voodoos and Starfighters were the mainstays of the Royal Canadian Air Force, and Aikenhead's first job at CAE was to make sure that the RCAF's Starfighter simulators were up to date with changes in the aircraft.

"I was doing that, and they were interested in my background in NASA," Aikenhead recalled. CAE was looking at getting into processing imagery, specifically photos from weather satellites, which were then being launched for the first time. "So they sent me down to NASA Headquarters to see about the automatic picture transmission program that was installed on TIROS and Nimbus satellites. This system would transmit a photo to the ground using a facsimile machine technique. In this way, you could get daily weather photos of the area where the station was set up. If we could set up stations, we could sell them to the Canadian Department of Transport. In addition, if we could enhance the photos and get it to work, we could sell them to all kinds of people, including NASA." Aikenhead worked on developing the antennas for the system, but while this work was going on, Aikenhead had a fateful encounter with another member of the Avro-NASA group who had returned to Canada after the Mercury program ended, Gene Duret.

The son of a carpenter, Duret was known as perhaps the most artistic and thoughtful of the Canadians who went to NASA. Born in Creelman, Saskatchewan, on May 25, 1924, Duret

Eugene L. Duret at Langley Research Center in 1960.
Courtesy NASA.

had an active interest in politics and inherited his father's enthusiasm for the Cooperative Commonwealth Federation, the democratic socialist party that later became the New Democratic Party. Duret, a lifelong bachelor, painted in his spare time. "He was a very widely read, intellectual kind of fellow. He used to come to the house frequently, and our kids would call him Uncle Gene," Aikenhead said. During the war, Duret served as a navigator on Lancaster bombers, and after the war he served as a meteorological officer. Duret's brother Maurice, who himself went on to a career in Canada's nuclear industry, said he persuaded Gene to join him after the war studying physics at Queen's University in Kingston, Ontario. Gene eventually got work at Avro, where he worked on the heat transfer aspects of the Arrow, and he continued this work at NASA, where he worked on the Mercury and Gemini heatshields, as well as becoming one of the capsule communicators in Mercury. Duret moved to Houston, but he decided that life there didn't agree with him. He then moved to Montreal, where he got work with one of the most enigmatic, charismatic and controversial figures in the history of Canadian aerospace, Gerald Bull.

In 1966, Bull was 38. After getting his Ph.D. at the University of Toronto, he had gone to work at the Canadian Armament and Research Development Establishment in Valcartier, Quebec, where he discovered his love for research into improving cannons. Bull left CARDE in 1961 and went to McGill University, where he set up the High Altitude Research Program to test cannons and projectiles in ever higher trajectories. Starting on property Bull acquired that straddled the border between Quebec and Vermont, HARP won support from the U.S. military and the Canadian government. Soon Bull and his team were launching projectiles from a giant cannon located in Barbados. The research had obvious military implications, but one of its aims was to launch a satellite using a cannon.

It was at this point in 1966 that Duret joined Bull's team. "Gene used to visit us quite regularly in Montreal and told us about the projects that Gerry had going," Aikenhead said. "CAE was a place where the morale pendulum swung back and forth. We hit a low spot when Gene came along. Gerry Bull was an absolutely enthusiastic super salesman. I talked to Gerry and he was really enthused about what he could do. He was interested in what I could do and it looked like a whole lot of fun. So I made the switch and I joined the HARP program as a systems engineer. I was put in charge of the Martlet 4 projectile.

"Most of the work they had done was with the Martlet 2, a projectile that could go to 100 miles and expel a payload, usually chemicals that would react with the atomic oxygen in the atmosphere. By photographing the way the trail dispersed from different angles, you could get an idea of the velocity and the direction of the winds at that altitude. They were getting information about upper atmospheric winds that had never been done before.

"Martlet 3 was a series of test vehicles related to ways of enhancing stability, particularly putting out folding fins. They had to survive launch acceleration and then extend to provide enhanced stability. They had another group working on how to get solid rockets to survive launch accelerations. The trick was to immerse the rocket in a liquid with the same specific gravity as the propellant itself. At launch it was neutrally buoyant and therefore supported, and the liquid would not crumble or crack. He had another group making a hollow steel projectile which contained another projectile. The projectile was in effect the first stage and then the nose would come off and the rocket inside - the second stage - would ignite. In the nose of the second rocket was the payload. This was the Martlet 4 concept. All our studies showed we could attain orbital velocity. The analysis showed you could put a 20-pound payload in orbit for a fraction of the normal launch cost.

"We took the Martlet 4 to Barbados and it was fitted with a dummy rocket with inert propellant. The thing broke up in the gun. The analysis showed that the engineer had taken a chance with a bit of steel that was questionable. We were about to do it again. The analyses showed we were going down the right track. It was a very nice swords-into-plowshares kind of thing. He and his people were taking weapons and converting them into research tools. But in the process they were learning how to make better weapons. That's why we had such great cooperation from the U.S. Navy. We were able to buy expired propellant and other things at bargain prices.

"It looked to be a promising thing, and if it had not been cancelled, I'm sure we would have made orbit. But Gerry had a tendency to jump from one project to another. The upper atmosphere winds project - instead of developing a good bread-and-butter project with it, he kind of got bored with it and left it for other things. He could have set up a division and marketed it. But research and development funds were drying up on both sides of the border, and it was the Avro Arrow all over again."

In 1967 the Canadian government withdrew support from HARP, followed by McGill and the U.S. military. Aikenhead and Duret left that year as Bull reconstituted his work on cannons through a private corporation and concentrated on doing research work for various military clients. Bull was jailed in 1980 for breaking the U.S. arms embargo against the apartheid regime in South Africa, and later he began doing work for Saddam Hussein's Iraqi dictatorship. He was murdered at his Brussels apartment in March 1990, a few months before the Gulf War. Although the killer has never been found, it is widely believed that Bull's death was ordered by the Israeli Mossad secret service.

After leaving HARP, Gene Duret moved to Toronto and got work teaching physics at Humber College until his death from cancer in 1985. Duret had enjoyed teaching, and he had been popular with his students, so a bursary in his name was established at the college when he died.

"There was a chance of staying with HARP when it was cancelled, but that would involve going to the States, and my son was draft age," Aikenhead said. "I heard that RCA was looking for people to work on a satellite." The satellite was called ISIS 2, and it was to be the fourth Canadian satellite.

Like its three predecessors, ISIS 2 was dedicated to research about the ionosphere, a layer of electrons in the upper atmosphere that make radio communications, particularly short-wave radio, possible because the signals bounce off it. But changes in Earth's magnetic field caused by changing levels of incoming solar and cosmic radiation have strong effects on the ionosphere. These effects lead to fluctuations and blackouts of radio communications. Canada had long been conducting soundings of the underside of the ionosphere, recording changes in the height of the ionosphere and changes in the concentration of electrons in this area. When satellites began to fly, Canadian scientists started to look at the idea of sounding the ionosphere from above. The result was Canada's first satellite, Alouette 1. Built by the Defence Research Telecommunications Establishment under the leadership of John H. Chapman, Alouette was launched atop an American Thor-Agena rocket on September 28, 1962, making Canada the third nation to have a satellite in space. Alouette was one of the most complex satellites built to that time, and it became the first to operate for longer than a decade.

One of Alouette's major innovations was the Storable Tubular Extendible Member (STEM) antenna, which was one of many inventions created by George Johnn Klein of the National Research Council of Canada (NRC). This technology allowed Alouette to deploy two antennas, one of them 23 m long and one 45 m long. STEM antennas were made of pre-stressed metal, rolled up like a carpenter's rule, that became rigid antennas when they were unrolled after launch. STEM antennas built by the special products and applied research (SPAR) division of de Havilland Canada were used as HF antennas in Mercury and as UHF recovery antennas in Gemini and early Apollo spacecraft as well as on many satellites, including the Agena target

Alouette, the first Canadian satellite.
Courtesy NASA.

vehicles for Gemini. The SPAR division started in 1960, and it was enlarged when de Havilland absorbed Canadian Applied Research, a company that had previously belonged to Avro Canada. As a result, de Havilland employed many former Avro hands who had gone there after the fall of the Arrow. The division became Spar Aerospace in 1968 when it was purchased by a group of employees.

In 1969, Spar won the contract to build booms based on the STEM concept for experiments that flew aboard the Apollo service module during the last three lunar landing missions, Apollo 15, 16 and 17. While these missions are best known for their extensive explorations of the lunar surface with the aid of the lunar roving vehicle, the command module pilots were kept busy operating sets of experiments in the service module's scientific instrument module while their colleagues were on the Moon. On Apollo 15 and 16, the module was equipped with a mass spectrometer that measured the density and composition of the Moon's extremely limited atmosphere, and with a gamma ray spectrometer that mapped radioactive sources on the lunar surface. Both instruments were set up at the end of 7.6 m STEM booms.

Apollo 17 flew a number of different experiments, including the lunar sounder, which probed lunar mascons in the Moon's crust using HF and VHF antennas. The HF antenna was a STEM dipole antenna 24 m long. The primary investigator on the lunar sounder was Stanley Ward, originally a Canadian, who had used similar techniques to search for ore bodies in Northern Ontario.

Spar's project manager for the Apollo STEM antennas was C. Malcolm Hinds, a British engineer who came to Canada in 1964. "Some of the tests we did were quite horrific," Hinds said. "I can't remember what it was, but we had to deploy the boom horizontally and hang a weight on the end, and the thing would go almost through a quarter of a circle in bending before it finally collapsed. For something that would deploy and retract into that relatively small space, it was

Artist's conception of the Apollo 15 and 16 command and service module with gamma ray and mass spectrometer experiments deployed on Canadian-built STEM booms from the scientific instrument module. Courtesy Malcolm Hinds.

Artist's conception of the Apollo 15 command and service module Endeavour with Canadian-built STEM antennas deployed as part of lunar sounder experiment. Courtesy Malcolm Hinds.

quite a feat to see that." The durable BI-STEM antennas were retracted for operations with the LM and for transearth injection, and could be ejected in case the retraction mechanism failed. "On Apollo 16, they took one of the booms and they actually did a jettison and fired it at some point well away from the [LM]. They fired it into the surface of the Moon just so they could get a seismic reading on the sensors to give them some idea of the composition of the Moon or the surface depth of things. They just fired the whole thing javelin style towards the Moon, and then recorded the impact on these seismographs," Hinds said. As a result of their work on Apollo, Hinds and 12 of his colleagues at Spar became the first resident Canadians to win the Silver Snoopy Award, a special award from the astronaut office at JSC for quality work done on human spacecraft.

•

De Havilland and Spar weren't the only Canadian firms that worked on Apollo. The most celebrated example of Canadian manufacturing in Apollo involves the lunar module legs, the bulk of which were manufactured in a plant in the Montreal suburb of Longueuil. Héroux Machine Parts Ltd. was building landing gear and other precision machined parts for many aircraft manufacturers in the 1960s, including Grumman Aircraft of Bethpage, New York, which was a major aircraft builder for the U.S. Navy.

Lionel Whyte, an engineer who was sales manager for Héroux, made a sales trip in 1964 to Grumman, which had recently won the contract to build the lunar module. When Whyte asked about bidding to subcontract some of Grumman's work on the LM, he was told that work was being restricted to U.S. firms. But when he got back to Longueuil, he got a call from someone in a California company with a question about milling a steel ingot for parts for the LM. Whyte was dumbfounded when he was told that the steel came from Belgium. The next day, Whyte was on a plane to New York, where he stormed into the Grumman plant and demanded to know why he couldn't bid on LM work if the Belgian foundry could. When the manager who had refused him realized this inconsistency, he offered to let Whyte bid on some work and immediately took him to a room where he could look through LM plans to choose what he wanted to bid on. "Now, Héroux at the time were designing and building landing gear for de Havilland and Canadair, for the water bomber and the Twin Otter. This would only complement the kind of business we were in. We make landing gear for everything, even the stuff that goes on the Moon. This is what prompted me to pick out all the components that made up the basic landing gear, the four legs, primary struts, the secondary support struts, and a crossover fitting in the middle, which was the most expensive part of the whole thing. I said, 'We can make all of them.' I put them in a briefcase, ran to the airport, caught the last flight out of New York Kennedy at 9:30 at night back to Montreal." Whyte and his group at Héroux bid on 10 parts for the LM legs, and won the work for nine of them. They lost the work for the piston that serves as the lower part of each LM leg, but milled the upper legs and the struts from American aluminum. Contrary to the legend that has built up since the lunar landings, Héroux did not bid for or make the footpads at the bottom of the legs.

Lunar module "Eagle" on the lunar surface showing the leg and struts. Courtesy NASA.

Whyte said that by 1967, Héroux had made 15 sets of four legs for lunar modules, and today six of those sets stand on the lunar surface. The manufacturing involved precision milling that required tolerances of thousandths of an inch. Because of the complicated shape of each part, including the legs, which also included lugs to fit the struts, the work was difficult. But Whyte said U.S. manufacturers were capable of doing the same work. "There are many companies in the United States, many companies right on Long Island, that could do this actual machining of these legs and do any of the processing that was required," he said. But he knew that Héroux could compete on this work just as it did on other aerospace work. Now Héroux-Devtek Inc., the firm continues to make landing gear for aircraft, along with other precision parts from metal and composites for a variety of high-tech industries. Héroux parts are in satellites, rockets and the Canadarm, and will fly on the International Space Station.

•

At first satellites formed the basis of Canada's space industry. After the success of Alouette 1 in 1962, Canada and the U.S. began the International Satellites for Ionospheric Studies (ISIS) program. The first satellite was Alouette 1's backup, which when launched in 1965 became known as Alouette 2. The next satellite, known as ISIS 1, was built at RCA in Montreal and was being prepared for its launch in 1969 when Bruce Aikenhead joined RCA to work as project engineer for ISIS 2. "This involved working out detailed specifications for the spacecraft itself and for the interfaces with the various experiments so that everything would work together properly. As well, there was the planning for testing the spacecraft itself. It was quite an intricate spacecraft. We needed to test all subsystems separately and collectively. We had to devise a test program to test for vibrations, changing atmospheric pressures, and temperature extremes, as well as compatibility under various operating modes. We had to make sure it was compatible with the ground support equipment."

Artist's conception of RADARSAT tracking over Canada. Courtesy CSA.

Like ISIS 1, ISIS 2 had an onboard computer, and Aikenhead said much of his work on the satellite involved programming the computer. As well, Aikenhead was part of the team that conducted prelaunch checkout on the satellite at its California launch site. ISIS 2 marked an advance over its predecessors in that it was equipped with a photometer to provide images of auroral activity near the Earth's poles. Auroras are caused by trapped particles of solar wind passing along the Earth's magnetic field into the Earth's atmosphere near the poles. After ISIS 2 was launched on March 31, 1971, it provided new data on the ionosphere and groundbreaking new information on auroras. The ISIS satellites operated for more than a decade and were given to Japanese scientists after the Canadian research work with them had ended. "It worked very well. Hundreds of scientific papers were written from the Alouette and ISIS programs."

There had been plans to build an ISIS 3 satellite, but the Canadian government decided to shift resources to communications research. A crown corporation, Telesat Canada, had been set up and was preparing to launch Anik A1, the world's first domestic communications satellite, into geosynchronous orbit in 1972. Telesat, which was later privatized, has since launched many satellites in the Anik series. But as Telesat was getting started in the early 1970s, the Canadian government wanted to build a satellite that would lead the way in new communications technologies. Most communications satellites at the time transmitted between large antennas connected to land-based networks. The Canadian government decided to build the Communications Technology Satellite, which would experiment with new frequencies and transmit to small, portable antennas that would open up satellite communications to small communities. "This was to look into the problems you would run into when you have a satellite with direct broadcast capability, with sensitive onboard receivers and a powerful onboard transmitter, so you could broadcast using very small one-metre antennas," said Aikenhead, who RCA assigned to work on the satellite before it was approved.

When the CTS program got the go-ahead, Aikenhead and his family moved to Ottawa, where he worked in the Communications Research Centre of the federal department of communications, which wanted to make use of the expertise of Aikenhead and other former members of the ISIS team. "My job on that was to be concerned with systems and interfaces. I wrote the countdown manual for CTS. My bosses had seen my handiwork with ISIS." CTS was a cooperative program with NASA and the European Space Agency, and when the satellite was launched from Cape Canaveral aboard a Delta rocket on January 17, 1976, it was renamed Hermes. The satellite operated for nearly four years, well beyond its designated two-year operational lifetime. Aikenhead said there was a moment of anxiety when, shortly after entering orbit, valves inside Hermes' attitude control system were released, and the fuel moving through the plumbing caused an unanticipated "water hammer" effect. But after an initial failure, the system worked properly. Hermes conducted a variety of direct communications tests, including direct-to-home broadcasting and telemedicine to remote places in the western hemisphere. Near the end of its lifetime, it was moved to a location above Australia. Today, direct-to-home broadcasting has become a major industry.

Hermes also marked a growing maturity for the private sector side of the Canadian space effort. RCA in Montreal had been responsible for the satellite's electronics. Spar Aerospace built the satellite's structure and electrical subsystems. Although the first Anik satellites were built by the American satellite manufacturer Hughes Aircraft, Spar obtained a great deal of subcontracting work and later became a manufacturer of communications satellites in its own right. Spar's growth included the 1977 takeover of RCA Canada's government and commercial division, where Bruce Aikenhead worked. Spar was to be best known, however, for another space device, an "arm" for the space shuttle, which Aikenhead was involved in from the start.

After Hermes was safely operational, Aikenhead and his family were preparing in 1976 to move back to Montreal when he got a phone call to see if he was interested in working with the NRC.

"They were setting up a program office to work with NASA on some kind of robotic

Hermes satellite. Courtesy CSA.

device to be used on the shuttle. It was to be a contract with Spar, RCA and a consortium. People who were used to working for those companies and NASA and were engineers were the sort of people they were looking for. So I said yes, I was interested. It would be better to remain in Ottawa than move to Montreal. So the next week I was across town to join what was called the teleoperator project office. It was run by the NRC. Ultimately Spar became the prime contractor and the others became secondary. My old colleagues at CAE were doing the display and control panel, and the hand controllers, and some of my old colleagues at RCA were doing the electronics systems to control each of the joints, and to do the clever logic to make it all work together. DSMA [Dilworth, Secord, Meagher and Associates, Ltd.] did the intricate things, notably the end effector, and some of the test and checkout equipment. NRC was right in the middle of all this, and it was our job to make sure that the people on the Canadian side did the job to design the hardware to meet the NASA requirements, and at the same time, protect them from growing requirements from NASA for performance or schedule or whatever. We were truly middlemen, and we were sometimes loved by all and sometimes hated by all. But it was great sport. We developed a team in the office and we were constantly travelling back and forth to Toronto. I spent a lot of time travelling back and forth to the Cape because among the many hats I was wearing was responsibility for the ground checkout equipment. This meant arranging participation by varying people at the Cape."

During this time, when Aikenhead was systems manager for the remote manipulator system (RMS), as the arm was officially called, he remained an employee of RCA and later of Spar, on loan to the National Aeronautical Establishment of the NRC, until 1981, when he formally became an employee of the NRC and deputy manager of the RMS program.

The arm, which became known in Canada as the Canadarm, is 15.2 m long and weighs 360 kg. Although it cannot support its own weight on the ground, in space it can move objects with a mass of up to 30 tonnes. The arm has a shoulder joint near the point where it is attached to the shuttle, which can be moved up, down and sideways. The elbow joint, halfway up the arm, can be moved in or out, much like the human elbow. The wrist joint, near the end of the arm, can be moved sideways, up and down, or rotated. This gives the arm six degrees of freedom. At the end of the arm is what is called the end effector. It is like a cup with three wire snares that, when closed, firmly attach the arm to a fixture on the object being held. The arm is equipped with television cameras that allow the operator to see what he or she is doing, and the arm uses computer software that permits the operator to direct the arm in a variety of automatic or manual modes. The arm is berthed in the shuttle along the side of the payload bay, with the shoulder attached at the front of the bay. Astronauts operate the arm from a control panel at the back of the shuttle's flight deck, below windows that look into the payload bay.

As part of the agreement the Canadian government reached with NASA in 1975, Canada financed the development of the RMS and gave the first Canadarm to NASA, which undertook to buy further arms as they were needed. The arrangement was part of NASA's effort at the start of the shuttle program to defray development costs and foster international cooperation in space flight. This also led to the building of the European Spacelab module.

Aikenhead and the project manager, Art Hunter, made a last-minute addition to the arm just before it was handed over to NASA in February 1981: a patch of beta cloth with the Canada wordmark that incorporates the Canadian flag. Although legend has it that NASA opposed the use of the wordmark, NASA authorized the addition before it was put on the arm. Once the Canadarm had been installed, American flags began to appear on equipment inside the payload bay.

While the Canadarm was being developed, astronauts came to the Spar plant in Toronto where a simulator had been built to teach them how to use this system. Two astronauts who spent a lot of time in Toronto were Sally Ride and Judith Resnick, who became the first and second American women to fly into space. NRC managers such as Aikenhead, Hunter and Garry

Canadarm aboard the space shuttle. Courtesy NASA.

Lindberg acted as liaisons between NASA and Spar, and Aikenhead remembers taking part in many long teleconferences between NASA centers on shuttle software issues. "Everyone needed more software, more control capability, and the remote manipulator system was no exception," said Aikenhead. Resnick, who later lost her life in the Challenger disaster, often joined these meetings to support the needs of the Canadians "because she was very very familiar with the whole thing and was quite a forceful arguer."

The arm's handover to NASA took place at a ceremony at Spar's plant in Toronto, and the arm was trucked to Cape Canaveral where it was prepared and tested for its first flight, which was scheduled for the second shuttle test flight in November 1981. Aikenhead said the first flight of the shuttle in April had caused concern for the arm because the noise and vibrations at launch were so great they had damaged a wing strut. "There was some anxiety that the shoulder joint would shake apart. The shoulder brace was looked at quite closely," he said. A beefed-up system to dampen sound and vibration was installed on the launch pad for Columbia's second flight, which began on November 12, 1981.

"To everybody's delight, the arm survived launch and the STS-2 mission moved the arm through all the tests we had planned for it," Aikenhead said. "The very last test was of a backup control mode. That was where we used a completely independent control mode for each joint. The last was the shoulder yaw mode and that didn't work. So we switched back to primary mode. We did a full checkout as soon as it returned to find that fault. It turned out to be a broken wire in a connector that we knew we had tested repeatedly. The control panel to which it was attached had been swung in and out repeatedly as other things were installed in the cabin. We're pretty sure the wear and tear on the cable was the cause. To my knowledge, that was the only problem we had with the Canadarm system."

Aikenhead was at the Cape for the launch, and then he flew to Houston to be present when astronauts Joe Engle and Richard Truly unberthed the arm for the first time. "My duties also included making sure that the performance was as advertised. So we had to do a lot of analysis of the telemetry data to make sure that the performance in flight was consistent with what had been predicted. We would see if the modelling in the simulator was incorrect or see if the dynamics of the arm in flight were different."

On this first flight the Canadarm did not lift any payloads, but its successful operation sparked a celebration in Houston the evening STS-2 landed. Malcolm Hinds, by then Spar's marketing director for the Canadarm, recalled that the party was enlivened by the presence of Truly, who had come to the celebration after flying home from the shuttle's California landing site. The success of that first test led to tests where the arm grappled and moved payloads under a variety of conditions during the third and fourth test flights of the shuttle in 1982. Said Aikenhead: "The arm flew, satisfied its requirements, and as a result was declared operational. We had an official operational readiness review with official presentations at NASA. So it was declared operational, and NRC and Spar hosted a whole bunch of NASA people at a shindig at a hotel in downtown Washington.

"All in all, it led to a substantial amount of business for Canada. We had spent about $113 million by 1983, and subsequently NASA has spent far more than that in buying new systems from Spar. So it was better than a break-even deal. Canada developed a world reputation for that kind of hardware and a sophisticated control system that has led to many other products and work by the team. And subsequently we began working on the follow-on system for the space station."

•

When the RMS was declared operational after passing its first tests in space, the project office at NRC was wound down, and soon Aikenhead was the only person left. But NRC was moving into another endeavour, which Aikenhead soon joined. In the late 1970s, NASA invited Canada to consider sending astronauts of its own aboard the space shuttle. As Aikenhead said, that invitation had "fallen down the cracks," but with the shuttle beginning to fly in 1981, NASA renewed its invitation. At the same time it was inviting the European Space Agency and other users to consider sending researchers into space aboard the shuttle. At a celebration in September 1982 marking the 20th anniversary of Alouette 1, the U.S. invitation was made public.

The following June 8, the space shuttle test vehicle Enterprise was flown to Ottawa atop its 747 carrier aircraft. Enterprise was on its way home from a tour of Europe as part of NASA's efforts to get international partners on board for the space station, and its visit to Canada included stops in Montreal and Toronto as well as the capital. In all three cities, large crowds came out to see Enterprise. When the shuttle landed in Ottawa, NASA administrator James Beggs alighted from the 747 carrier aircraft and joined Canadian science minister Don Johnston for the announcement that Canadians would fly aboard shuttles as part of the newly formed Canadian Astronaut Program. Advertisements for astronauts appeared in Canadian newspapers the following month, and at NRC, Aikenhead joined Karl Doetsch and Garry Lindberg, with whom he had worked on the Canadarm, to select Canadian astronauts for the shuttle.

"We had done enough homework to know that we needed two types - medical specialists and engineers. I was still looking after the [Canadarm] office, while Karl and Gary were running the recruitment campaign with another branch of NRC. I only got involved at the last to meet the 20 finalists. They had something like 4,900 applicants, and I was involved in reviewing the summary sheets. They had quite a few people at NRC reviewing the applications. The logistics of the operation necessitated successive screenings. They had regional screenings, leading to the final 20 going to Ottawa. One dropped out at that stage. The 19 were subjected to quite intensive medical examinations and an exposure to a variety of people to evaluate them. I remember being quite impressed by Marc Garneau. He was very sharp and very pleasant to talk to. A couple of days later I got a call from Karl Doetsch, who was working through his list, and he said he noticed me chatting with Garneau and what did I think of him?"

On December 5, the names of Canada's first six astronaut trainees were announced. They included Garneau, a naval officer who was an expert in electronic systems and the only French Canadian in the group; physicist Steve MacLean; engineer Bjarni Tryggvason; Ken Money, a physiologist who belonged to a group of Canadian researchers that led the world in research on space sickness; and two physicians, Bob Thirsk and Roberta Bondar, the only woman in the group. Money, Bondar and Tryggvason were also pilots.

"They were chosen in December, and they reported in January. I was still wearing two hats, still looking after the remnants of the NRC involvement in the Canadarm program and also helping Karl look after the astronaut program. We grouped them, a medical type and an engineering type, to share offices. We thought a little cross pollination wouldn't hurt. We put Bondar and Garneau in one office, Tryggvason and Money in the next office, and Thirsk and MacLean in the third office. We had offices for public relations people and the training people and the secretarial staff. We had a conference room. We managed to squeeze it all in and get it ready by the time they reported for duty in January."

Aikenhead, with his experience at NASA training the Mercury astronauts, was one of two people training the new recruits at NRC in Ottawa and arranging work with NASA in Houston. The first Canadian astronaut was expected to fly in late 1985, a year and a half away. The Canadian astronauts would fly as payload specialists, a category of shuttle crew with specific assignments and limited training.

"They came on board in January, and about the middle of March we got a call from NASA headquarters asking if we could have somebody ready to fly by that summer. Good grief, no. Then they said, how about October. We said yeah, October would be fine. We had been promised two flights, one in life sciences and one in engineering, and this would be a third flight, an extra flight. They had decided to go ahead with payload specialists from other countries and they thought they would start with Canada. We didn't want to send someone along on a joyride. We wanted them to have something useful to do for Canadian scientists, and therefore we needed to organize some experiments in a hurry, and train the person."

Marc Garneau, Canada's first astronaut. Courtesy NASA.

The person chosen was Marc Garneau, with Bob Thirsk as his backup. Aikenhead and Doetsch went to NASA to work out the details of the flight and arrange training in Houston and the placement of experiments aboard a locker in the shuttle. "We were able to get Marc and Bob down to Houston for training, and I had to scramble to organize all the equipment to be stowed on board the locker. We had to get it all flight qualified and get it ready through the summer and fall." Equipment for the 10 Canadian experiments, known collectively as Canex, had to fit into a locker no bigger than a breadbox, be able to withstand the stresses of launch, weightlessness and re-entry, and meet NASA's strict materials and safety standards.

Just before dawn on October 5, 1984, the shuttle Challenger lifted off on its 6th flight and the 13th mission of the shuttle program with Marc Garneau and six other astronauts aboard. The flight, commanded by veteran Bob Crippen, included the first space walk by a U.S. woman, Kathy Sullivan, and the deployment of an Earth research satellite. Aikenhead, by then program manager for the Canadian Astronaut Program, spent the flight in a back room near the Houston mission control room, ready in case the crew or the flight director had questions or problems. "We had a small group set up and I was in charge of it. We manned it around the clock, and we had representatives from each of the investigative teams. We were also in a position to answer questions put to us by the public relations people."

During the flight, Aikenhead noted, Garneau was so silent that "we wondered at one point if he'd really got on board." As a last-minute addition to the crew, Garneau was conscious of the fact that he should do his tasks without interfering with other work and be available to help the others when possible. "So Marc tried, as a good Navy man, to stay out of the way. We actually had to nudge the system a bit to get a daily voice report from him," Aikenhead said. His 10 experiments included two technology tests, including preparatory work for the Canadian-developed space vision system, which was to be used in later shuttle flights, two space science experiments, and six medical experiments looking into reflexes, perception and space sickness. Garneau's work, and the whole mission, ended successfully after eight days in orbit.

After helping Garneau deal with his postflight tour of Canada, Aikenhead and the others in the astronaut program set to work preparing for further launches, including a flight featuring medical tests and a flight in which Steve MacLean would begin testing the space vision system. But the astronaut program came to a halt when Challenger exploded on its 10th launch in January 1986. "It did mean our astronauts were grounded, and the flights we planned for '86 didn't happen until '91 or '92." For much of that time, the six astronauts left the program to pursue their research while they waited to see what would happen. "We had hired them because of their skills, and if we wanted them to keep those skills, we had to let them keep their skills by doing research. They continued to do PR work and keep abreast of the shuttle program and do all those other things." NASA was reluctant to launch payload specialists after the loss of Challenger, and this put the Canadian Astronaut Program into some question.

But at the same time, NASA was asking Canada to take part in the space station program, and when Canada agreed to provide an advanced version of the Canadarm for the station, Aikenhead's bosses moved on to take over this work. Aikenhead became manager of the Canadian Astronaut Program. When the Canadian Space Agency (CSA) was formed in 1989, Aikenhead's title was changed, according to civil service rules, to director-general of the program. Not long before the new agency was formed, Aikenhead reached the age of 65. He was busy preparing for the next two flights of Canadian astronauts, so he decided to stay because "I was up to my ears in things and was having fun." By 1991, Aikenhead was busier than ever. Two Canadian astronauts, Roberta Bondar and Steve MacLean, were preparing to fly in space, and Aikenhead had special responsibilities for the space vision system, which would be tested on MacLean's flight.

Bondar was slated to fly aboard the first mission of the international microgravity laboratory, which used the Spacelab module for a set of space science and medical experiments. With Ken Money serving as her backup, Bondar became Canada's second astronaut when Discovery lifted off on January 22, 1992. During her eight days in space, Bondar carried out a set of medical experiments that, among other things, looked into space sickness that strikes about half the astronauts who go into space during the first few days of flight. The seven-year gap between her flight and Garneau's, and the fact that Bondar was the first Canadian woman to fly into space, led to the flight being very highly publicized in Canada.

Less publicized was the 10-day flight of MacLean, which took place aboard Columbia starting October 22. Although MacLean had a whole set of Canadian-developed experiments to work with, his major task was to test the space vision system. The SVS uses video images of targets placed on objects to tell a computer, and ultimately the operator of the Canadarm, the precise position of the payload and changes in that position. The SVS is designed to help arm operators on the shuttle and space station when they can't see what they are moving, and ultimately it may allow computers to do the work of human operators. During his flight, MacLean worked with the arm operators on Columbia to put the SVS through its paces for evaluation on the ground. During Bondar's flight, Aikenhead worked in the science control room in Huntsville, Alabama, and during MacLean's flight, he worked out of Houston, much as he did during Garneau's flight.

About the time of Bondar's flight, another advertisement soliciting applications for the Canadian Astronaut Program appeared in Canadian newspapers. Aikenhead was in charge of selecting new astronauts to beef up the ranks of Canadian astronauts in anticipation of the space station, and to replace Bondar and Money, who left the program. There were 5,000 applications, and Aikenhead, Garneau and Thirsk were among the group that interviewed the finalists. In June, the agency announced that test pilot Chris Hadfield, engineer Julie Payette, physician Dafydd

Canadian Astronaut Team. From left to right, back row: Bjarni Tryggvason, Robert Thirsk, Julie Payette, Steve MacLean; front row: Dave Williams, Marc Garneau, Chris Hadfield.
Courtesy CSA.

(Dave) Williams and geophysicist Robert Stewart would join the program. Within days, Stewart dropped out when he reconsidered his scientific career, and his replacement, Mike McKay, an engineer in the Canadian Forces, also had to drop out due to vision problems.

The Canadian astronaut group now had seven members. As soon as the new astronauts were selected, Aikenhead had to decide who would get a year's training in Houston that would lead to designation as a full-fledged mission specialist (the category for most American astronauts). Aikenhead chose Garneau, who was already well respected at NASA from his previous flight, and Hadfield, who was also known in Houston because of his work as a test pilot. Aikenhead admitted that his decision to designate Hadfield was not popular with members of the 1983 astronaut intake who were still waiting to fly, but within five years, all seven of Canada's astronauts had received mission specialist training and were living in Houston as full-fledged astronauts. As well, by 1999 all seven had flown in space at least once. The Canadian flights included a 1995 journey by Hadfield to the Mir space station, and flights to the International Space Station by Payette in 1999 and Garneau in 2000.

After MacLean's flight, Aikenhead began to prepare for retirement. His wife Helen had already set up a home in Salmon Arm, B.C., and the CSA was moving from Ottawa to a new headquarters at St. Hubert, Quebec, just south of Montreal. Aikenhead got what he called a "big sendoff" when he retired on March 12, 1993, and the next day he flew out to his new home in the British Columbia interior. Four years later, Aikenhead was given one of Canada's highest honours when he was named an officer of the Order of Canada. "I was lucky to always have an interesting job, usually working in a program that had never existed before," Aikenhead reflected in an interview. "There was always some element of pioneering involved, or there were engineering problems where there were no solutions readily available."

•

The Canadian space program today consists of far more than the astronaut program, the Canadarm and the space station. It also includes its old mainstay of satellites, notably Radarsat 1, which was launched in November 1995. Using a sophisticated and powerful radar system, Radarsat images the Earth, providing information that can be used for a variety of purposes, including mapping, geological research, mineral prospecting, crop identification, oil spill tracking and iceberg detection. Radarsat's radar provides regular coverage of most of the Earth, except for Antarctica.

When the agency needed some help to handle the intricate and difficult operation of turning the satellite around so that it could map Antarctica, it turned to George Harris, a veteran of tricky operations with satellites. Since he had helped rescue TDRS in 1983, Harris had helped modernize the U.S. Information Agency's Voice of America network and helped launch communications satellites on Ariane rockets at the French launching site at Kourou in French Guiana. Hired through a contractor, the ex-Avro engineer moved with his wife Martha to the CSA headquarters at St. Hubert from their home in New Mexico to serve as Radarsat's control facility manager in 1997 and 1998. "What made me come here was the challenge again," Harris said during his stint at St. Hubert. "We've taken this Radarsat, turned it around so it's flying backwards, and it's mapping Antarctica. That had never been done before. Nobody's ever taken an Earth resources satellite and done this kind of maneuver. That's all the challenge right now." Once the mapping had been completed and Radarsat turned back around to its usual orientation, Harris returned home to New Mexico.

Harris and Aikenhead weren't the only members of the Avro-NASA group who returned to work on the Canadian space program. In 1986, Bryan Erb, who had retired from NASA, joined the National Research Council of Canada as its liaison at the Johnson Space Center for the space station program. To Erb, this turn of events was not without its ironies. When Erb had joined NASA in 1959, he had turned down a job offer from NRC, but he told them that he might be

George Harris inside Radarsat Control Room at CSA Headquarters under "George's clock," which operates on MET (Mission Elapsed Time) rather than GMT (Greenwich Mean Time), October 8, 1997. Photo by Chris Gainor.

interested after a couple of years at NASA. Two years later, a letter duly arrived from NRC, but Erb, by then immersed in developing the Apollo heatshield, was not interested. Finally, 35 years later, he was ready to join NRC.

"In the fall of 1985, Canada's involvement in the space station was pretty well settled, and Canada came looking for a Houston liaison person," Erb said. "They were recruiting all across Canada, and this came to my attention when my son saw the advertisement in the *Globe and Mail* in Calgary and sent a copy to me. I was having a good time consulting. I had all the work I cared to do, and I thought, oh well, I'm working close to full time, and threw my hat in the ring. I went up to Ottawa that fall and met with Karl Doetsch and Mac Evans, and it turned out they liked the idea. It took awhile to negotiate the terms and the start time, but I joined the space division of NRC in March of 1986. Due to the 'obsolete' fingerprints, I was still sitting with Canadian citizenship when Karl Doetsch came looking for a liaison person. You wonder about these twists of fate that change your career."

Just two years out of JSC, Erb was familiar with all the players there. And his consulting work after NASA had included marketing for Spar Aerospace. "I got to know some of the players on the Canadian scene. I knew enough about robotics and automation so I could communicate with members of that community. So I've been representing Canada on the space station," Erb said in 1995.

Erb's new job was not his first involvement with the space station. In 1984, a powerful U.S. senator persuaded NASA to join in his effort to promote automation and robotics by dedicating a major part of station funding for robotics. NASA formed a committee and called on Erb, who was still with NASA, to serve as a technical writer. By the time Erb returned to the station effort in his new job, U.S. funding problems meant Canada had been handed responsibility for the station's robotic equipment.

Canada's contribution to the International Space Station is known as the mobile servicing system, a sophisticated system that will be able to move along the main truss of the ISS using a mobile base so it can transfer and assemble equipment almost anywhere on the space station. The MSS includes the space station remote manipulator system or Canadarm2, a more sophisticated version of the Canadarm, which is 17 m long and has seven joints where the Canadarm had six. Erb said Canadarm2 can "walk" around the station by grappling station parts using both of its ends, and can move up to 100 tonnes, the mass of a fully loaded space shuttle. As well, the MSS will be equipped with what is called the special purpose dexterous manipulator or the "Canada hand," a set of small manipulators that will be equipped with an artificial sense of touch to carry out the finer work that the large arms cannot do. The MSS will relieve astronauts of space walking duty. The MSS's prime contractor is MD Robotics, a subsidiary of MacDonald Detwiler and Associates, which bought the Spar space robotics operation in 1999. Installation of the MSS began in April 2001 when Chris Hadfield helped set up Canadarm2 on the ISS during the first space walk by a Canadian astronaut.

"The robotic systems that we will be supplying are absolutely crucial to the success of the space station venture," Erb told a 1998 briefing on the station. "What Canada does for the International Space Station is, we put it together and we keep it working."

During his time at NRC and the Canadian Space Agency, Erb had to deal with the crises and redesigns that have bedevilled the space station, along with one near-cancellation of the Canadian government's participation in the space station. As assistant director of the Canadian space station program, Erb was involved in many design issues as he represented Canada in international negotiations and a variety of bodies from technical working groups to blue ribbon review panels. As the Houston representative for the CSA, he hosted prominent Canadian visitors, and he and Dona provided help and local orientation for Canadians assigned to JSC, including flight controller trainees and Canadian astronauts.

The systems Canada is contributing to the ISS will keep Canada at the cutting edge of technology and will provide research opportunities for Canadian scientists, Erb said. As well, the astronauts Canada is sending to the station will inspire younger Canadians to venture into space and conduct scientific research.

"We can build space hardware that is as sophisticated as any in the world, and we can train our own and other countries' astronauts to use it," Erb said. "The International Space Station is a facility which will extend the intellectual reach and habitat of humanity – people living in space and doing useful things."

Bryan Erb with test article of the SSRMS in stowed position on the Mobile Servicing System (MSS) made for use in the JSC Weightless Environment Training Facility (WETF) circa 1989 or 1990. Courtesy Bryan Erb.

Karl H. Doetsch, a key driver behind the Canadian space station program, formerly vice-president of Canadian Space Agency and now President of the International Space University, with Bryan Erb in front of ISS model. Courtesy Bryan Erb.

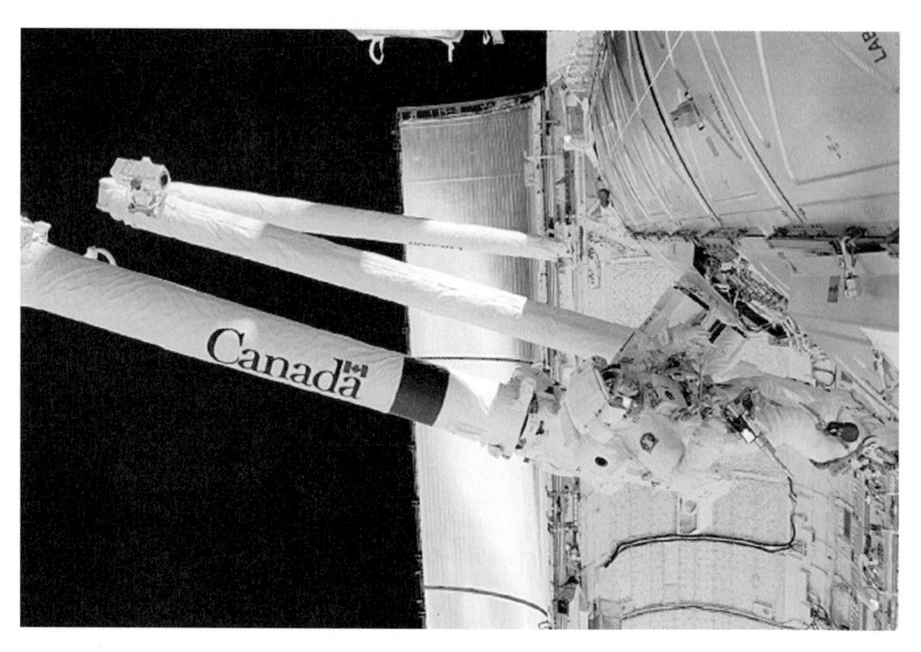

Astronaut Scott E. Parazynski, mission specialist for STS 100, works with cables associated with the Canadarm2 during one of two days of extravehicular activity. Parazynski shared both space walks with astronaut Chris A. Hadfield of the Canadian Space Agency. Courtesy NASA.

Conclusion: A Godsend to NASA

On the 20th anniversary of the Apollo 11 Moon landing, NASA held a week of celebrations at the Johnson Space Center, including an abbreviated re-enactment of the Apollo lunar landing symposium that Owen Maynard organized in 1966. Maynard had left NASA early in 1970, true to his promise that he would remain for two lunar landing attempts, and went to the large aerospace firm Raytheon, near Boston, where he remained for the rest of his career. In 1989, as Maynard returned to JSC, he was nearing retirement age but still working at Raytheon. He strode onto the same stage in the center's auditorium that he had taken 23 years before, and with six other participants from that symposium he told a new audience what the original had meant. "My intent at the time [was] to orient us all, and no one needed more orientation than I," Maynard said.

When asked what lessons he drew from Apollo, Maynard recalled Bob Gilruth's dictum that the program should be able to be carried out by ordinary people. "Don't ever plan for heroics at any level," Maynard went on. "I've discovered personally that there's no such thing as uninvited constructive criticism. If the conceiver doesn't invite it, they won't accept it." He urged the spectators to help new people along. "Be very encouraging and don't be negative unless the person is dead wrong. I discovered that I needed around me a lot of people assuming a lot of responsibility, and I discovered that people can assume tremendous responsibility if one assigns that responsibility to match their potential capability."

Maynard had gone to Raytheon to work with his former boss, Joe Shea. During the 21 years he spent there, Maynard worked on many projects in the defence field, including the Patriot missile, and on some schemes involving space. Maynard was space systems manager and project manager at Raytheon for a proposed power satellite program. He had hoped to return to Canada in the 1970s and continue his work in aerospace, but he was kept too busy at Raytheon to move back home until his retirement. From his retirement home in Waterloo, Ontario, where he and his wife Helen moved in 1992, Maynard worked with young members of the Canadian Space Society, providing advice on spacecraft design. Maynard died in Waterloo on July 15, 2000.

Like Maynard, many of the Avro-NASA engineers moved into work in the private sector when they left NASA. Many went into consulting for aerospace firms. Fred Matthews, who went to the RCA Aerospace division in Boston in 1962, worked for nearly 30 years on tactical command and control systems, including airborne command posts and army battle management systems, for the U.S. armed forces. After leaving NASA in 1964, George Watts had a long career at Lockheed, working on its proposal for a supersonic transport, then on the L-1011 airliner and then on advanced helicopters and other aircraft.

Peter Armitage was reunited in 1988 with former astronaut Deke Slayton, with whom he had worked on the troublesome lunar landing training vehicle. "Slayton was now president of Space Services, Inc. [SSI], a small company trying to move into the potential area of commercial launch vehicle design and operations. When Deke heard that I might be available, he made contact with me about joining SSI. I took Deke's old position as program manager of the Conestoga launch vehicle, which the company was designing. Later, I was given the position of vice-president for programs with responsibility for launch vehicle design and operations. It was a very small company, with most of our engineering work being done by various outside contracts. So it was a management job in one sense, but very rewarding to be one of the leading companies pursuing the commercialization of launch-to-orbit vehicles and payloads. We pursued many contracts with NASA and the [defense department] and other potential launch vehicle users. We designed and launched several small payloads out of White Sands Missile Range in New Mexico with our 'Starfire' sub-orbital launch vehicle. This was a great job! A small group of good engineers doing great things, very reminiscent of my days with Avro flight test, and the early days with landing and recovery at NASA. The contracts to move from government-launched spacecraft and payloads to commercial companies were, however, very slow in coming. Too slow, as it turned out, for our major investor, and our main operating funds were withdrawn in 1990." After some further consulting work, Armitage settled into retirement at his home near JSC, where he restores vintage British cars.

•

Maynard spent much of his career at Raytheon advocating satellites as a source of power. Solar cells on the satellites would collect the power, which would then be converted into microwaves and sent to receiving stations on Earth. Raytheon was one of several contractors supporting NASA and the U.S. Department of Energy in studying this concept and conducting demonstrations. Despite technical promise, such systems were not implemented, in part because key people in the energy department believed that electricity from nuclear fusion would soon be available. Another technology advocated by Maynard and his colleague Bill Brown of Raytheon was to beam power generated on the ground up to distant users, to keep aircraft flying or even to boost satellites from low Earth orbit to higher orbits. This concept was later demonstrated on a small scale in Canada in 1986 by the SHARP microwave-powered aircraft project.

Around 1991, concerns over the environmental impact of energy use caused a resurgence of interest in space solar power. While Brown and Maynard continued their advocacy, others became interested, and Bryan Erb, always interested in new technological challenges, began his own research as a side interest to his work on the International Space Station. "I've long been interested in global ecosystem conditions and the survivability of the human species on the planet," Erb said in 1995. "I'm persuaded that the notion of importing energy from space is going to come around again. So for the past five years now, I've been conducting a little side study on the notion of solar power imported from space. There's increasing interest worldwide in this. I would say the energy establishment of the world only barely recognizes renewable resources on the surface of the planet. They haven't yet begun to think about extraterrestrial resources." Erb, who formed and now manages the Canadian Space Power Initiative, said current energy practices are creating serious problems on Earth. "We are crudding up our environment. We will get a substantial shift in the mindset of the energy establishment. And companies who are at present purveyors of oil and coal will come to the realization that they can be broader based energy companies. That they have skills they can bring to broader resources."

Four years later, Erb said he believes that space power could be providing a significant portion of the Earth's electricity needs within 50 years. "It's impossible to predict what the future will be. A lot of the [new energy sources] that are considered, such as tidal and wave and ocean thermal - there's just not enough energy there and it's difficult to capture. My feeling is that in about 20 years we will see the beginnings of a space power industry." There remain several important challenges, including bringing down the cost of putting payloads into orbit. Canada, with its well-developed electrical industry, will likely play a role in space power, perhaps in developing the equipment needed to receive the microwaved energy on Earth. Both Maynard and Erb learned that the success of endeavours such as space power depend on how effective they are at improving life on Earth.

•

To many people, the next logical target for human space exploration is Mars. Erb and an associate were awarded a patent in 1964 on a heatshield for a spacecraft returning directly from Mars. But the member of the Avro group most strongly associated with a flight to the red planet is Morris Jenkins, who led an effort at the Manned Spacecraft Center to draw up plans for a flight to Mars. Jenkins was working for John Hodge and Dennis Fielder on future programs at the time of Apollo 11 when NASA Headquarters asked both MSC and Marshall to submit studies for human flights to Mars, and Bob Gilruth designated Jenkins as head of the effort at MSC. Jenkins was able to call on help from many of his colleagues from MSC and from the aerospace contractor TRW. "Even with a realistic perspective on the whole thing, we put out an energetic effort on it. It was a good first draft and sent to headquarters," he said.

The Jenkins report, as it is often known, was completed in February 1971 and called for an "austere" low-budget program sending an initial expedition of 570 days to Mars in 1987 and 1988 after 11 years of development. Jenkins' plan assumed that NASA had already developed a shuttle and that components of the Mars vehicle would be assembled in Earth orbit following seven launches using shuttle booster vehicles. As the solar-powered vehicle with a crew of five flew to Mars, chemical propulsion stages would be jettisoned after they were used to boost or slow down the vehicle. After 15 days in orbit around Mars, three of the five astronauts would descend to the surface in a Mars exploration module for 45 days of exploration using two roving vehicles similar to those used in Apollo. On its way back to Earth, the spacecraft would swing by Venus and enter Earth orbit for pickup by shuttle at the end of its trip.

Said Jenkins of the report: "Just as it was completed, NASA became officially committed to the shuttle and, understandably, headquarters terminated all further discussion on the Mars mission, and I'm sure that was the right decision at the time." Although his report was not widely circulated at the time, it has been referred to often by present-day advocates of Mars missions because of its cost-effective ideas on crew safety and crew composition.

When Jenkins began to draw up his Mars plans, their realization seemed imminent. Apollo's voyages to the Moon had laid to rest the old cliché comparing any seemingly impossible effort to going to the Moon. Even before humans retreated back to the environs of Earth, a new cliché was born, which has sadly been appropriated by those who want to pick up where Apollo left off: If we can go to the Moon, why can't we go to the Moon? More than a generation has passed without a serious effort to take humans beyond Earth orbit again, as the resources that were briefly diverted to take astronauts to the Moon are used for military or social security programs, or put into tax cuts. At the beginning of a new century, Apollo seems to be an achievement of greater audacity than was even acknowledged in the superlatives of the time.

•

Although those who went from Malton to NASA continue to look to the future with projects such as power satellites and private launch vehicles, they have much in their past to celebrate.

One of Maynard's most treasured keepsakes was a copy of a 1969 letter from Bob Gilruth, for whom he worked during his career at NASA. Joseph Hobin of Sarnia, perhaps responding to some of the publicity in Canada about Maynard's role in putting men on the Moon, wrote Gilruth asking for his appraisal of Owen Maynard. Gilruth wrote:

> Maynard was involved in Apollo from the very beginning. When he first came to our organization from Canada, he made many of the preliminary plans for the Apollo command module. The space ship Columbia, which flew to the Moon in July, 1969, resembled in many ways the layouts that Mr. Maynard first made in 1960. Later on, we had to decide not only how to get to the Moon but also how to land on the Moon. Again, Owen Maynard played an important role. He had a major part in the design of the lunar module, the Eagle, that successfully landed on the Moon in July 1969. But Owen Maynard is not only a planner, he is also a doer. He stayed with Apollo from the very beginning and helped carry through to the lunar landing. As chief of our systems engineering division he directed the team that was responsible for the many design details that evolved in the intervening years. He was also responsible for the detailed planning of each flight mission. Under his direction the flight test program, including first Earth orbital flights, then lunar orbital flights, and finally the lunar landing, was evolved. Owen Maynard's role in the greatest technological achievement of mankind, man's first landing on the Moon, was a most significant one. It has given me a great deal of pleasure to be associated with him through these many years.

A few days after Apollo 11 returned to Earth, Rod Rose got a similar letter from his boss, Chris Kraft:

> Dear Rod: I don't believe that there is anyone at MSC who has contributed more to the success of Apollo than you have. It is the attention you give to detail that assures a successful flight. Every organization at MSC is aware of the important work that you have done. You should be very proud of the success we have had in the program. I personally appreciated the fact that you were involved in all of the planning, and it gave me a great deal of confidence that things were being done properly. One of the greatest assets we have is the motivation and dedication of people like you.

At the time of Apollo 11, NASA honoured Jim Chamberlin, John Hodge, Maynard and Rose with the NASA Exceptional Service Medal. As well, Dr. Bill Carpentier, Tom Chambers, Bryan Erb, Len Packham and Bob Vale were awarded Manned Spacecraft Center Certificates of Commendation, and Burt Cour-Palais and Norm Farmer won MSC Superior Achievement Awards.

Maynard had already won the Exceptional Service Medal for his work on Apollo 8, and Hodge for his service as a flight director in Gemini. Rose later won the medal again for his work on shuttle. Chamberlin had been given NASA's Exceptional Scientific Achievement Medal for his work on Gemini. In 1973, Armitage and Jenkins won the Exceptional Service Medal, and Erb won the same medal five years later for his remote sensing work. Robert Lindley was given the Exceptional Service Medal in 1981 for his role in the shuttle, Chamberlin won a similar medal for shuttle work, and David Ewart was awarded the medal in 1985. Tecwyn Roberts won NASA's Outstanding Leadership Medal on his retirement in 1984. John Hodge was given a NASA Outstanding Leadership Medal in 1985 for his work on the space station, and his other honours include an honorary degree from the University of London in 1966. In 1990, Erb was awarded an honorary doctor of science degree by the University of Alberta.

The University of Toronto awarded Maynard an honorary doctor of engineering degree in 1996. In his address to a crowd of graduating engineers, most of whom weren't yet alive when Apollo went to the Moon, Maynard emphasized that inside nine years, "the extraordinary had been accomplished by quite ordinary people." Invoking the view of the Earth as seen from Apollo 8, Maynard said that view has given engineers and others the challenge of dealing with environmental degradation caused by the Earth's human population.

In 1987, Bruce Aikenhead was awarded the Order of Canada. An unusual honour belongs to Maynard and to Bryan and Dona Erb. Asteroid 1990 SG4, which was discovered in 1990, was named Erb in honour of both Erbs and their work on space travel. Asteroid 1990 ME, discovered the same year, is named after Maynard. Chamberlin and the Avro Arrow were immortalized in 1996 on a Canadian $20 silver coin. As part of the same series, Avro test pilot Jan Zurakowski's image appeared on a coin with the CF-100.

In June 2001, Chamberlin was inducted into Canada's Aviation Hall of Fame, joining others from Avro including Zurakowski and Jim Floyd. "His engineering genius, technical direction and leadership have been of significant benefit to Canada," Chamberlin's induction citation reads. "Further, his contributions to the United States space programs have given much credit to his home country of Canada."

•

The best-known group of foreign engineers that worked for NASA was the team of 118 German engineers who came to the U.S. after World War II under the leadership of Wernher von Braun. After 14 years working for the U.S. Army, most of that group followed von Braun to NASA, where they formed the nucleus of the Marshall Space Flight Center in Huntsville, Alabama. They worked together, and many lived near each other in a suburb of Huntsville. Von Braun's group had to live down the fact that it had been formed in service to Hitler's Reich and that its work involved direction of concentration camp workers who produced V-2 rockets.

Unlike the Germans, the Avro group never held meetings after it settled into work at NASA. Avro was such a large establishment that not all the men who went to NASA knew each

Owen E. Maynard (right) with James C. Floyd (left),who was in charge of the Avro Arrow Program, after Maynard received an honorary degree from University of Toronto. June 12, 1996. Photo by Chris Gainor.

other. The group had been chosen by NASA officials who were looking for specific skills. Jim Chamberlin probably had a voice in NASA's selections, but many of his close associates from Avro didn't go south with him. Many of the Avro group maintained friendships with former colleagues over the years and some took part in Avro reunions, but the only formal reunion of the NASA Avro engineers took place in Toronto in 1994. Only nine people made it, and two of them were the two Canadian doctors, Owen Coons and Bill Carpentier, who weren't at Avro but were considered part of the group just the same.

More than 20 of the 32 Avro engineers who went to NASA remained in the United States and became U.S. citizens. Maynard held dual citizenship and was proud of both countries. When Maynard and his wife returned to Canada for retirement, their children remained in the United States. The children of some of the others who remained in the U.S. returned to Canada, but most have set down roots in the United States.

The members of the Avro group are generally remembered for the contributions they made to NASA as individuals. On the occasions when the Avro engineers were referred to by NASA as a group, they were generally called the "Canadians." On the 25th anniversary of Apollo 11, the members of this group were hailed as the "NASA Canadians" and presented with special certificates by the Canadian Space Society and the U.S. National Space Society. While many of the Avro engineers, including many of British origin, embrace this title, not all of them do. Seventeen of the 32 who went from Avro to NASA were from England. Frank Chalmers was Scottish, Tecwyn Roberts was Welsh, and John Shoosmith was born in England but partly raised in Canada. The remaining 12 were born and raised in Canada.

At the time the Avro Arrow was cancelled in 1959, Canada had much closer familial, political, trade and corporate links to the United Kingdom than it does today. Outside the province of Quebec, Great Britain was considered Canada's mother country. The demise of Avro Canada, one of the largest corporations in Canada and one that was under British control, was one of the major steps moving Canada out of the British orbit and more firmly into the American one. Today, Canada's links to Great Britain outside of the monarchy are only a dimming memory for middle-aged and older Canadians.

The dual nature of the Avro group was inadvertently exposed at a panel discussion during NASA's 20th anniversary celebrations for Apollo 11. In one of his last public appearances, Bob Gilruth spoke about his early days at NASA: "They shut down a lot of things in England, and we were able to get a good number of aeronautical engineers with English backgrounds ..."

George Mueller, sitting nearby, interjected, "Canada."

"... that were very bright," Gilruth continued, "and they didn't have to unlearn a lot of things that they otherwise have had to do."

Chris Kraft added: "We tried to hire engineers that had five to 10 years experience. Max [Faget] had the foresight to get those kind of people around him to do that kind of work, and along with the people that we had from Canada, we had a good nucleus."

Because the Germans were concentrated in one place, their legacy to Apollo is easy to see: the Saturn rockets that propelled Apollo into orbit and to the Moon. Because the Avro engineers were spread out in NASA, their own legacy hasn't been seen as clearly. But it can be found in the details of the Mercury capsule, in Jim Chamberlin's design of Gemini, in Owen Maynard's and Bryan Erb's work on the Apollo spacecraft, in Hodge's work as a flight director and manager, in Rose's mission plans, in Jenkins' trajectory and software efforts, in Armitage's recovery designs, and in Hodge's, Roberts' and Fielder's efforts to design the tracking network for Mercury, Gemini and Apollo, and their work in charting the future of NASA's human space exploration. The other Avro engineers made their contributions to Mercury, Gemini and Apollo as well, and their handiwork can be found in Skylab, the shuttle and the International Space Station. Avro engineers made their mark in building the ground and now satellite networks that keep the Earth in touch with spacecraft, while the two physicians, Coons and Carpentier, made large contributions to the new field of space medicine.

Everyone in that group agrees that the U.S. would have made it to the Moon had the Avro engineers never come to NASA. There were many equally talented engineers in the U.S. who didn't go to NASA because of concerns that the Mercury program would be just a short-term dead-end program, or because of the better money in private industry. The Avro group proved that Canadian and British engineers could work with the best America had to offer. "I think there was a feeling that people from Canada can do anything anybody else can do," Erb once put it.

Without Jim Chamberlin, Owen Maynard, John Hodge, Rod Rose and their colleagues, the road to Tranquillity Base may have been different and more bumpy, just as the lack of a German rocket team at Huntsville would have rendered the task more difficult. In particular, the work of Jim Chamberlin was crucial in making Mercury, Gemini, Apollo and the space shuttle the successes they were.

In separate interviews, both Faget and Gilruth called the Avro group a "godsend" for NASA. Gilruth's former deputy, James C. Elms, said the Avro group was "one of the greatest things that ever happened" to NASA.

In 1994, Owen Maynard was speaking to a group in Kitchener, Ontario, a Canadian community with a large German population. An elderly man of German descent stood up and said: "Now we've heard a lot about the Americans did this great Apollo spacecraft program and landed people on the Moon, but we all know they really didn't do it themselves. How much of it was done by the Germans? How much did the Germans contribute to it?" Maynard got a standing ovation that not even he expected when he responded, "Almost as much as the Canadians."

Left to right - Bryan Erb, Stanley Galezowski, Bruce Aikenhead, Owen Maynard, David Ewart and Owen Coons May 1994 reunion in Toronto. Courtesy Toronto Star/J. Wilkes

Avro Canada Engineers at NASA

Bruce Alexander Aikenhead — Born in Didsbury Alberta in 1923, he grew up in London, Ont. After wartime service as a radar mechanic with the Royal Canadian Air Force and the Royal Air Force, he obtained an honours BSc at the University of Western Ontario in physics and mathematics. After working in radio and television design and production for five years, he joined CAE Industries in Montreal in 1955 and worked on aircraft simulators. He worked on simulators for Avro Canada in 1958 and 1959. At NASA, he helped set up astronaut training in Mercury, and then returned to Canada in 1962, rejoining CAE Industries. He worked for Gerry Bull on his advanced cannon program in the 1960s. He later worked on Canadian satellite programs, the Canadarm on the Space Shuttle, and finally was director of the Canadian Astronaut Program. Upon his retirement, he and his wife Helen moved to Salmon Arm, B.C. Aikenhead was awarded the Order of Canada in 1997.

Peter J. Armitage, who worked on recovery systems during Mercury, Gemini and Apollo, managed the Lunar Receiving Laboratory in 1971 for Apollos 14 and 15. He directed earth observations programs on Skylab in the 1970s and Science payloads on the early Shuttle missions. Born in Leeds, Yorkshire, England, in 1929, he spent much of his childhood, including most of the war years, on the south coast of England. He studied engineering at Southampton University as part of his apprenticeship in local aircraft companies, including Air Service Training. Subsequently he became a designer for the Cierva Autogyro Company. After two years as a flight engineer in the Royal Air Force, he joined Avro Canada in 1952, where he worked in flight test. While at Avro, he was awarded a scholarship to the College of Aeronautics at Cranfield, England, now Cranfield University, where he received a Masters degree in Aeronautical Engineering. Joining NASA in 1959 he worked in recovery operations, and in 1969 was awarded an Alfred P. Sloan Executive Fellowship at Stanford University. Returning to NASA he became the manager of the Lunar Receiving laboratory and then served as a director in the space sciences area. After NASA, he worked with astronaut Deke Slayton with responsibilities for the engineering and operations of commercial Space flight boosters. He is now retired in Houston, Texas with his wife June, also from the UK.

David Brown — Born in South Croydon, Surrey, England, in 1927, he worked in the engineering directorate while at NASA, which he left around 1970. He took another job in the U.S. aerospace industry.

Richard R. Carley played a leading role in developing guidance and navigation systems in Mercury, Gemini, Apollo and shuttle. He was born in Saskatoon, Saskatchewan, in 1927 and earned his engineering degree at the University of Saskatchewan before working for two years with Canadian National Telegraph and five years at Avro. He helped develop the Arrow's semi-automatic fly-by-wire control system and worked on a similar system in Mercury. He had responsibility for Gemini's guidance and control system, and drew up specifications for the shuttle's control systems. He remained with NASA until 1981, spending much of his career working out of NASA headquarters in Washington after a stint at the NASA Electronics Research Center in Cambridge, Mass.

Frank J. Chalmers — Born in 1927 in Wishaw, Scotland, he studied engineering at the University of Strathclyde in Glasgow. After working for Vickers, he came to Canada in 1955 and worked for Westinghouse in Hamilton before joining Avro Canada, where he was flight test engineer on the first CF-105 to fly. He left NASA after only a few months in 1959. He moved to Canada where he worked in a printing business until he died in 1975 in Toronto. He and his wife June had three children.

James Arthur Chamberlin — Born in Kamloops, B.C., in 1915, he began his education B.C. but had most of his schooling in Toronto. He studied mechanical engineering at the University of Toronto and earned a Diploma at the Imperial College of Science and Technology in London, England. In England, he worked briefly for Martin-Baker, the ejection seat manufacturers. Back in Canada during the war, he worked for Clark Ruse Aircraft in Dartmouth, N.S., and Noorduyn Aircraft in Montreal. In 1946, he joined Avro Canada, and was chief aerodynamicist on the CF-100 Canuck and the C-102 Jetliner. During the development of the CF-105 Arrow, he was chief of technical design. He became head of engineering for Project Mercury, where he acted as project manager and played a key role in the success of Mercury. Then as the first project manager for Gemini, he designed this spacecraft. Chamberlin played a key role in NASA's decision to use lunar orbit rendezvous to go to the moon, and when he left Gemini in 1963, he became a technical advisor to Manned Spacecraft Center director Bob Gilruth, and troubleshooter on Apollo. He worked on concepts for the space shuttle at NASA and later at McDonnell-Douglas, which he joined in 1970. He worked for the contractor for four years in St. Louis and then in Houston helping direct development of the shuttle until his death in Houston in 1981. He and his wife Ella, who he married in 1942, had a daughter and a son. Ella Chamberlin died in 1992.

Thomas V. Chambers worked for NASA from 1959 to 1980. After 13 years in the Royal Navy during and after World War Two, he worked in the aircraft industry in England before joining Avro Canada in 1957. After being responsible for the Mercury attitude control system, he worked on on-board computers for Apollo and the shuttle. He was born in 1920 in Yorkshire, England and was trained as an electrical engineer. He lives near Houston with his wife Doreen.

Jack Cohen — A computer expert from England who left NASA in the early 1960s.

Stanley H. Cohn was one of the computer experts at Avro and NASA. Educated at the University of Toronto and Indiana University, he was a senior computing specialist at Avro and head of the digital computer group at STG. He returned to Canada in 1962 where he had a successful career in computers at IBM and later at the U of T. He was born in Toronto in 1926 and lives with his wife Joyce in Oakville, Ontario.

Burton G. Cour-Palais worked in engineering in Mercury and then turned to work on the issue of micrometeoroid impacts on spacecraft. He was one of the first to raise the problem of orbital debris left behind by satellites and rockets. He was born in Najpur, India in 1925 of English parents, and grew up in India. He was educated at St. Francis Xavier College in Calcutta and the College of Aeronautical Engineering in London, England. He worked at Vickers-Armstrong Ltd and Bristol Aircraft Ltd., and on his own as an aircraft structures analyst before joining Avro Canada in 1957. After retiring from NASA in 1989, he continued his work on orbital debris in private industry and at the Southwest Research Institute in San Antonio, near his retirement home in Texas, where he lives with his wife Audrey.

Eugene L. Duret left NASA in 1965 after helping to set up the tracking network and working as a capsule communicator during Mercury flights. He was born in Creelman, Saskatchewan, in 1925 and served in World War Two in the Royal Canadian Air Force as a navigator in Lancaster bombers. After the war, he continued in the RCAF as a meteorological officer. He studied physics at Queen's University in Kingston, and joined Avro Canada shortly after graduation, specializing in heat transfer. After he returned to Canada from NASA and working for Gerry Bull in his cannon research program, he became a physics instructor at Humber College in Toronto and died in 1985. A bachelor, Duret was interested in painting and sailing.

R. Bryan Erb — Born in Calgary, Alberta, in 1931, he studied at the University of Alberta and at Cranfield in the U.K. He joined Avro Canada in 1955, where he conducted aerothermodynamic analysis on the Arrow. During Project Mercury, he predicted the performance of the heatshield, and in 1960 became a member of the 8-person advanced vehicles team that laid the foundations for Apollo. Later he managed the development of the Apollo heatshield and oversaw that of other Apollo subsystems. After spending a year at MIT on a Sloan Fellowship, Erb became deputy manager and then manager of the Lunar Receiving Laboratory. He joined NASA's remote sensing program in 1971, and later became Chief of the Earth Observations Division. After retiring from NASA in 1985 and working as a consultant, Erb became Canada's representative in Houston and Assistant Director of the Canadian Space Station Program. He is now manager of the Canadian Space Power Initiative.

Dona M. Erb — Born Dona Marie German, she married Bryan Erb in 1955. Dona Erb, a Calgary native who graduated in education at the University of Alberta, worked as a mathematics teacher, and lectured in computer science at the University of Houston until the early 1970s, when she began work as a computer programmer and manager working for Lockheed on the Shuttle Program and later for The MITRE Corporation, from which she retired as Lead Scientist in 1993. She and her husband have two children, Richard - a rancher in Saskatchewan - and Evelyn - a bond and equity investor for an insurance company in Connecticut.

David D. Ewart spent most of his 35 years with NASA as the NASA office at North American (and its successors) in Downey, Cal., working on Apollo, and the space shuttle orbiter. In 1971 he represented NASA at McDonnell Douglas at Huntington, Beach, Cal., where Skylab was built. He was born in 1927 in Stoke-on-Trent, Staffordshire, England, and served in the Royal Air Force and the Royal Canadian Air Force reserve. At Avro Canada, he did wind tunnel work on the Avro Arrow. In 1994, he retired after having spent the previous nine years as the NASA resident manager at Downey. He lives with his wife Sally in Nevada.

Joseph E. Farbridge, was born in 1931 in South Shields, Durham, England. He studied engineering at the Royal Aircraft Establishment at Farnborough, and joined Avro Canada in 1953. Avro sent him to Cranfield for further studies. He left NASA after only a few months, in 1959, and returned to England, where he worked for Hawker Siddeley and then in tourism for two years. He returned to Canada in 1964. He died in 1993 after a successful career with de Havilland Canada in the Toronto area.

Norman B. Farmer worked for NASA for about 20 years in electrical engineering, specializing in development flight instrumentation for Apollo and shuttle. In Mercury, he worked as a subsystem engineer, and in Apollo, he helped develop the communications system for the Lunar Roving Vehicle. He was born in London, England, in 1927, and studied at Southall Technical College in England. Trained in electronics, he worked for EMI, English Electric, and a small aircraft firm, ML Aviation. He joined Avro Canada in 1957. He now is retired in Oregon with his wife Helen.

Dennis E. Fielder was born in London, England, in 1930 and took a five-year apprenticeship at the Royal Aircraft Establishment in Farnborough, England. After working on guided missiles at the RAE, he came to Canada in 1954 and worked on radar systems at General Electric in Toronto. In 1959 he joined Avro and worked on fire control systems and flight test instrumentation in the CF-100 and CF-105. After joining NASA in 1959, Fielder worked on setting up the mission control center and the worldwide tracking network. In the late 1960s, he joined John Hodge's group that planned future programs for NASA. He was one of the first to propose the use of satellites to communicate with piloted spacecraft, an idea that was adopted in the shuttle program. He worked on mission planning for Skylab, and on shuttle and space station design. He left NASA in 1985, and now lives in Houston.

Stanley H. Galezowski was born in 1933 in Toronto, Ontario, and earned an engineering degree at the University of Toronto. He worked on control systems at Avro, and at NASA, he worked with analog computers to simulate space flight conditions. He left NASA in 1962 and went into business in Canada. He and his wife Emily are now retired in the Bahamas.

George Harris Jr., born in 1929 in Willenhall, England, he attended Wolverhampton Technical College and then apprenticed at the Midlands Electricity Board. After coming to Canada, he joined Avro Canada in 1954, where he worked in flight test. After a brief stint at North American Aviation in Ohio, he joined NASA in 1960 and helped set up the worldwide tracking network for Mercury, Gemini and Apollo. He left NASA in 1968 and worked for the European Space Agency, setting up their first control center in Germany. While working with a contractor, he was involved with the rescue of TDRS-1 following a booster failure after its deployment on a 1983 shuttle flight. He also worked on launching communications satellites with Ariane rockets and also worked with the U.S. Information Agency. He returned to Canada in 1997 and 1998 on a contract managing the Canadian Radarsat spacecraft. He and his wife Martha live in New Mexico.

John Dennis Hodge — Born in Leigh-on-Sea, England in 1929, Hodge was educated at the University of London and worked at Vickers Armstrong, before moving to Toronto and Avro Canada in 1952. At Avro, he worked on loads and flight test for the Arrow. At NASA, he became one of the first flight directors, along with Chris Kraft and Gene Kranz of Apollo 13 fame, working as flight director in Mercury, Gemini and early Apollo missions. He was the lead flight director for Gemini 8, the first emergency in space. Hodge worked in operations, helping set up the network aspects of Mercury and the control center for Mercury and Gemini. He became chief of the flight control division at MSC and was responsible for the hiring and training of flight controllers. In 1968, Hodge went to plan the later lunar missions, and then NASA's piloted space program after Apollo. In 1970, he went to the U.S. Department of Transportation. He returned to NASA in 1982 to get the space station program started at NASA Headquarters, became deputy associate administrator for space station, and remained there until 1987. Hodge is now a consultant living near Washington, D.C. He and his wife Audrey have four children.

John K. Hughes — Of English descent, he worked in the Flight Operations Directorate in Mercury.

Morris V. Jenkins was a trajectory expert whose group drew up the initial lunar trajectories for Apollo. He was born in Southampton, England, in 1923. After a wartime stint as a navigator in the RAF, which included training in Canada, he worked for Vickers Armstrong in England and moved to Avro in Canada in 1956, working in the stability and control group. After Apollo and a brief time working in the Skylab program, he worked on mission planning and analysis in the shuttle program. He left NASA in 1984 and lives in Texas. His wife Joan was a school teacher who later studied engineering and worked at McDonnell Douglas. They have two children.

Robert N. Lindley worked at A.V. Roe in England, where he became a protégé of famed designer Roy Chadwick and worked on such aircraft as the Lancaster and the York. Before leaving for Canada in 1947, he did preliminary design work on the Vulcan. After a year and a half at Canadair, he became one of the top people at Avro Canada, where he worked for 10 years. During the Arrow program, he was chief engineer. Lindley was born in England in 1921. He helped organize the hiring of the Avro engineers by NASA but then he went to McDonnell Aircraft rather than NASA. He was McDonnell's engineering manager for Gemini. Lindley joined NASA in 1969 and worked in headquarters. In 1972, he moved to the Goddard Space Flight Center where he worked until 1979. During this time, he was involved in the early shuttle program and in Spacelab. He lived in Maryland with his wife Chris until his death in May, 1999.

C. Frederick Matthews trained Mercury flight controllers and helped set up the Mercury Control Center. Along with other Avro engineers, he had a key role in setting up mission control concepts. He left NASA in 1962 and went to work for RCA in Boston for 29 years on various military and space projects. He lives with his wife Fran in Lexington, Mass. He was born in 1922 in Guelph, Ontario, and was raised and educated in Toronto. He served in the RCAF during the Second World War, and after earning his engineering degree in 1948, he joined Avro Canada as a stress engineer, although he spent most of his time at Avro in flight test.

Owen Eugene Maynard — Born in 1924 in Sarnia, Ont., he worked as a boatbuilder and machinist before joining the Royal Canadian Air Force in 1942. He flew a number of aircraft, including the Mosquito, in Canada and overseas. After the war, he got work at Avro Canada while he earned his aeronautical engineering degree at the University of Toronto. He held a number of jobs at Avro, including layout of the Jetliner and working on the design and testing of the Avro Arrow weapons pack and landing gear. He joined NASA in 1959 and after a brief period working in Mercury, he moved to advanced projects and became involved in the early design of the Apollo command and service modules, and was the first person in NASA to begin working on the design of the lunar module. He helped sell the concept of lunar orbit rendezvous to NASA. In 1964, he became chief of systems engineering in the Apollo program office, where he was in charge of the integration of Apollo systems. He was chief of mission operations in 1966 and 1967 before returning to systems engineering, Maynard organized the Apollo Lunar Landing Symposium in June, 1966, that helped set out the requirements for the first lunar landing flight. In 1970, Maynard left NASA and went to Raytheon in the Boston area, and worked on many programs, including solar power satellites. He retired in 1992, and he and his wife Helen moved to Waterloo, Ont. Maynard died there on July 15, 2000. They had four children.

John K. Meson — He worked for NASA from 1960 to 1984, spending much of his career at NASA Headquarters. He later worked for the Defence Advanced Research Projects Agency, and died in 1996.

Leonard E. Packham worked on communications, tracking and radar systems in Mercury, Gemini and Apollo. He retired in 1974 after 15 years with NASA and died in 1979. He was born in 1922 in Saskatoon, Saskatchewan, and educated at the University of Saskatchewan. During the war, he was an aircraft mechanic in the RCAF. After his service and university, he went to work for Avro Canada, where he was involved in communications as part of flight testing. He and his wife Wilda had a son.

Tecwyn Roberts was the original Flight Dynamics Officer in mission control and was one of those most responsible for the design of the Mercury Control Center at Cape Canaveral and the Misson Control Center in Houston. In 1965, he moved to Goddard Space Center and became the director of networks there. He left NASA in 1984 and died in 1988. He and his wife Doris had a son. Roberts was born in 1925 in Liverpool, England, of Welsh parents. He studied at the Beaumaris Grammar School and attended Southampton and Isle of Wight Technical College. After working with Saunders, Roe, on the Isle of Wight, Roberts joined Avro Canada in 1952.

Rodney G. Rose — He was born in Huntingdon Hunts, England, in 1927, and completed a five-year apprenticeship in aeronautical engineering with A.V. Roe in Manchester, England. He obtained a Master of Science degree at the College of Aeronautics at Cranfield, now Cranfield University, in England. Rose came to NASA after six years of work at Vickers Armstrong Ltd. in England and two years at Avro Canada, where he worked in flight test analysis and as an aerodynamicist. He worked on Little Joe rockets during Mercury and the paraglider program in Gemini. Rose then moved to operations, where he worked as assistant to Chris Kraft, planning missions in Apollo and shuttle until 1984. Rose helped organize the Christmas prayer from lunar orbit during the historic Apollo 8 mission, and was responsible for monitoring radiation dangers during Apollo flights. Rose helped set up mission planning and communications for the space shuttle, and after leaving NASA in 1984, he worked for Rockwell for the following five years. Now retired in Wimberley, Texas, with his wife Leila, Rose is the unofficial historian among the Avro group.

Leslie G. St. Leger was born in 1922 in Calcutta, India, to English parents. He moved to England at the outbreak of the Second World War, and served in the Army as a glider pilot. After working for de Havilland in England, he joined Avro Canada in 1950, working as a stress engineer. After a brief stint at General Dynamics following the cancellation of the Arrow, he joined NASA and continued his work as a stress engineer in Mercury and Apollo. During the 1970s and 1980s, he joined the structures test laboratory at JSC, and at the time of his retirement in 1989, he was manager of the laboratory. He lives in Houston with his wife Emar.

John N. Shoosmith worked on computers in Mercury and Gemini, and moved to Langley Research Center in 1964, where he remained for the rest of his career as a computer specialist. He was born in London, England, in 1934, but his family moved to Ottawa, Ontario, when he was 14. He studied engineering physics at Queen's University in Kingston, Ont., before joining Avro Canada. The youngest of the Avro group, he was the last to leave NASA when he retired on April 1, 1995 after nearly 36 years there. He lives in Williamsburg, Virginia, with his wife Carolin.

Robert E. Vale joined NASA in 1959 and worked mainly in structures and mechanical engineering in the engineering directorate at STG and Houston during his career at NASA. In the mid-1960s, he took time out from his work in structures to supervise the preparation of Apollo lunar surface experiments. He was born in Toronto, Ontario in 1922, studied engineering at the University of Toronto, and served in the RCAF during the Second World War. At Avro Canada, where he worked from 1947 to 1960, he was responsible for stress analysis and structural analysis. He died in 1980, shortly after retiring from NASA. He and his wife Marguerite had five children.

George A. Watts was born in Trail, B.C., in 1928 and studied at the Spartan School of Aeronautical Engineering in Tulsa, Oklahoma, and the University of Toronto. He worked at Avro Canada for 10 years, starting in 1949. At NASA, he worked in engineering on loads for Mercury, Gemini and Apollo. He left NASA in 1964 and joined Lockheed in California. He worked on Lockheed's proposal for a supersonic transport aircraft, on the L-1011 and on rotary winged aircraft. He has worked as a teacher and consultant since his retirement from Lockheed in 1991. He remarried after the death of his first wife Norma.

Other noteworthy Canadian and British experts

Others who came from Canada and the U.K. to work in the U.S. Space Program:

Dr. Dwight Owen Coons was born and raised in Hamilton, Ontario. He studied medicine at the University of Toronto, but was unable to qualify as a pilot because of colour blindness. Nevertheless, he became a flight surgeon with the Royal Canadian Air Force for 15 years, where he used his training as a para-rescue specialist and consulted on the design of the Avro Arrow. In 1963, he left NASA with the rank of Wing Commander and joined NASA's Manned Spaceflight Center in Houston, where he worked as a flight surgeon on Gemini and early Apollo missions and became chief of the medical office. He left NASA in early 1969 and opened an aviation medicine practice in Dallas, Texas, which he continued until his death in December, 1997 at age 72.

Dr. William R. Carpentier was born in Edmonton, Alberta, and raised there and on Vancouver Island in B.C. After studies at Victoria College, the University of B.C. and Ohio State University, he worked on medical programs at MSC in Gemini and Apollo. Dr. Carpentier examined many returning Gemini and Apollo crews on the recovery ships, and was placed in quarantine in the trailer on the Hornet with the Apollo 11 astronauts following their historic mission. He left NASA in 1971. He lives in Texas and continues his work in nuclear medicine at the Scott and White Clinic in Temple, Texas. He and his wife Willie have two sons.

C. Malcolm Hinds was project manager for Spar Aerospace in Toronto on the STEM antennas used for experiments in the service modules of Apollos 15, 16 and 17. Along with 12 of his colleagues, he was the first in Canada to receive the Silver Snoopy Award given to people who make outstanding contributions to NASA human spaceflight programs. Born and educated in England, Hinds came to Canada in 1964. After working for de Havilland in Toronto and Rohr Industries in San Diego, Hinds joined Spar Aerospace in 1968. During his 16 years there, he worked on STEM antennas for Apollo and for various other satellite and earthbound applications. He also was sales manager for Spar's role as builder of the Canadarm. Since leaving Spar, Hinds has worked for other aerospace companies and is now a consultant based in Winnipeg, Manitoba. The other Spar employees who received the Silver Snoopy are: Robert Alexander, Harry Brown, James Crawford, Helmut Dieners, Kenneth Doughty, Sidney Herbage, Phillip Ibbotsson, Lawrence Lebland, William McMillan, Douglas MacQueen, William Morgan and Terence Ussher.

Carl V. Lindow was a top manager at Boeing for launch systems and propulsion systems from 1961 to 1966. In this job, he was involved with Boeing's unsuccessful bids for the first stages of the Saturn I and IB rockets, and for the lunar module. His work in this job also included various other launch systems using liquid, solid and nuclear fuels at a time when Boeing was building the first stage of the Saturn V rocket. Born in Nelson, B.C. in 1924, he attended school in Salmo and Nelson, and studied at the Spartan School of Aeronautical Engineering in Tulsa, Oklahoma. After working at Noorduyn Aviation in Montreal for two years, he moved to Avro Canada in 1946. He became chief stress engineer, engineering project manager, and after the cancellation of the Arrow, chief design engineer. He left Avro in 1961 for Boeing. After 1967, he worked on the Boeing 747 and on other aircraft and space programs in Boeing. He died in 1986.

Mario A. Pesando — After officer training and work at National Steel Car during the war, he joined Avro Canada when it started up in 1945 and worked there until 1959. During this time, he also earned a degree in engineering physics at the University of Toronto. At Avro, Pesando worked as chief project research engineer and later as chief flight test engineer. He worked on the Jetliner, the CF-100 and the Arrow, developing a data handling system for the Arrow. He moved to the U.S. and went to work for RCA, where he worked on various military programs and on RCA's proposal for Operational Flight Control for the Saturn rockets. From 1963 to 1967, he worked for Avco on the Minuteman ICBM, and then rejoined RCA. Pesando later returned to Canada and worked for Indac Technologies in Mississauga on helicopter systems, including a system that helps draw helicopters onto the decks of small, unstable warships. He and his wife Dorothy are now retired near Victoria, B.C.

Keith A. Richardson is a Toronto native born in 1934 who studied geology at the University of Toronto and pursued graduate studies at Rice University in Houston. A specialist in radioactivity in rocks, he was hired by NASA at Houston in 1964 and worked on training astronauts for their missions. He was also involved in designing radiation testing facilities in the Lunar Receiving Laboratory, and was principal investigator for lunar geology experiments in Apollos 11 and 12. He left NASA in 1971, and came to the Geological Survey of Canada in Ottawa, where he became head of the exploration geophysics group. He retired in 1996.

John W. Sandford began his career as an apprentice at Westland Aircraft in England and took a master's degree in aeronautical engineering from the College of Aeronautics at Cranfield. He moved to Canada in 1957, where he joined Avro Canada. In 1959, he helped form a small company that built auto gyros, and then worked for Stanley Aviation. In 1961 he joined North American Aviation, where he became director of engineering for advanced launch systems, working on the Saturn V rocket. He was also involved in North American Rockwell's winning bid for the space shuttle, but his subsequent career has focussed on aircraft. After work in Rockwell's General Aviation Division, he returned to Canada in 1975 as president of Canadian Admiral Corp. In 1978, he became president of de Havilland Aircraft of Canada, where he remained until 1985. He has since held top positions at Fairchild Republic, Gulfstream Aerospace, Rolls Royce North America Inc., Cade Industries, and Avcorp Industries Inc.

David W. Strangway — A geophysicist who became involved in Apollo as one of many investigators from around the world who conducted research on lunar samples. From 1970 to 1973, Strangway worked for NASA as the chief of the geophysics branch, where he was responsible for geophysical aspects of Apollo missions. Strangway was born in Simcoe, Ont., in 1934, and was brought up in southern Africa by his missionary parents. He earned degrees, including a Ph.D., at the University of Toronto in physics and geology, and worked for two mining companies before stints on the faculty of the University of Colorado in Boulder, and the Massachusetts Institute of Technology. In 1968, he joined the physics department at the U of T and returned there after his stint at NASA as chairman of the geology department. From 1980 to 1985, he served as vice president and acting president of the U of T, and from 1985 to 1997, he was president of the University of British Columbia. He is now working on developing a private university in British Columbia.

R. Lionel Whyte was responsible for his firm, Héroux Ltd. of the Montreal suburb of Longueuil, winning the contract to fabricate the most of the leg assemblies for the Apollo lunar module. Born in Lachine, Quebec, in 1918, Whyte lived most of his life in the Montreal area. He worked for Dominion engineering for many years and studied engineering at night at the Montreal Technical Institute. In the 1960s, he joined Héroux and became general sales manager. Héroux builds landing gear and other precision parts as a subcontractor with U.S. and Canadian aerospace companies, including Grumman, the prime contractor on the LM. Whyte and his wife Eleanor retired to Brockville, Ont., where Whyte died in 1998.

Sources

NASA Histories

Bilstein, Roger E. *Stages to Saturn: A Technological History of the Apollo/Saturn Launch Vehicles.* Washington D.C.: National Aeronautics and Space Administration, 1980.

Brooks, Courtney G., James M. Grimwood and Loyd S. Swenson, Jr. *Chariots For Apollo: A History of Manned Lunar Spacecraft.* Washington D.C.: National Aeronautics and Space Administration, 1979.

Brooks, Courtney G., and Ivan D. Ertel. *The Apollo Spacecraft: A Chronology, Volume III.* Washington D.C.: National Aeronautics and Space Administration, 1976.

Compton, William David. *Where No Man Has Gone Before: A History of Apollo Lunar Exploration Missions.* Washington D.C.: National Aeronautics and Space Administration, 1989.

Compton, W. David, and Charles D. Benson. *Living and Working in Space: A History of Skylab.* Washington, D.C.: National Aeronautics and Space Administration, 1983.

Dethloff, Henry C. *Suddenly Tomorrow Came...., A History of the Johnson Space Center.* Houston, Texas: National Aeronautics and Space Administration, 1993.

Ertel, Ivan D., and Mary Louise Morse. *The Apollo Spacecraft: A Chronology, Volume I.* Washington D.C.: National Aeronautics and Space Administration, 1969.

Ertel, Ivan D., and Roland W. Newkirk, with Courtney G. Brooks. *The Apollo Spacecraft: A Chronology, Volume IV.* Washington D.C.: National Aeronautics and Space Administration, 1978.

Gawdiak, Ihor, with Helen Fedor. *NASA Historical Data Book, Volume IV, Resources 1969-78.* Washington, D.C.: National Aeronautics and Space Administration, 1994.

Grimwood, James M. *Project Mercury: A Chronology.* Washington D.C.: National Aeronautics and Space Administration, 1963.

Grimwood, James C., Barton C. Hacker and Peter J. Vorzimmer. *Project Gemini Technology and Operations: A Chronology.* Washington, D.C.: National Aeronautics and Space Administration, 1969.

Hacker, Barton C., and James M. Grimwood. *On the Shoulders of Titans: A History of Project Gemini.* Washington D.C.: National Aeronautics and Space Administration, 1977.

Hansen, James R. *Spaceflight Revolution, NASA Langley Research Center from Sputnik to Apollo.* Washington, D.C.: National Aeronautics and Space Administration, 1995.

Hansen, James R. *Enchanted Rendezvous: John C. Houbolt and the Genesis of the Lunar Orbit Rendezvous Concept, Monographs in Aerospace History Series #4.* Washington, D.C.: NASA History Office, 1995.

Heppenheimer, T.A. *The Space Shuttle Decision: NASA's Search for a Reusable Space Vehicle.* Washington, D.C.: National Aeronautics and Space Administration, 1999.

Logsdon, John M., ed. *Exploring the Unknown: Selected Documents in the History of the U.S. Civil Space Program.* Washington, D.C.: NASA History Office, National Aeronautics and Space Administration, 1995.

Logsdon, John. *Together In Orbit: The Origins of International Participation in the Space Station, Monographs in Aerospace History #11.* Washington, D.C.: NASA History Division, 1998.

Lord, Douglas R. *Spacelab: An International Success Story.* Washington, D.C.: National Aeronautics and Space Administration, 1987.

Mogan, Kathleen M. and Frank P. Mintz, editors. *Keeping Track: A History of the GSFC Tracking and Data Acquisition Networks: 1957 to 1991*. Greenbelt, Md.: NASA Goddard Space Flight Center, 1992.

Morse, Mary Louise, and Jean Kernahan Bays. *The Apollo Spacecraft: A Chronology, Volume II.* Washington D.C.: National Aeronautics and Space Administration, 1973.

NASA History Office. *The Mission Transcript Collection: U.S. Human Spaceflight Missions From Mercury Redstone 3 to Apollo 17.* CD ROM, NASA History Office, Washington, D.C., 2000.

Newkirk, Roland W., Ivan D. Ertel and Courtney G. Brooks. *Skylab: A Chronology.* Washington, D.C.: National Aeronautics and Space Administration, 1977.

Portree, David S.F. *Humans to Mars: Fifty Years of Mission Planning, 1950-2000, Monographs in Aerospace History #21.* Washington, D.C.: NASA History Office, 2001.

Siddiqi, Asif A. *Challenge to Apollo: The Soviet Union and the Space Race, 1945-1974.* Washington, D.C.: National Aeronautics and Space Administration, 2000.

Swanson, Glen E., ed. *Before This Decade is Out: Personal Reflections on the Apollo Program.* Washington, D.C.: National Aeronautics and Space Administration, 1999.

Swenson, Loyd S., James M. Grimwood, and Charles C. Alexander. *This New Ocean: A History of Project Mercury.* Washington D.C.: National Aeronautics and Space Administration, 1966.
Tomayko, James E., *Computers Take Flight: A History of NASA's Pioneering Digital Fly-By-Wire Project,* Washington D.C.: National Aeronautics and Space Administration, 1999.

Van Nimmen, Jane, Leonard C. Bruno, Robert L. Rosholt. *NASA Historical Data Book, Volume I, Resources 1958-68.* Washington, D.C.: National Aeronautics and Space Administration, 1988.

Books

Aldrin, Col. Edwin E., and Wayne Warga. *Return to Earth.* New York:Random House, 1973.

Armstrong, Neil, Michael Collins, Edwin E. Aldrin Jr. *First on the Moon.* Boston: Little, Brown, 1970.

Arnold, H.J.P., ed. *Man In Space: An Illustrated History of Space Flight.* New York: Smithmark Publishers, 1993.

Baker, David. *The History of Manned Space Flight.* New York: Crown Publishers Inc., 1982.

Berton, Pierre. *Vimy.* Toronto: McClelland & Stewart, 1986.

Bizony, M.T., General Editor. *The Space Encyclopedia.* New York: E.P. Dutton & Co., 1964.

Bizony, Piers. *Island In The Sky: Building the International Space Station.* London: Aurum Press Ltd., 1996.

Campagna, Palmiro. *Storms of Controversy: The Secret Avro Arrow Files Revealed.* Toronto: Stoddart Publishing Company Ltd., 1992.

Chaikin, Andrew. *A Man on the Moon: The Voyages of the Apollo Astronauts.* New York: Viking, 1994.

Collins, Michael. *Carrying the Fire.* New York: Ballantyne Books, 1975.

Collins, Michael. *Liftoff: The Story of America's Adventure in Space.* New York: Grove Press, 1988.

Cooper, Henry S.F. *Moon Rocks.* New York: The Dial Press, 1970.

Divine, Robert A. *The Sputnik Challenge.* New York: Oxford University Press, 1993.

Dotto, Lydia. *Canada in Space.* Toronto: Irwin Publishing, 1987.

Dotto, Lydia. *The Astronauts: Canada's Voyageurs in Space.* Toronto: Stoddart Publishing Co., 1993.

Dow, James. *The Arrow*. Toronto:James Lorimer & Company, 1979.

Floyd, Jim. *The Avro Canada C102 Jetliner*. Erin, Ontario: Boston Mills Press, 1986.

Gatland, Kenneth. *The Illustrated Encyclopedia of Space Technology.* New York: Harmony Books, 1981.

Godwin, Robert, editor. *Friendship 7: The First Flight of John Glenn: The NASA Mission Reports.* Burlington, Ont.: Apogee Books, 1999.

Godwin, Robert, editor. *Apollo 8: The NASA Mission Reports*. Burlington, Ont.: Apogee Books, 1998.

Godwin, Robert, editor. *Apollo 9: The NASA Mission Reports.* Burlington, Ont.: Apogee Books, 1999.

Godwin, Robert, editor. *Apollo 11: The NASA Mission Reports, Volume 1.* Burlington, Ont.: Apogee Books, 1999.

Harford, James. *Korolev.* New York: John Wiley & Sons, Inc., 1997.

Harz, Theodore R. and Irvine Paghis. *Spacebound.* Ottawa: Canadian Government Publishing Centre, Supply and Services Canada, 1982.

Heppenheimer, T.A. *Countdown: A History of Space Flight.* New York: John Wiley & Sons, 1997.

Hodge, Paul W. *Meteorite Craters and Impact Structures of the Earth.* Cambridge University Press, 1994.

Hotson, Fred W. *de Havilland in Canada.* Toronto: Canav Books, 1999.

Jelly, Doris H. *Canada: 25 Years in Space.* Montreal: Polyscience Publications Inc., 1988.

Jenkins, Dennis R. *Space Shuttle: The History of Developing the National Space Transportation System.* Osceola Wisconsin: Motorbooks International, 1992, 1993.

Johnson, Nicholas L. *The Soviet Reach for the Moon, Second Edition.* Washington, D.C.: Cosmos Books, 1995.

Jones, Eric M., editor. *The Apollo Lunar Surface Journal.* http://www.hq.nasa.gov/alsj, 1995 - 1999.

Kelly, Thomas J. *Moon Lander: How We Developed The Apollo Lunar Module.* Washington, D.C.: Smithsonian Institution Press, 2001.

Kirton, John, ed. *Canada, the United States, and Space.* Toronto: Canadian Institute of International Affairs, 1986.

Kraft, Christopher C. *Flight: My Life in Mission Control.* New York: Dutton, 2001.

Kranz, Gene. *Failure is Not an Option: Mission Control from Mercury to Apollo 13 and Beyond.* New York: Simon & Schuster, 2000.

Launius, Roger D., and Howard E. McCurdy, eds. *Spaceflight and the Myth of Presidential Leadership.* Chicago: University of Chicago Press, 1997.

Lindsay, Hamish, *Tracking Apollo to the Moon.* London: Springer-Verlag, 2001.

Logsdon, John M. *The Decision to Go to the Moon.* Chicago: University of Chicago Press, 1970.

Lowther, William. *Arms and the Man: Dr. Gerald Bull, Iraq and the Supergun.* Toronto: Seal Books, 1991.

Mark, Hans. *The Space Station: A Personal Journey.* Durham: Duke University Press, 1987.

McConnell, Malcolm. *Challenger: A Major Malfunction.* Garden City, NY: Doubleday & Co., 1987.

McCurdy, Howard E. *The Space Station Decision: Incremental Politics and Technological Choice.* Baltimore: Johns Hopkins University Press, 1990.

Milberry, Larry. *Air Transport in Canada.* Toronto: Canav Books, 1997.

Milberry, Larry. *The Avro CF-100.* Toronto: Canav Books, 1981.

Murray, Charles and Catherine Bly Cox. *Apollo: The Race to the Moon.* New York: Simon and Shuster, 1989.

Newkirk, Dennis. *Almanac of Soviet Manned Space Flight.* Houston: Gulf Publishing Co., 1990.

Oberg, James E. *Red Star in Orbit.* New York: Random House, 1981.

Oberg, James E. *Mission to Mars.* Harrisburg, PA: Stackpole Books, 1982.

Organ, Richard, Ron Page, Don Watson, Les Wilkinson. *Avro Arrow, the Story of the Avro Arrow from its Evolution to its Extinction.* Erin, Ont.: Boston Mills Press, 1980. Revised edition published 1992 by Stoddart Publishing.

Peden, Murray. *Fall of an Arrow.* Toronto: Stoddart Publishing Company Ltd., 1987.

Powell, Joel W. and Lee Robert Caldwell. *Space Shuttle Almanac.* Calgary, Alta.: Microgravity Press, 1992-1999.

Powell, Joel W. and Lee Robert Caldwell. *Soviet/Russian Manned Space Program.* Calgary, Alta.: Microgravity Press, 1994-1999.

Purser, Paul E., Maxime A. Faget and Norman F. Smith, eds. *Manned Spacecraft: Engineering Design and Operation.* New York: Fairchild Publications, 1964

Reeves, Robert. *The Superpower Space Race.* New York: Plenum Press, 1994.

Roads to Space: An Oral History of the Soviet Space Program. Trans. Peter Berlin. The McGraw-Hill Companies, 1995.

Roy, Reginald H. *For Most Conspicuous Bravery: A Biography of Major-General George R. Pearkes, V.C., Through Two World Wars.* Vancouver: University of British Columbia Press, 1977.

Seitzen, Frank, and Aerospace F.Y.I. *Apollo 11: America on the Moon.* Arlington, Va.: Aerospace F.Y.I., 1994.

Shaw, E.K. *There Never Was an Arrow, Second Edition.* Ottawa: Steel Rail Publishing, 1981.

Simpson, Theodore R., ed. *The Space Station: An Idea Whose Time Has Come.* New York: IEEE Press, 1985.

Smith, Denis. *Rogue Tory: The Life and Legend of John G. Diefenbaker.* Toronto: Macfarlane Walter and Ross, 1995.

Société Europééne de Propulsion. *Baikonur, La Porte des Etoiles.* Armand Colin, 1994.

Stewart, Greig. *Shutting Down the National Dream.* Scarborough, Ont.: McGraw-Hill Ryerson, 1988.

Stewart, Grieg. *Arrow Through the Heart: The Life and Times of Crawford Gordon and the Avro Arrow.* Toronto: McGraw-Hill Ryerson, 1998.

Stockton, William, and John Noble Wilford. *Spaceliner.* New York: Times Books, 1981.

Stursberg, Peter. *Diefenbaker: Leadership Gained 1956-62.* Toronto: University of Toronto Press, 1975.

Whitaker, Reg and Gary Marcuse. *Cold War Canada.* Toronto: University of Toronto Press, 1994.

Tomayko, James E. *Computers in Space: Journeys with NASA.* Indianapolis: Alpha Books, 1994.

Wilhelms, Don E. *To a Rocky Moon: A Geologist's History of Lunar Exploration.* Tucson: University of Arizona Press, 1993.
Zaslow, Morris. *Reading the Rocks: The Story of the Geological Survey of Canada 1842-1972.* Toronto: Macmillan Company of Canada, 1975.

Zimmerman, Robert. *Genesis: The Story of Apollo 8.* New York: Four Walls Eight Windows, 1998.

Articles

Agle, D.C. "Flying the Gusmobile." *Air & Space Smithsonian*, September, 1998, pp. 46 – 55.

Alderman, Tom. "Canada's Men on the Moon Shot." *The Canadian Magazine*, June 21, 1969, pp. 2 – 9.

"Armitage to leave JSC," *JSC Space News Roundup*, Johnson Space Center, May 16, 1986.

"Asteroids with Canadian Connections." *Journal of the Royal Astronomical Society of Canada,* Vol. 89, No. 6, p. 267, 1995.

"Apollo 11's Engineer Returns Here For Visit," *Sarnia Observer,* Aug. 21, 1969, p. 15.

Baker, David. "Reaching Those Parts That Others Can't." *Flight International,* March 2000, Vol. 58 No. 3, pp. 180-184.

Berry, Charles A., et. al. "Man's Response to Long-Duration Flight in the Gemini Spacecraft." *Gemini Midprogram Conference*. National Aeronautics and Space Administration, Houston, Texas, 1966, p. 235.

Burke, Walter F. "How Company Met Gemini Challenge," *McDonnell Airscoop,* McDonnell Company, St. Louis, MO, vol. XXV, no. 10, Dec. 1966, p. 3.

"Canadians to work in space, Alouette ceremony told." *Science Notes*, Fall 1982, v. 1, no. 3, Ministry of State for Science and Technology Canada, Ottawa.

Carson, Catherine. "Bringing them back alive: Bryan and Dona Erb, space pioneers." *New Trail,* Winter, 1998/99, p. 44.

Chaikin, Andrew. "Fallen Arrow." *Air & Space Smithsonian,* April/May 1998, pp.33 -41.

Chamberlin, James A. "Project Gemini Design Integration." From *Manned Spacecraft: Engineering Design and Operation*, Eds Paul E. Purser, Maxime A. Faget, and Norman F. Smith. Fairchild Publications, New York, 1964, pp. 365 — 374.

Chamberlin, James A. and Andre J. Meyer, Jr. "Project Gemini design philosophy." *Astronautics and Aerospace Engineering*, AIAA, no date, p. 35.

Chamberlin, James A. "Orbital Space Station Design for Permanent Residence." Delivered at American Institute of Aeronautics and Astronautics Third Annual Meeting, Boston, Nov. 29 – Dec. 2, 1966.

Cooper, Henry S.F., Jr. "Annals of Space: We Don't Have to Prove Ourselves," *The New Yorker,* September 2, 1991.

"Eight MSC Employees Become U.S. Citizens." *Space News Roundup,* Manned Spacecraft Center, Houston, Texas, Nov. 11, 1964.

Falkner, Mrs. John William. "Chamberlin's fame comes as no surprise to mother." *The Province,* Vancouver, B.C., August 14, 1962.

Floyd, Jim. "Canada's Gift to NASA: The Maple Leaf in Orbit." *Wayfarers,* Heirloom Publishers, pp. 340 – 345.

Goudy, Donald V. "The Canadian Who Will Put 2 Yanks Into Space." *Canadian Weekly,* Date unknown, about August, 1962.

Hall, Joseph. "The Canadians behind NASA's space success." *Toronto Star,* May 28, 1994, A2.

Halvorson, Todd. "NASA places a massive bet on city in sky: United venture in space holds many big rewards, big risks." *Florida Today*, Sept. 20, 1998.

Hengeveld, Ed. "Land Landings for Gemini." *Quest*, Fall 1993, pp. 4 – 12.

Hengeveld, Ed. "Training for a Lunar Landing: The LLRV and LLTV." *Quest: The History of Spaceflight Quarterly,* Summer 1998, pp. 50 – 54.

Hodge, John D. and Tecwyn Roberts, "Flight operations facilities," *Astronautics and Aerospace Engineering,* 1962.

Hodge, John D. and Jones W. Roach. "Flight Control Operations." *Gemini Midprogram Conference,* National Aeronautics and Space Administration, Houston, Texas, 1966.

"Hodge says modest space station can be built for $4-$6 billion." *Defence Daily,* March 3, 1983, p. 21.

Humphries, Kelly. "Last Canadians break 'pact' to leave together." *JSC Space News Roundup,* Johnson Space Center, June 2, 1989, p. 4.

Ingram, Jay. "How U of T helped save Apollo 13 astronauts." *Toronto Star,* July 2, 1995.

Isinger, Russell. "Flying Blind: The Politics of the CF-105 Avro Arrow Program." *Western Canada Aviation Museum Aviation Review,* December 1996, pp. 23 - 30.

Karasiuk, Marion. "Flying High: Canadian Engineers and the Apollo Moon Landings." *Engineering Dimensions,* July/August, 1994, pp. 44-46.

Karasiuk, Marion. "The NASA Canadians." article draft for *Engineering Dimensions*, June, 1994.

Lafferty, Michael. "For the record: He keeps track of shuttle firsts." *Star-Advocate,* June 24, 1984.

Lindley, Robert N. "The Economics of a New Space Transportation System." AIAA paper 71-806, presented at the AIAA Space Systems Meeting, Denver, Colorado, July 19-20, 1971.

Morrison, Harold. "Gemini Project Big Step to Moon, Engineer Says." *Newport News Times-Herald*, Feb. 28, 1962.

Maynard, Owen E., and Rene A. Berglund. "Space Stations: Configurations and Design Considerations." *Astronautics and Aerospace Engineering*, American Institute of Aeronautics and Astronautics, New York, February, 1963.

Maynard, Owen. Convocation Address, University of Toronto, June 12, 1996.

McLean, Alasdair, and Michael Sheehan. "A Hare Turned Tortoise: 40 Years of UK Space

Policy." *Quest, The History of Spaceflight Quarterly,* v. 6, no. 4, Winter, 1998.

"NASA announces baseline configuration of space station." *NASA News*, Washington, D.C., May 14, 1986.

"No Moonbugs Found So Far." *Victoria Daily Times*, July 25, 1969, p. 1.

Pealer, Donald. "Manned Orbiting Laboratory (MOL) Part 3." *Quest,* Vol. 5, No. 2, 1996, p. 16.

Pesando, Mario. "The CF-100 goes Supersonic with help from gravity." unpublished and undated.

Portree, David S. F., and Joseph P. Loftus, Jr. *Orbital Debris: A Chronology, NASA/TP-1999-208856.* Johnson Space Center, January 1999.

Quig, James. "The Canadians and the Moon." *Weekend Magazine*, Vol. 16, No. 28, July 9, 1966, p.2.

Reguly, Robert. "Canadians who helped Gemini 'Go.'" *Toronto Star,* Sept. 3, 1965.

Rittinger, Carolynne. "Astronauts here for blast." *Medicine Hat News*, July 22, 1970.

"Scientists asked to consider space station transportation base." *Aerospace Daily,* Nov. 1, 1983, p. 5.

"Spacecrafts' Rubber Models Prepared by Sarnian, Space Official's Mother." *London Free Press*, July 9, 1963, p. 32.

"Space Decision-Maker: John Dennis Hodge." *New York Times,* March 18, 1966, p. 20.

Space News Roundup, Nov. 1, 1961, Manned Spacecraft Center, Langley AFB, Virginia.

Space News Roundup, Manned Spacecraft Center, June 9, 1972.

"Spar at 15 – a retrospective." *Infospar,* v. 15, no. 2, June 1983, Spar Aerospace Ltd., Mississauga, Ont.

The Avro News, Oct. 18, 1957

Varfolomeyev, Timothy, "Sputnik Era Launches," *Spaceflight,* Vol. 39, pp.331-2, October, 1997.

Wooten, James T. "Flight Surgeon Reports." *New York Times*, July 26, 1969, p. 1.

Wade, Mark. "Gemini – Landing on the Moon Faster, Better, Cheaper." *Encyclopedia Astronautica,* http://solar.rtd.utk.edu/~mwade/craft/gemcraft.htm, 1998.

Woodman, Jack. "Flying the Avro Arrow." presented to the Canadian Aeronautics and Space Symposium, Winnipeg, May 16, 1978.

Young, S. "Forty Years of Work at the McDonnell Douglas Canada Ltd. Plant, Malton." *Canadian Aeronautics and Space Journal*, Vol. 25, No. 2, Second Quarter 1979, pp. 128-132.

Memos and Documents

A Shuttle Chronology 1964 – 1973, Abstract Concepts to Letter Contracts, Volume 1: Abstract Concepts to Engineering Data; Defining the Operational Potential of the Shuttle. Houston Texas: Management Analysis Office, Administration Directorate, Johnson Space Center, December, 1988.

"A Background Review, Spar Aerospace Ltd.," February, 1982.

"Alouette – Canada's First Satellite." *NASA Facts,* National Aeronautics and Space Administration, Washington D.C., 1962.

"An Arm in Space," Public Information Branch, National Research Council of Canada, Ottawa, 1980.

"Apollo Lunar Landing Mission Symposium, Manned Spacecraft Center, Houston, Texas, June 25-27, 1966." Houston Texas: NASA Technical Memorandum, NASA TM X-58006, NASA, 1966.
Apollo Site Selection Board Meeting, minutes, June 3, 1969.

Apollo 8 Mission Commentary, December 24, 1968.

"Canada and the Space Station: Gateway to the 21st Century," Space Station Program Office, National Research Council of Canada, Ottawa, 1988.

"Canadian Astronaut Program Announced," press release, Ministry of State for Economic Development, Science and Technology, June 8, 1983.

Chamberlin, J.A. Avro Aircraft inter-departmental memo to J.G. Floyd, "Project Research Group," Dec. 28, 1956.

Chamberlin, James A. Letter to James T. Grimwood, March 26, 1974.

Chalmers, Frank. Letters to June Chalmers, April and May, 1959, supplied by June Chalmers.

Ertel, Ivan D. Manned Spacecraft Center Fact Sheet 291, Gemini Program Series, Houston Texas, 1965-66.

Glennan, T. Keith. Memorandum, "Approval to employ alien scientists from Canada," March 18, 1959.

Grimwood, James T. Memo from to Ivan D. Ertel regarding interview with Caldwell Johnson on March 10, 1966, undated.

Hodge, John D. "Statement of John D. Hodge, Director, Space Station Task Force, NASA, before the Subcommittee on Science, Technology and Space," U.S. Senate, Nov. 15, 1983.

"IML-1 Mission Media Kit," Canadian Astronaut Program, 1992.

"International Space Station Assembly Sequence (Rev. E, June, 1999)." NASA Facts, Johnson Space Center, Houston.

Jenkins, Morris. Memorandum to D.C. Cheatham, "Apollo studies to be completed or started by the Operational Analysis Section during the 12 months from the above date," Manned Spacecraft Center, Nov. 13, 1961.

Johnson, C.C. "Rough Draft." Sept. 18, 1959.

Johnson, C.C. "Mercury Optimization - first look." Dec. 7, 1959.

Matthews, C. Fred. and Christopher C. Kraft Jr. "Project Mercury Flight Control Facilities and Operation." Draft of a paper for NASA-Industry Apollo Technical Conference, Washington D.C., July 18-20, 1961.

Matthews, C.F. "Summary of an Operational Flight Control System (OFC) for Saturn." Nov. 13, 1962, Radio Corporation of America, Burlington, Mass.

Maynard, Owen E. "Preliminary Evaluation." Sept. 27, 1961,

Maynard, Owen E. Letter to Frederick J. Lees, Chairman, Inventions and Contributions Board, NASA, November 13, 1982.

Maynard, Owen E., et. al. "Radial Module Space Station," United States Patent 3,300,162, Patented Jan. 24, 1967.

Maynard, Owen E. Papers from Owen Maynard collection including job descriptions, biographical sketches, management charts, sketches and descriptions of early Apollo concepts, and notes and itinerary of April 16, 1962 presentation at Marshall Space Flight Center, and "A Brief Introductory Description of Apollo" presented to the Lunar Sciences Subcommittee of the NASA Space Sciences Steering Committee, Washington, D.C., Nov. 14, 15, 1961, "A Brief Introductory Description of Apollo" presented to the Lunar Sciences Subcommittee of the NASA Space Sciences Steering Committee, Washington, D.C., Nov. 14, 15, 1961, and memo, "Apollo principal technical problems," Sept. 10, 1965.

"Mercury Capsule Water Stability," memo by Peter Armitage and E. H. Harrin, Oct. 31, 1960.

"NASA-STG Concept of Operational control for Project Mercury with Particular Reference to Abort Responsibilities," F.J. Chalmers and C.F. Matthews, July 21, 1959.

NASA biographical data sheets on Burton Cour-Palais, David Ewart, Tecwyn Roberts, and Robert E. Vale, undated.

NASA personnel announcements, biographical data sheet on John Hodge.

"Organization of Space Task Group," Memorandum, Aug. 10, 1959.

"Original layout drawing for the Interim Mission Control Center in a trailer for the Mercury-Redstone flights," sketch by C. Fred Matthews, late summer, 1959.

"Plan for Project Mercury Remote Site Flight Controllers Training Equipment," probable author Bruce Aikenhead, April 8, 1960.

"Radarsat: Canada's Earth Observation Satellite," Radarsat Canada, Richmond, B.C., 1996.

Rose, Rod. "Summary of air drop and fatigue program with production capsule no. 5," memo, May 4, 1961.

"Saturn Operational Flight Control," Radio Corporation of America, 1962, and associated documents.

"Shuttle mission 41-G Media Kit," Canadian Astronaut Program, 1984.

Slayton, Donald K. "LLTV Review," memo, Jan. 27, 1969.

"Spar's Part in Apollo Lunar Science: An Outline of the Apollo Lunar Orbit Science Program and the Part Played by Spar Aerospace Products Ltd." Spar Aerospace Products Ltd., Toronto, Ont., December, 1972.

"Some Operational Aspects of Project Mercury," Christopher C. Kraft Jr., presented at the annual meeting of the Society of Experimental Test Pilots, Los Angeles, Oct. 9, 1959.

"Space Task Group: Canadian Personnel, Duty Assignments, need to know, and travel requirements," NASA STG, April, 1959.

"Specific Problems," memo, Owen Maynard, Oct. 7, 1959.

"Typical Down Range Station," sketch by C. Fred Matthews, Dec. 4, 1959.

Documents from the National Archives of Canada

Office of the Prime Minister, Canada, "Revision of the Canadian Air Defence Programme," Sept. 23, 1958.

Gordon, Crawford, President, A.V. Roe Canada Ltd. "Statement to Annual Meeting of Shareholders: The Arrow and Iroquois Programs," Oct. 27, 1958.

O'Hurley, Raymond, Minister of Defence Production. Letter to John Drysdale, M.P., June 6, 1960.

Campbell, Hugh. Memorandum to Minister of National Defence, April 24, 1959.

Cabinet Conclusions, various dates between October 24, 1957 to February 27, 1959.

Belyea, A.D. Cable to D.L. Thompson, Department of Defence Production, March 6, 1959. From National Archives of Canada, through Palmiro Campagna and Owen Maynard.

Speech texts and other documents from the National Archives of Canada.

Interviews

Interviews by author:

Bruce Aikenhead, Peter Armitage, Richard Carley, Dr. William Carpentier, Thomas Chambers, Robert Cockfield, Stanley Cohn, Dr. D. Owen Coons, Burton Cour-Palais, Bryan Erb, Dona Erb, Dave Ewart, Norman Farmer, Dennis Fielder, Stanley and Emily Galezowski, George Harris, C. Malcolm Hinds, John Hodge, Morris Jenkins, Robert Lindley, C. Frederick Matthews, Owen Maynard, Mario Pesando, Rod Rose, John Shoosmith, Leslie St. Leger, David Strangway, George Watts, R. Lionel Whyte, Keith Richardson, John Sandford, Thomas Kelly, Christopher C. Kraft Jr., Charles W. Mathews, Paul Purser, James T. Rose, Bob Thompson, John Yardley, Bernard Etkin, J. Barry French, Peter Hughes, James Floyd, Peter Beck, Hal Smith, Frank Brame, Helen Maynard, Ross Maynard, Merrill Marshall, Beth Devlin, Annette Maynard Franklin, Arthur Chamberlin, Shirley Ditloff, Isabel Rowe, Peter Wilson, June Chalmers, Ian Chalmers, Joan Farbridge, Kevin Farbridge, Maurice Duret, Marguerite Vale, Mark Vale, Jo Lindow, and Leslie Lindow.

Interview of Richard R. Carley by Rebecca Wright, NASA Johnson Space Center Oral History Project, March 2, 1999.

Interview of James A. Chamberlin by James M. Grimwood, NASA Historian, June 9, 1966.

Interview of Robert G. Chilton, by Loyd S. Swenson, NASA Historian, March 30, 1970.

Interview of Stanley Cohn by Summer Chick Bergen and Carol Butler, Johnson Space Center Oral History Project, October 19, 1998.

Interview of James C. Elms by Martin Collins, Sept. 16, 1988, National Air and Space Museum, Glennan-Webb-Seamans Project for Research in Space History.

Interview of David Ewart by Summer Chick Bergen, Carol Butler and Rebecca Wright, NASA Johnson Space Center Oral History Project, March 6, 1999.

Interview of Dr. Robert R. Gilruth by James M. Grimwood, NASA Historian, March 21, 1968.

Interview of Dr. Robert Gilruth by Drs. David DeVorkin and John Mauer, March 2, 1987, National Air and Space Museum, Glennan-Webb-Seamans Project for Research in Space History.

Interview of John D. Hodge by Ivan D. Ertel, NASA Historian, Houston, March 12, 1968.

Interview of John D. Hodge by Robert B. Merrifield, NASA Historian, March 15 and 18, 1968.

Interview of John D. Hodge by Ivan D. Ertel, NASA Historian, December 17, 1969.

Interview of John Hodge by Rebecca Wright, NASA Johnson Space Center Oral History Project, April 18, 1999.

Interview of Robert Lindley by James M. Grimwood, NASA Historian, April 13, 1966.

Interview of C. Fred Matthews by Rebecca Wright, NASA Johnson Space Center Oral History Project, June 18, 1999.

Interview of Owen and Helen Maynard by Charles Murray and Catherine Bly Cox, July 1988.
Interview of Owen Maynard by Loyd Swenson, NASA Historian, January 9, 1970.

Interview of Owen Maynard by Loyd Swenson and Courtney Brooks, NASA Historians, Feb. 18, 1970.

Interview of Tecwyn Roberts by Robert B. Merrifield, NASA Historian, July 26, 1968.

Interview of James T. Rose by James M. Grimwood, NASA Historian, April 13, 1966.

Interview of Rodney G. Rose by Loyd Swenson, NASA Historian, May 6, 1970.

Interview of David Strangway by Robert Sherrod, July 27, 1972, Courtesy of NASA Headquarters History Office.

Interview of George A. Watts by Carol Butler, NASA Johnson Space Center Oral History Project, March 7, 1999.

Interview of Walter C. Williams by James M. Grimwood, NASA Historian, May 15, 1967.

Films

Apollo Program Oral History. Videotape, July 21, 1989, Gilruth Recreation Center, Johnson Space Center, Houston.

James Lloyd. *Arrow: From Dream to Destruction.* Aviation Videos, 1992.

Fly Me to the Moon And Back – Mission Analysis and Planning Division. Film, Manned Spacecraft Center, National Aeronautics and Space Administration, Houston, 1966.

Friendship 7. NASA film, 1962.

Planning the Apollo Missions. Videotape, July 18, 1989, Teague Auditorium, Johnson Space Center, Houston.

Remarks of Bryan Erb from International Space Station briefing, Johnson Space Center, Houston, Texas, May 12, 1998.

Audio

Audio tape of MA-8 Air-to-Ground conversations. Supplied by Maurice Duret.

Audio tapes of Apollo Lunar Landing Mission Symposium, Manned Spacecraft Center, Houston, Texas, June 25-27, 1966. Supplied by Owen Maynard.

Audio Tape of Apollo 8 Countdown Demonstration Test, December 11, 1968, supplied by Owen Maynard.

Gemini Highlights. A Sounds from Space cassette, Space Frontiers Ltd., Havant, Hampshire, England.

INDEX

About the Author

Chris Gainor, the author of *Arrows to the Moon*, has had a lifelong interest in space flight and is a Fellow of the British Interplanetary Society. An award-winning journalist who has written about space exploration for many publications in Canada, the U.S. and the U.K., he is now a communications consultant in Victoria, British Columbia.

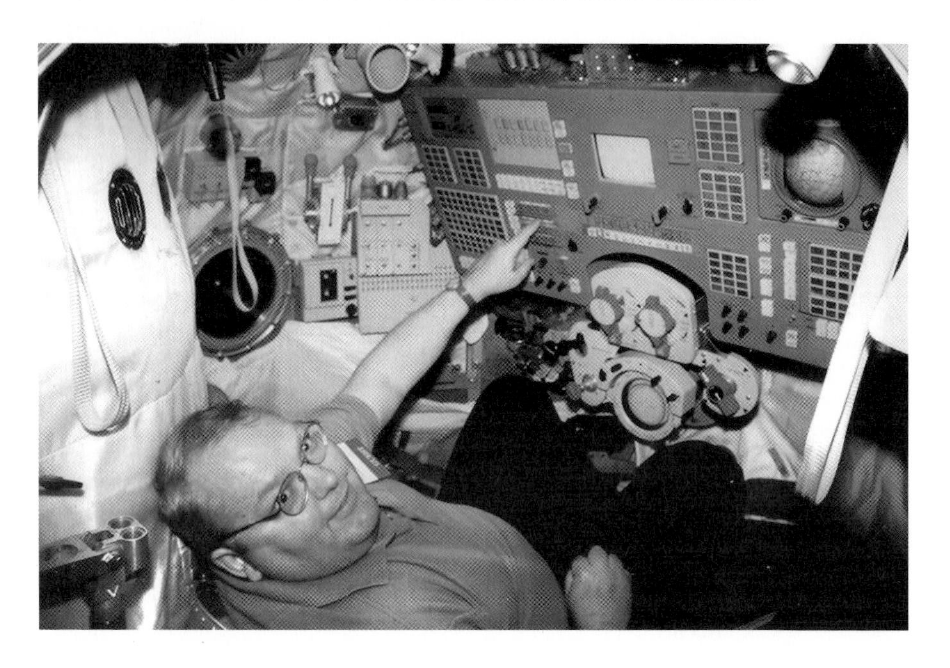

The author inside Soyuz trainer at Johnson Space Center, 1999.